Universitext

Series Editors
Nathanaël Berestycki, Universität Wien, Vienna, Austria
Carles Casacuberta, Universitat de Barcelona, Barcelona, Spain
John Greenlees, University of Warwick, Coventry, UK
Angus MacIntyre, Queen Mary University of London, London, UK
Claude Sabbah, École Polytechnique, CNRS, Université Paris-Saclay, Palaiseau, France
Endre Süli, University of Oxford, Oxford, UK

Universitext is a series of textbooks that presents material from a wide variety of mathematical disciplines at master's level and beyond. The books, often well class-tested by their author, may have an informal, personal, or even experimental approach to their subject matter. Some of the most successful and established books in the series have evolved through several editions, always following the evolution of teaching curricula, into very polished texts.

Thus as research topics trickle down into graduate-level teaching, first textbooks written for new, cutting-edge courses may find their way into *Universitext*

Chris Bowman

Diagrammatic Algebra

Chris Bowman
Department of Mathematics
University of York
York, UK

ISSN 0172-5939　　　　　　　ISSN 2191-6675　(electronic)
Universitext
ISBN 978-3-031-88800-7　　　ISBN 978-3-031-88801-4　(eBook)
https://doi.org/10.1007/978-3-031-88801-4

Mathematics Subject Classification (2020): 05E10, 20C08, 81R05, 20C30, 20G43

© The Editor(s) (if applicable) and The Author(s), under exclusive license to Springer Nature Switzerland AG 2025

This work is subject to copyright. All rights are solely and exclusively licensed by the Publisher, whether the whole or part of the material is concerned, specifically the rights of translation, reprinting, reuse of illustrations, recitation, broadcasting, reproduction on microfilms or in any other physical way, and transmission or information storage and retrieval, electronic adaptation, computer software, or by similar or dissimilar methodology now known or hereafter developed.
The use of general descriptive names, registered names, trademarks, service marks, etc. in this publication does not imply, even in the absence of a specific statement, that such names are exempt from the relevant protective laws and regulations and therefore free for general use.
The publisher, the authors and the editors are safe to assume that the advice and information in this book are believed to be true and accurate at the date of publication. Neither the publisher nor the authors or the editors give a warranty, expressed or implied, with respect to the material contained herein or for any errors or omissions that may have been made. The publisher remains neutral with regard to jurisdictional claims in published maps and institutional affiliations.

This Springer imprint is published by the registered company Springer Nature Switzerland AG
The registered company address is: Gewerbestrasse 11, 6330 Cham, Switzerland

If disposing of this product, please recycle the paper.

"The problem with the symmetric group is there's far too much symmetry and not enough group" – Gordon James.

Foreword

One of the most beautiful aspects of mathematics is the diversity of methods and points of view on a subject. Representation theory is no exception. Ever since Frobenius' discovery of the character table of a finite group in 1894, Representation Theory has influenced, and been influenced by, a remarkable array of fields. Weyl was led to ground-breaking work on the representation theory of compact Lie groups by his attempts to understand Einstein's work in general relativity, and Serre's famous book on representations of finite groups contains a part for chemists studying crystallography.

Modern representation theory continues to be influenced by neighbouring veins of mathematical development. Looking at the last decades one can identify two major themes. The first is the influence of algebro-geometric techniques, pioneered by Lusztig and others. Remarkably, calculations possible in sheaf theory provide the answer to questions which appeared intractable in representation theory.

The second major theme is the theory of categorification and higher representation theory. Here algebra provides answers about algebra, but the new algebras involved are different and unexpected. For example, Khovanov–Lauda–Rouquier algebras have told us new and exciting things about the group algebra of the symmetric group, one of representation theory's oldest and most fundamental objects. The "new" algebras that arise in categorification are often diagrammatic in nature, and it is these algebras which are the focus of this book.

This book builds up enough theory in order to explain my counter-examples to the expected bounds in Lusztig's conjecture. My path began in geometry, and slowly became more algebraic as I realised the computational power of diagrammatic algebra. After our work, several authors (including the author of this book!) have done beautiful work which explains that much of what we did can be recast in much simpler algebraic terms. It is this new language that is explained here. I hope that the reader is inspired by this treatment, and is able to find their own new ways of thinking about the fundamental objects of representation theory. Many tantalising mysteries remain!

Sydney, Australia *Geordie Williamson*
January 2025

Preface

Over the past twenty years, the field of representation theory has undergone a "diagrammatic revolution" whereby age-old questions have been reimagined using visual calculus borrowed from physics, knot theory, and categorical Lie theory. This diagrammatics has provided the intuition necessary to cut through some of the most famous conjectures and questions in modern algebra. This book is an attempt to make this diagrammatic insight available to everyone: we build-up from elementary group theory, to representation theory, Lie theory, and finally categorification. We blithely dispense with the usual formalism and encyclopaedic approach to these topics, in favour of a light-touch perspective which we hope will accelerate the reader's understanding. By the end of this book, the reader should be able to understand Williamson's counterexamples to arguably the most famous conjecture in the history of modular representation theory: Lusztig's conjecture.

The best way to illustrate our diagrammatic philosophy is with an example. For $i \geqslant 1$ we let f_i be the ith Fibonacci number in the sequence $1, 1, 2, 3, 5, 8, 13, 21, \ldots$ where $f_i = f_{i-1} + f_{i-2}$. We will prove the following equation

$$\sum_{1 \leqslant i \leqslant k} f_i^2 = f_k f_{k+1} \qquad (\dagger)$$

by replacing these products with richer diagrammatic structures. The proof is by induction. We first check that the base case holds for $k = 3$. One can easily check that

$$f_1^2 + f_2^2 + f_3^2 = 1^2 + 1^2 + 2^2 = 2 \times 3 = f_3 \times f_4$$

by hand, but we note that a more pleasing visualisation is given by interpreting the terms of this equation as the areas of certain squares and rectangles where the f_4 along the bottom of the rectangle comes from the sum of the widths of the two smaller squares of width f_2 and f_3 (where $f_4 = f_2 + f_3$ by the definition of the Fibonacci numbers). We continue by induction, adding an $(f_k \times f_k)$-square to the longest edge of the previous $(f_{k-1} \times f_k)$-rectangle at each stage. For example, when $k = 7$ we obtain

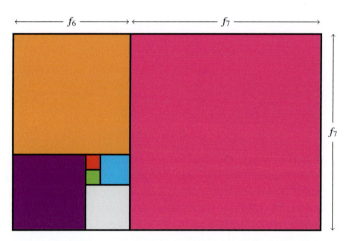

If one looks at this picture and equation (†) one sees that they are expressing the same thing (for $k = 7$). Not only does this diagrammatic approach provide the desired proof of (†), but the visualisation adds a deeper understanding and has its own physical beauty.

The above picture is often augmented with the "golden spiral", drawn so as to pass through the diagonally-opposite corners of every constituent square, in order. This is a logarithmic spiral whose growth factor is $\frac{1+\sqrt{5}}{2}$, the famous golden ratio. This gives a concrete visualisation of the the famous limiting behaviour or Fibonacci numbers:
$$\lim_{k \to \infty} (f_{k+1}/f_k) = \frac{1+\sqrt{5}}{2}.$$
The golden ratio also appears in Binet's closed formula for calculating the kth Fibonacci number:
$$f_k = \frac{(\frac{1+\sqrt{5}}{2})^k - (\frac{1-\sqrt{5}}{2})^k}{\sqrt{5}}.$$
This formula will make a surprise appearance later in the book in the counterexamples to Lusztig's conjecture. The importance of the Fibonacci numbers in our context is derived from our understanding (and lack thereof!) of their prime divisors. For instance, it is believed that there are infinitely many prime Fibonacci numbers, and yet at present we can only find fifty of them! On the other hand, it is known that each f_k (for $k > 12$) has at least one prime divisor p such that p does not divide any earlier Fibonacci number f_j for $j < k$, this theorem is over 100 years old and was an integral part of the disproof of Lusztig's conjecture.

Preface

What you will have learnt by the end of this book

A prime focus of twentieth century representation theory was the (doomed!) attempt to prove Lusztig's conjecture for p-modular representations. Loosely speaking, this conjecture stated that the most important problems in representation theory should admit combinatorial solutions via an efficient algorithm for calculating "p-Kazhdan–Lusztig polynomials" when p is any *reasonable* prime number. This was too optimistic... indeed, Geordie Williamson showed that calculating these polynomials subsumes the problem of determining the prime divisors of Fibonacci numbers — a notoriously difficult problem for which a combinatorial solution is almost certainly impossible. This was done by re-imagining Lusztig's conjecture in terms of calculating "p-torsion" within the "intersections forms" of the "anti-spherical Hecke category". By the end of this book, the reader should be intimately familiar with the inner workings of these Hecke categories and have a complete understanding of Williamson's examples of p-torsion. We expect the ambitious reader to be in a position, by the end of the book, where they would be able to conduct research and begin publishing papers on the topic of Hecke categories.

As $p \to \infty$ the situation outlined above simplifies dramatically and we encounter the classical Kazhdan–Lusztig polynomials; these are important *combinatorial objects* which can be calculated via a recursive algorithm. In this book, we gradually build-up towards the (initially, rather daunting) anti-spherical Hecke categories by first restricting our attention to the most combinatorially well-understood infinite family (those of maximal parabolics of finite symmetric groups). By restricting to this family first, we can provide complete and elementary proofs of all the major theorems in the literature (which would normally be way beyond what could possibly be covered in a single book) and then explain how the ideas generalise to arbitrary parabolic Coxeter systems (without proof!). We then present Williamson's examples of p-torsion, which we calculate in complete and glorious detail. Formulating these counterexamples undoubtably required startling insight, however the proof itself is actually an elementary exercise in diagrammatic algebra — this is why we chose this result as the centrepiece of this book.

Before discussing the details of the book in more detail, we take a moment to discuss the implications of Williamson's examples of p-torsion in modular representation theory. As mentioned above, an explicit understanding of the prime divisors of Fibonacci numbers (and therefore of p-Kazhdan–Lusztig polynomials in full generality) is a question whose answer appears utterly beyond the realms of hope. One could view this as a depressing conclusion to thirty years of research in modular Lie theory. However, we instead choose to marvel at the fact that modular representation theory is rich and varied enough to admit "generic solutions" in the form of beautiful combinatorial algorithms (as $p \to \infty$) and yet complete solutions are out of the question — this is not unlike the big questions in statistical mechanics and theoretical physics. Understanding this gulf between the combinatorial and non-combinatorial realms within p-Kazhdan–Lusztig theory will no doubt be one of the prime focuses of representation theory in the twenty-first century.

Part I: Groups

Part 1 of this book gives a clear and accessible introduction to group theory, emphasising fundamental concepts and building intuition through numerous examples. The first chapter slowly builds up to the definition of a group, first via the familiar idea of "symmetries" of a polygon, then via matrix groups, before axiomatising an abstract group and proving the Sudoku property. We also discuss the notation of "sameness" for groups and give examples of substructures. The second chapter is built around the concrete "real world" problem of Samuel Lloyd's famous 15 puzzle. This chapter introduces the symmetric group, both in terms of permutation matrices and string diagrams, and utilises the length function to resolve Samuel Lloyd's question. We then extend our strand diagrams to encompass hyper-octahedral groups and wreath products, thereby providing an introduction to Coxeter groups (these groups will also be important later in the book). In Chapter 3 we discuss one of the most universal ideas in algebra: that of simplicity. We begin by visualising how cosets partition a group into copies of smaller (sub)groups, and we prove Lagrange's theorem. We then introduce normal subgroups and quotient groups; we prove that any finite group can be broken down in a *canonical fashion* into constituent simple parts. This allows us to speak of the *composition factors* of a finite group, a notion that is essential in various guises throughout the book. Finally, in Chapter 4 we discuss the symmetry groups of the Platonic and Archimedean solids; of which, the permutahedron (and its symmetry group) will make another appearance later in the book when we define the *permutahedron relation* for the Hecke category. We discuss group actions on concrete objects such as dice, necklaces, and ID cards and play various counting games with these objects by way of the Burnside's counting theorem and the orbit-stabiliser theorem.

For those wanting to delve deeper into the topics of Part 1, we recommend the books of Mark Armstrong [Arm88] and James Humphreys [Hum90].

Part II: Algebras and Representation theory

In Part 2, we cast off the yoke of invertible symmetry and encounter our first algebraic structures that are not bound by group axioms. Chapter 5 provides a gradual development of the notion of a $\Bbbk$-algebra, guided by concrete examples and a preliminary examination of matrix algebras. We introduce the Temperley–Lieb, zig-zag, and binary Schur algebras, which will provide us with a concrete foundation for our exploration of representation theory in the following chapter.

We begin Chapter 6 with examples chosen to illustrate the manner in which we can glue together simple modules to obtain non-semisimple representations. We use the Temperley–Lieb algebra as a prototypical example of a *cellular algebra*, allowing us to formally axiomatise whilst encountering the ideas through concrete diagrammatics. We introduce the all-important *intersection forms* and use these to give a theoretical construction of simple modules of arbitrary cellular algebras using only elementary linear algebra. We put this cellular framework to use in concrete examples of small rank symmetric groups and binary Schur algebras. From this point onwards, we start to slowly incorporate the ideas of *highest weight theory* into our cellular algebras, axiomatising in stages as we go, and highlighting these axiomatic features in the illustrative cases of the zigzag and binary Schur algebras. We emphasise the powerful grading and idempotent tricks for understanding the Gram matrices of intersection forms (which will be pivotal later on in the book, when we come to disproving Lusztig's conjecture). We introduce Alperin diagrams in order to give the reader a visual way of understanding submodule structures. The main result of Chapter 6 is the determination of the simple modules and the full submodule structures of cell modules for the binary Schur algebras over arbitrary fields — we do this in terms of modular colourings on Pascal's triangle. We conclude the chapter with some technical points regarding truncations and quotients which will be needed later in the book.

All of the tools and ideas of Part 2 will be integral to the disproof of Lusztig's conjecture and therefore this chapter serves two audiences: those who are entirely new to representation theory, as well as experts who have perhaps not used these tools before (for example, this is very different to the Brauer-character-theoretic approach often favoured by finite group theorists).

The style of this chapter was inspired by the books of Gordon James [Jam78] and Andrew Mathas [Mat99], which we heartily recommend for those who wish to delve further into the representation theoretic and group theoretic ideas of this book. In particular, we have tried to emulate James' elementary style by suppressing any unnecessary machinery and using the bare minimum of abstraction. Our idempotent-heavy approach to representation theory was inspired by Stephen Donkin's treatment of quasi-hereditary algebras [Don98, Appendix] and Paul Martin's approach to diagrammatic algebras arising in statistical mechanics (see, for example [MW00b]).

Part III: Combinatorics

The purpose of Part 3 of this book is to motivate the study of Kazhdan–Lusztig polynomials solely through their beautiful combinatorial properties. In Chapter 7, we approach the Kazhdan–Lusztig polynomials side-on: we start by considering the quantum binomial coefficients, proving that they satisfy a beautiful palindromic symmetry by interpreting them as the rank generating functions of posets of partitions inside rectangles $\mathscr{P}_{m,n}$. We discuss how this palindromic property fails for subposets of $\mathscr{P}_{m,n}$. Our search for a "correction factor" for this palindromic-failure leads us to define the "Kazhdan–Lusztig polynomials of $\mathscr{P}_{m,n}$" in terms of path-counting polynomials within this poset. We then unveil that $\mathscr{P}_{m,n}$ is in fact the Bruhat graph for the maximal parabolic subgroups of finite symmetric groups, which was our hidden motivation all along. We conclude this chapter by calculating these Kazhdan–Lusztig polynomials explicitly and by proving that they do, indeed, satisfy beautiful palindromy and unimodality properties.

In Chapter 8 we generalise the path-counting definition of Kazhdan–Lusztig polynomials to arbitrary Bruhat graphs. In our first, naive, attempt at generalisation we quickly come up against the well-definedness problem; we overcome this by a simple matrix factorisation. We further motivate our study of Kazhdan–Lusztig polynomials through the striking combinatorial invariance conjecture and take a quick tour through some of the other beautiful combinatorial properties they exhibit.

With hindsight, one can see that the two chapters of Part 3 are chosen so as to factorise the complexity of defining/calculating Kazhdan–Lusztig polynomials into two stages: in the first chapter we consider Bruhat graphs which contain only lines (2-gons) and squares (4-gons); in the second chapter we allow arbitrary polygons in our Bruhat graphs. This will guide our treatment of the Hecke categories in Part 4.

For further discussion of the ideas of Part 3, we again refer to the seminal book of James Humphreys [Hum90] and the famous paper of Wolfgang Soergel [Soe97]. In particular, like many before us in the literature we have chosen not to prove the well-definedness of Kazhdan–Lusztig polynomials (which requires the theory of canonical bases, not developed here).

Part IV: Categorification

In Part 5 we define the anti-spherical Hecke categories, $\mathcal{H}_{(W,P)}$, for parabolic Coxeter systems. We restrict our attention to those categories whose underlying Bruhat graphs contain only squares and hexagons. Again we make this complex construction more tractable by factorising the material into two stages of difficulty, as follows.

In Chapter 9 we consider the $\mathcal{H}_{(W,P)}$ whose underlying Bruhat graphs contain only lines (2-gons) and squares (4-gons). We first introduce the monochrome generators and relations and provide the basis and multiplication table of $\mathcal{H}_W$ for $W = \mathfrak{S}_2$; we construct an isomorphism from this algebra to a matrix algebra we encountered in Chapter 6. We then introduce the multi-colour relations for $\mathcal{H}_{(W,P)}$ and give a "how-to guide" for manipulating diagrams, emphasising how to obtain the trapezoidal shapes that we desire. We then take a short detour in order to prove that the simplest of the algebras $\mathcal{H}_{(W,P)}$ are actually isomorphic to the zigzag algebras from Chapter 6 (thus grounding the reader in the familiar). Finally, we conclude the chapter by constructing the "light leaves" cellular basis of $\mathcal{H}_{(W,P)}$ and providing a complete proof. This allows us to define and compute the $(p\text{-})$Kazhdan–Lusztig polynomials, determine the radical layers of the cell modules, and discuss the Elias–Williamson categorification theorem in this context.

In Chapter 10 we gently introduce the additional generators and relations required for defining the algebras $\mathcal{H}_{(W,P)}$ in the more complicated case that the Bruhat graph of $^P W$ contains hexagons. We experiment with the newly developed braid generators and provide concrete examples of intersection forms for $\mathcal{H}_{(W,P)}$ which do depend on the characteristic of the field. The bulk of Chapter 12 is dedicated to the construction and proof of an infinite family of intersection forms, indexed by $k \in \mathbb{Z}_{\geq 0}$, whose rank is zero if and only if the characteristic of the field is a prime divisor of the

Preface

kth Fibonacci number (and this is done in such a way as to contradict Lusztig's conjecture, phrased in terms of the Hecke category). We conclude the chapter by discussing "what happens next", in the wake of these explosive examples of p-torsion; presenting He–Williamson's hope for finite $\mathcal{H}_W$, and work of Lusztig–Williamson and Hazi for affine Weyl groups.

We strongly recommend Elias–Makisumi–Thiel–Williamson's book [EMTW20] — which is the only other existing book on the topic of Hecke categories. Part 4 of this book differs from their treatment of this topic in both the the material covered (for example, they do not discuss "torsion explosion" whilst we do not focus on the Kazhdan–Lusztig positivity conjecture) and the assumed background knowledge of the reader.

Part V: Group theory versus diagrammatic algebra

Part 5 of this book might well be of interest only to those who are intent on pursuing new research in this field. The purpose of this chapter is to rephrase Lusztig's conjecture (originally posed in the context of group theory) entirely within the language of Hecke categories. We do this "after the fact" because it preserves the flow of the book and allows us to first understand everything in terms of Hecke categories. This section is more of a high-level review of material and, as such, requires that we be less self-contained.

Chapter 11 briskly develops the representations of symmetric groups from the point-of-view of Jucys–Murphy elements, cellularity, and the semi-normal form (we will reap the benefits of this extra work in Chapter 12). We then introduce the Schur algebras and construct their cellular bases. This allows us to state Lusztig's and Andersen's conjectures in their original, group-theoretic, settings. Chapter 12 begins on the combinatorial level, where we introduce the LLT algorithm in terms of graded tableaux combinatorics. We strongly emphasise the connections between the LLT theory and Kazhdan–Lusztig theory. We illustrate the back-and-forth between graded tableaux combinatorics and that of paths in Bruhat graphs, thus equating the Kazhdan–Lusztig and LLT polynomials. We then lift this graded tableaux combinatorics to a structural level by introducing the quiver Hecke algebra of the symmetric group and constructing its graded cellular basis. Chapter 13 bridges the gap between the original group-theoretic statements of the Lusztig and Andersen conjectures and our reformulations of these conjectures within the Hecke category. This is done by establishing isomorphisms between $\mathcal{H}_{(W,P)}$ and the quiver Hecke algebra of the symmetric group. In this way, we see that the modular simple representations of symmetric groups can be calculated in terms of p-Kazhdan–Lusztig polynomials for almost tautological reasons.

Rings and things

Throughout this book, $\Bbbk$ will always be a commutative integral domain. If this means nothing to you, the only examples we will consider are the following. We have the integers
$$\mathbb{Z} = \{\ldots, -2, -1, 0, 1, 2, \ldots\}$$
together with usual addition and multiplication. The integer-valued Laurent polynomials
$$\mathbb{Z}[q, q^{-1}] = \{\cdots + a_{-2}q^{-2} + a_{-1}q^{-1} + a_0 + a_1 q + a_2 q^2 + \cdots \mid a_i \in \mathbb{Z} \text{ for } i \in \mathbb{Z}\}\}$$
(with only finitely many non-zero coefficients) with the usual addition and multiplication of polynomials. The rational field $\mathbb{Q} = \{a/b \mid a, b \in \mathbb{Z}, b \neq 0\}$ with the usual addition and multiplication of rational numbers, and the real and complex fields $\mathbb{R}$ and $\mathbb{C}$. Finally, a starring role will be played by the modular field of p elements $\mathbb{F}_p = \{0, 1, 2, \ldots, p-1\}$ with addition and multiplication modulo the prime p, for example the addition and multiplication tables for $\mathbb{F}_5$ are as follows,

+	0	1	2	3	4
0	0	1	2	3	4
1	1	2	3	4	0
2	2	3	4	0	1
3	3	4	0	1	2
4	4	0	1	2	3

×	0	1	2	3	4
0	0	0	0	0	0
1	0	1	2	3	4
2	0	2	4	1	3
3	0	3	1	4	2
4	0	4	3	2	1

(for example $3 + 4 = 7 = 5 + 2 \equiv 2$ modulo 5).

Categorification without categories

This is an aside for the expert reader. In this book we will discuss the category algebra of the Hecke category and truncate this to be finite dimensional (by only considering "morphisms between reduced words"). This allows us to keep the amount of background formalism to a minimum, instead allowing the diagrammatics to do the heavy lifting for us. For example, we will not introduce "monoidal categories", instead we will simply speak of "horizontal concatenation of diagrams" (which we denote ⊗). For more details on this back-and-forth between a category and its category algebra we refer to [BHN22, Remark 1.1] and [BS17, Section 2.2 & Remark 2.3].

As a further aside for the expert reader, we remark that we will entirely suppress any mention of the realisation underlying our diagrammatic algebras $\mathcal{H}_{(W,P)}$. For those in the know, we state here that we will use the (dual) geometric realisation.

Too much symmetry and not enough group?

We now take a moment to reflect on the quotation of Gordon James, above. For the purposes of modular representation theory, our philosophy is to forget entirely about the group-theoretic structure and focus solely on hidden, richer "categorical" structures which afford us new tools such as gradings and efficient diagrammatic presentations. This viewpoint sees James' over-abundance of "symmetry" being explained via its origins in categorical Lie theory, rather than finite group theory. Indeed over the past 20 years, many of the most spectacular results concerning the structure of the symmetric group have been proven using categorical ideas [CR08, KL09, Rou, BK09, Wil17, CMT17, EK18, GKM22].

Further reading

Diagrammatic algebra is now a sprawling area of mathematics, the boundaries of which are quite fuzzy. We certainly do not attempt a comprehensive treatment of this theory here. Interesting diagrammatic constructions which are not discussed here (for purposes of brevity), but which are very enjoyable include: the Brauer algebras, first defined in [Bra37] but whose structure remained very mysterious for over seventy years [CDM09b, CDM09a, CDM11, Mar15, DM17]; the walled, cyclotomic, and periplectic Brauer algebras defined in [Tur89, Koi89, HO01, Moo03] and studied in [CDDM08, CD11, BC14, BCD13, BDEA$^+$20, BCD19, Cou18]; the partition algebras of statistical mechanics [Mar91] whose representation theory was developed in [Mar96, BDK15, BEG18a, BDE19]; the spider/web algebras of [Kup96]; and the modular and super Schur–Weyl dualities involving these diagram algebras [DH09, DDH08, ES16c, BDM22a, BDM22b, CKM14]. From a more categorical perspective, we mention here the Khovanov arc algebras of [Kho00, Str09] which have been of much interest in representation theory [BS10, BS11b, BS12a, BS12b, BW] and which continue to have spectacular applications in knot theory [Ras10, KM11, Pic20]; Webster's beautiful theory of weighted quiver Hecke algebras [Web19, MT24] with its connections to knot theory [Web17a], Coulomb branches [LW23], and Cherednik algebras and their Schur–Weyl dualities [Web20, Web17b, BCS17, BS18, BC18, BS19]; and the theory of Kac–Moody and Heisenberg categorification [CR08, Rou, KL09, LS12, BSW20b, BSW21, BSW22, BSW20a].

York, England
January 2025

Chris Bowman

Acknowledgements

First and foremost, this book owes a special debt of gratitude to Maud De Visscher — I have come to understand much of the material of this book over our past decade of collaborating and talking about mathematics together. I would also like to thank Amit Hazi for teaching me what the Hecke category *actually is*, as well as Christine Bessenrodt, Joe Chuang, Anton Cox, Stephen Donkin, Stephen Doty, Harry Geranios, Paul Martin, Stuart Martin, Andrew Mathas, Rob Muth, Emily Norton, Liron Speyer, David Stewart, Catharina Stroppel, Jay Taylor, Ben Webster, and Alexander Woo for teaching me various bits and pieces of representation theory and combinatorics which have made it into this book in one form or another. Finally, I would like to thank Ben Elias, David Kenepp, Cailan Li, Evuilynn Nguyen, Alexis Langlois-Rémillard, Dani Tubbenhauer, Geordie Williamson, Keke Zhang, and Regina Zhou for suggestions and comments which no doubt improved the exposition of the book immeasurably.

Contents

Part I Groups

1 Symmetries 3
   1.1 Size isn't everything 3
   1.2 Matrix groups 6
   1.3 Abstract groups and the Sudoku property 11
   1.4 Presenting groups 14
   1.5 Isomorphisms 15
   1.6 Subgroups and direct products of groups 17

2 Coxeter groups and the 15-puzzle 21
   2.1 The symmetric group 21
   2.2 The length function and the alternating group 26
   2.3 The 15-puzzle 28
   2.4 The hyper-octahedral groups 34
   2.5 Groups of decorated strand diagrams 35
   2.6 Coxeter groups 37

3 Composition series 39
   3.1 Cosets and Lagrange's theorem 39
   3.2 Simple groups and normal subgroups 43
   3.3 Homomorphisms 47
   3.4 The Jordan–Hölder Theorem 48

4 Platonic and Archimedean solids and special orthogonal groups 55
   4.1 The platonic solids 56
   4.2 Group actions 57

4.3	The symmetry group of the tetrahedron	62
4.4	The symmetry group of the cube	65
4.5	The symmetry group of the dodecahedron	67
4.6	The classification of 3-dimensional rotational symmetry groups	69
4.7	Archimedean solids and their symmetry groups	74

Part II Algebras and representation theory

5 Non-invertible symmetry ... 81
- 5.1 Algebras ... 81
- 5.2 The Temperley–Lieb algebras ... 84
- 5.3 Idempotents and things ... 87
- 5.4 Gradings and zig-zag algebras ... 90
- 5.5 Oriented Temperley–Lieb algebras ... 94
- 5.6 The binary Schur algebra ... 96
- 5.7 Hidden gradings on Schur algebras in positive characteristic ... 103

6 Representation theory ... 105
- 6.1 Examples of simple modules and representations ... 105
- 6.2 Temperley–Lieb algebras as the prototypical cellular algebras ... 115
- 6.3 The smallest non-semisimple Schur algebra ... 123
- 6.4 The symmetric group on 3 letters ... 126
- 6.5 Semisimple filtrations of non-semisimple modules ... 129
- 6.6 Weighted cellular algebras and gradings ... 131
- 6.7 The grading and idempotent tricks for intersection forms ... 134
- 6.8 Simple modules of zig-zag algebras ... 135
- 6.9 Simple modules of the binary Schur algebra ... 137
- 6.10 Decomposition numbers via highest weight theory ... 142
- 6.11 Alperin diagrams and submodule structures ... 144
- 6.12 Truncations, quotients, and saturation. ... 150

Part III Combinatorics

7 Catalan combinatorics within Kazhdan–Lusztig theory ... 161
- 7.1 Quantum binomial coefficients ... 161
- 7.2 The poset of partitions in a rectangle ... 163
- 7.3 Kazhdan–Lusztig polynomials ... 166
- 7.4 Tile-partitions, weights, and Bruhat graphs ... 169
- 7.5 Oriented Temperley–Lieb combinatorics ... 172

Contents   xxiii

8   General Kazhdan–Lusztig theory 181
    8.1   Weak Bruhat graphs of parabolic Coxeter systems 181
    8.2   Kazhdan–Lusztig polynomials in full generality 184
    8.3   The strong Bruhat order and combinatorial-invariance 191
    8.4   The infinite dihedral group 198
    8.5   Kazhdan–Lusztig combinatorics? 201

Part IV   Categorification

9   The diagrammatic algebra for $\mathfrak{S}_m \times \mathfrak{S}_n \leqslant \mathfrak{S}_{m+n}$ 209
    9.1   From paths to diagrammatic algebras 210
    9.2   The one colour relations 214
    9.3   The multi-colour relations for $\mathfrak{S}_m \times \mathfrak{S}_n \leqslant \mathfrak{S}_{m+n}$ 219
    9.4   How to manipulate diagrams 220
    9.5   Return of the zig-zag algebras 228
    9.6   Cellularity of $\mathcal{H}_{(W,P)}$ for $(W, P) = (\mathfrak{S}_{m+n}, \mathfrak{S}_m \times \mathfrak{S}_n)$ 229
    9.7   The categorification theorem 234

10   Lusztig's conjecture in the diagrammatic algebra $\mathcal{H}_{(W,P)}$ 239
    10.1   From paths to diagrammatic algebras (again) 240
    10.2   Multi-colour relations 244
    10.3   Fun with braids 247
    10.4   Light leaves and the Kazhdan–Lusztig positivity conjecture 250
    10.5   Calculating intersection forms 252
    10.6   Counterexamples to Soergel's conjecture in positive characteristic .. 261
         10.6.1   Picking pairs of permutations 261
         10.6.2   Fibonacci numbers as values of intersection forms 268
         10.6.3   The base case of the proof 271
         10.6.4   An observation and inductive reformulation 274
         10.6.5   Counterexamples to "Soergel's conjecture" 279
    10.7   Affinization: from Soergel's to Lusztig's conjecture 281
    10.8   What might the future hold? 284

Part V   Group theory versus diagrammatic algebra

11   Reformulating Lusztig's and Andersen's conjectures 295
    11.1   The symmetric group and its representation theory 295
    11.2   Generalised Temperley–Lieb algebras 301
    11.3   Schur algebras 303
    11.4   Lusztig's and Andersen's conjectures 306

12 Hidden gradings on symmetric groups ... 313
12.1 The tableaux-theoretic LLT algorithm ... 313
12.2 Bringing together LLT and Kazhdan–Lusztig polynomials ... 325
12.3 The quiver Hecke algebra ... 329
12.4 The Brundan–Kleshchev isomorphism theorem ... 337
12.5 The Hu–Mathas cellular basis ... 338
12.6 The quiver Temperley–Lieb algebra ... 343

13 The p-Kazhdan–Lusztig theory for Temperley–Lieb algebras ... 345
13.1 Graded path combinatorics of the Temperley–Lieb algebra ... 345
13.2 Hyperplane-coloured residue sequences ... 349
13.3 Recolouring the quiver Temperley–Lieb algebra ... 350
13.4 The isomorphism theorem ... 354
    13.4.1 Fork-annihilation in the KLR algebra ... 356
    13.4.2 Fork-spot contraction in the KLR algebra ... 357
    13.4.3 Isotopy in the KLR algebra ... 358
    13.4.4 The monochrome barbell relation in the KLR algebra ... 359
    13.4.5 Bijectivity and bases ... 361

References ... 363

Index ... 375

Part I
Groups

> *"The mathematician has, above all things, an eye for symmetry."*
>
> James Clerk Maxwell

Group theory is the mathematical study of "symmetry". The notion of "symmetry" is all-pervasive in the natural sciences — thus the pure mathematical questions we ask about groups have manifold connections to physics, quantum theory, chemistry, cryptography... For example, The Standard Model of particle physics uses group theory to describe the fundamental forces (electromagnetism, weak and strong nuclear force — excluding gravity); this model predicted the existence of the Higgs boson long before it was physically observed.

Group theory is at the heart of modern algebra, and it's where many interesting questions both begin and end. No self-respecting introduction to group theory can ignore the Classification Theorem of Finite Simple Groups. This isn't just a theorem; it is a magnum opus, the result of a mathematical relay race spanning decades, involving hundreds of mathematicians and thousands of pages. The Classification provides a complete list of all the simple "atoms" from which one can construct a general finite group. It is a true testament to collaborative effort and a cornerstone of modern mathematics.

This book is less concerned with "all possible groups" and more focussed on our favourite group: the symmetric group of all permutations of n objects. Gordon James lamented that the symmetric group has "too much symmetry and not enough group", and he was not wrong — Cayley's theorem tells us that every finite group appears as a subgroup of some symmetric group. Of course, this just tells us that trying to study symmetric groups via their subgroup structure is essentially meaningless. In Part 1 of this book we study symmetric groups, and more generally Coxeter groups and their subgroups, in terms of their actions on discrete combinatorial objects such as puzzles, necklaces, ID cards, and Platonic solids.

Conjectures of Lusztig, James, and Andersen laid the foundation for a major mathematical programme dedicated to revealing the structure and symmetries of the symmetric groups, via a mixture of representation theory, combinatorics, and categorification. These conjectures forms the guiding principle of this book.

Chapter 1
Symmetries

Groups provide the mathematical formalism for understanding invertible symmetries. In this chapter we approach this idea through a series of steps of abstraction: we begin by considering concrete examples of 2 and 3 dimensional symmetries of polygons, we build up to matrix groups, and then finally to the abstract formalism of a group. We also consider what it means for two groups to be "the same" and when and how one group can "sit inside" another.

1.1 Size isn't everything

One of the principal ideas of this section is that an object cannot be effectively encapsulated by considering only the *set* of symmetries it exhibits, a far better way of understanding an object is to study the extra structure that comes from taking into account the *composition* of these symmetries.

We first consider the planar symmetries of the triangle. Consider the rotation of the real plane through 120° clockwise, which permutes the three vertices of the equilateral triangle as follows

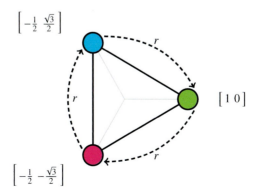

© The Author(s), under exclusive license to Springer Nature Switzerland AG 2025
C. Bowman, *Diagrammatic Algebra*, Universitext,
https://doi.org/10.1007/978-3-031-88801-4_1

We have that $r^3 = r * r * r$ is the rotation through $3 \times 120° = 360°$ and so r^3 takes us back to where we started. Thus there are three rotational symmetries of the triangle $\{\text{id}, r, r^2\}$ which can be represented by the matrices:

$$\text{id} = \begin{bmatrix} 1 & 0 \\ 0 & 1 \end{bmatrix} \quad r = \begin{bmatrix} \cos(120) & -\sin(120) \\ \sin(120) & \cos(120) \end{bmatrix} \quad r^2 = \begin{bmatrix} \cos(240) & -\sin(240) \\ \sin(240) & \cos(240) \end{bmatrix}$$

where we note that $\cos(360) = 1$ and $\pm \sin(360) = 0$ (and so $r^3 = e$, as required). Applying these symmetries to the above triangle (or equivalently, right-multiplying the (x, y)-coordinates of the vertices by the rotational matrices) we obtain the following distinct triangles,

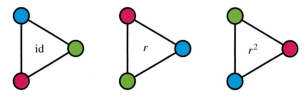

The composition of these symmetries (or if you prefer, the multiplication rule for these matrices) is completely encapsulated in the following table

$*$	id	r	r^2
id	id	r	r^2
r	r	r^2	id
r^2	r^2	id	r

Notice that this table satisfies a Sudoku-like property: every symmetry appears precisely once in each row and each column. Moreover, this matrix is symmetric with respect to reflection through the north-west to south-easterly diagonal.

The 2-dimensional rotational symmetries of a regular n-sided shape (also known as an n-gon) are the same as the 3-dimensional symmetries of a n-gon based pyramid (for $n > 3$). For example, there are 6 symmetries of the hexagonal pyramid, $\{\text{id}, r, r^2, r^3, r^4, r^5\}$, where r is the rotation by $60°$ through the xy-plane and the composition of these symmetries is recorded in the table

$*$	id	r	r^2	r^3	r^4	r^5
id	id	r	r^2	r^3	r^4	r^5
r	r	r^2	r^3	r^4	r^5	id
r^2	r^2	r^3	r^4	r^5	id	r
r^3	r^3	r^4	r^5	id	r	r^2
r^4	r^4	r^5	id	r	r^2	r^3
r^5	r^5	id	r	r^2	r^3	r^4

(1.1.1)

1.1 Size isn't everything

We now consider the 3-dimensional symmetries of the triangle. We have already considered the rotation, r, of the triangle through 120° clockwise within a plane. However, in 3-dimensional space we have the room to rotate the piece of paper on which we drew the triangle in the first place through 180° (of if you prefer to work in 2-dimensions, we can consider the 2-dimensional reflection of the triangle through a mirror). Either way, we picture this "flip" as follows:

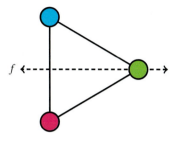

These symmetries ("rotate" and "flip") of the triangle can be represented by the matrices

$$r = \begin{bmatrix} \cos(120) & -\sin(120) \\ \sin(120) & \cos(120) \end{bmatrix} \qquad f = \begin{bmatrix} 1 & 0 \\ 0 & -1 \end{bmatrix}.$$

We let fr^2 be the element obtained by first flipping and then rotating the triangle (clockwise 240°). In other words we read from left-to-right as follows

There are 6 symmetries of the triangle (as there are 3 vertices and therefore $3 \times 2 \times 1$ choices for permuting these vertices) and they can be pictured as follows:

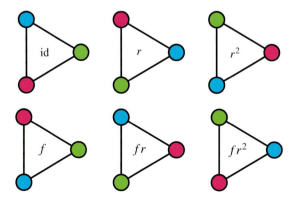

These are also the 6 3-dimensional symmetries of a triangular prism.

The composition rule for these symmetries is recorded in the following table:

$$\begin{array}{c|cccccc} * & \mathrm{id} & r & r^2 & f & fr & fr^2 \\ \hline \mathrm{id} & \mathrm{id} & r & r^2 & f & fr & fr^2 \\ r & r & r^2 & \mathrm{id} & fr & fr^2 & f \\ r^2 & r^2 & \mathrm{id} & r & fr^2 & f & fr \\ f & f & fr^2 & fr & \mathrm{id} & r^2 & r \\ fr & fr & f & fr^2 & r & \mathrm{id} & r^2 \\ fr^2 & fr^2 & fr & f & r^2 & r & \mathrm{id} \end{array} \qquad (1.1.2)$$

We have now seen two groups of size 6: the symmetries of the triangular prism form one such group and the symmetries of the hexagonal pyramid form another such group. Looking at the *number* of symmetries of these two shapes is not enough to distinguish them. However, the composition tables in (1.1.1) and (1.1.2) are completely different. In other words, we can distinguish between the algebraic constructions underlying these shapes, but only if we look at the *extra structure* offered by their composition rules.

This is a recurring theme in this book: we consider richer structures (in this case we are considering the composition of the symmetries, rather than merely the number of such symmetries) in order to better understand the objects that interest us.

1.2 Matrix groups

Let $\Bbbk$ be a field, for example the real or complex numbers. We set $\mathrm{GL}_n(\Bbbk)$ to be the set of all $(n \times n)$-matrices with entries from $\Bbbk$ and non-zero determinant. We let A^{-1} denote the inverse of the matrix $A \in \mathrm{GL}_n(\Bbbk)$. We let $*$ denote the usual matrix multiplication. For example if $n = 2$ we have that

$$\begin{bmatrix} a & b \\ c & d \end{bmatrix} * \begin{bmatrix} p & q \\ r & s \end{bmatrix} = \begin{bmatrix} ap + br & aq + bs \\ cp + dr & cq + ds \end{bmatrix} \qquad (1.2.1)$$

and (providing $ad - bc \neq 0$) we have that

$$\begin{bmatrix} a & b \\ c & d \end{bmatrix}^{-1} = \frac{1}{ad - bc} \begin{bmatrix} d & -b \\ -c & a \end{bmatrix} \qquad (1.2.2)$$

We set id_n to be the $(n \times n)$-identity matrix, that is the matrix with all entries equal to zero, except along the northwest-to-southeasterly diagonal, for which all entries are equal to 1. For example,

$$\mathrm{id}_2 = \begin{bmatrix} 1 & 0 \\ 0 & 1 \end{bmatrix}.$$

1.2 Matrix groups

Definition 1.2.3. Let $G \subseteq \mathrm{GL}_n(\Bbbk)$ be a collection of matrices. We say that G is a **matrix group** if it satisfies the axioms:

- $\mathrm{id}_n \in G$;
- $a, b \in G$ implies $a * b \in G$;
- $a \in G$ implies $a^{-1} \in G$.

We refer to these as the **identity**, **closure**, and **inverse** axioms, respectively.

Example 1.2.4. The set, $\mathrm{GL}_2(\mathbb{R})$, of (2×2)-invertible matrices with real entries forms a matrix group. To see this, note that we already verified the closure axiom in (1.2.1), the inverse axiom in (1.2.2), and it is clear that $\mathrm{id}_2 \in \mathrm{GL}_2(\mathbb{R})$.

Example 1.2.5. The set, $\mathrm{U}_2(\mathbb{R})$, of (2×2)-upper uni-triangular matrices (that is, the matrices with ones on the diagonal, a zero entry below the diagonal, and an arbitrary real number above the diagonal) forms a matrix group. To check the identity axiom, we simply note that the identity matrix is upper uni-triangular. We next check the closure axiom; for $a, b \in \mathbb{R}$ we have that

$$\begin{bmatrix} 1 & a \\ 0 & 1 \end{bmatrix} \begin{bmatrix} 1 & b \\ 0 & 1 \end{bmatrix} = \begin{bmatrix} 1 & a+b \\ 0 & 1 \end{bmatrix}$$

and we note that $a + b \in \mathbb{R}$ (and so the product does indeed belong to $\mathrm{U}_2(\mathbb{R})$, as required). For any $a \in \mathbb{R}$, we note that

$$\begin{bmatrix} 1 & a \\ 0 & 1 \end{bmatrix} \begin{bmatrix} 1 & -a \\ 0 & 1 \end{bmatrix} = \begin{bmatrix} 1 & a-a \\ 0 & 1 \end{bmatrix} = \begin{bmatrix} 1 & 0 \\ 0 & 1 \end{bmatrix}$$

and so any matrix in $\mathrm{U}_2(\mathbb{R})$ has inverse given by

$$\begin{bmatrix} 1 & a \\ 0 & 1 \end{bmatrix}^{-1} = \begin{bmatrix} 1 & -a \\ 0 & 1 \end{bmatrix}$$

where we note that $-a \in \mathbb{R}$ (and so the inverse matrix does indeed belong to $\mathrm{U}_2(\mathbb{R})$, as required).

Example 1.2.6. We set $C_\infty = \{\cos(\vartheta) + i\sin(\vartheta) \mid 0 \leq \vartheta < 2\pi\}$ to be the set of all complex numbers of modulus 1. These complex numbers form a matrix group (of (1×1)-matrices!) under the usual complex multiplication. To see this, note that expanding out the brackets for $(\cos(\vartheta) + i\sin(\vartheta))(\cos(\psi) + i\sin(\psi))$ we obtain

$$\cos(\vartheta)\cos(\psi) + i\cos(\vartheta)\sin(\psi) + i\sin(\vartheta)\cos(\psi) - \sin(\vartheta)\sin(\psi)$$

and hence using double-angle formulae we obtain

$$\cos(\vartheta + \psi) + i\sin(\vartheta + \psi) \in C_\infty$$

and so the closure axiom holds. The inverse axiom can be deduced using the same double-angle formulae. The identity axiom is trivial.

Exercise 1.2.7. We let $SL_n(\Bbbk) \subseteq GL_n(\Bbbk)$ be the subset of matrices with determinant equal to 1. Verify that $SL_n(\Bbbk)$ forms a matrix group.

Given $n \in \mathbb{Z}_{\geqslant 2}$, we consider the 2-dimensional rotational symmetries of a regular n-gon. For $1 \leqslant k \leqslant n$, we let r^k be the rotation given by the (2×2)-matrix

$$r^k = \begin{bmatrix} \cos(k\vartheta) & -\sin(k\vartheta) \\ \sin(k\vartheta) & \cos(k\vartheta) \end{bmatrix}$$

where ϑ is equal to $360/n$ degrees or equivalently $2\pi/n$ radians. In order to see that this is, indeed, the claimed rotation we let

$$x = \begin{bmatrix} 1 & 0 \end{bmatrix} \qquad y = \begin{bmatrix} 0 & 1 \end{bmatrix}$$

be the basis of the real plane $\mathbb{R}^2$. We have that

$$(x)r = \begin{bmatrix} 1 & 0 \end{bmatrix} \begin{bmatrix} \cos(\vartheta) & -\sin(\vartheta) \\ \sin(\vartheta) & \cos(\vartheta) \end{bmatrix} = \begin{bmatrix} \cos(\vartheta) & -\sin(\vartheta) \end{bmatrix}$$

$$(y)r = \begin{bmatrix} 0 & 1 \end{bmatrix} \begin{bmatrix} \cos(\vartheta) & -\sin(\vartheta) \\ \sin(\vartheta) & \cos(\vartheta) \end{bmatrix} = \begin{bmatrix} \sin(\vartheta) & \cos(\vartheta) \end{bmatrix}$$

which (with only some elementary trigonometry) can be pictured as follows

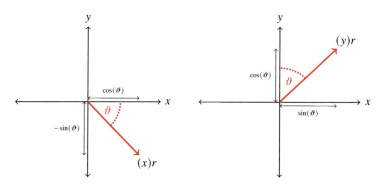

Multiplying the matrix r by itself n times, we obtain the identity matrix (since $\pm\sin(2\pi) = 0$ and $\cos(2\pi) = 1$), or equivalently, rotating an arrow through a full 2π radians the arrow comes back to its starting position. Therefore multiplying the matrix r by itself $n + 1$ times, we obtain the original matrix r. Carrying on in this manner, we realise that there are precisely n distinct matrices which can be obtained by repeated applications of this rotation, namely

$$C_n = \{\mathrm{id}_2, r, r^2, \ldots, r^{n-1}\} \tag{1.2.8}$$

and the composition of these symmetries is encoded in the rule

$$r^i \times r^j = r^{i+j \,(\mathrm{mod}\, n)}. \tag{1.2.9}$$

1.2 Matrix groups

In particular, the set of rotation matrices in (1.2.8) satisfies the closure axiom of a matrix group. For example, we now consider the effect of the rotation pictured above through $\pi/4$. We label the vertices of an "initial octagon" in a natural fashion, starting with the point $(1, 0) \in \mathbb{R}^2$. We now picture the initial octagon and its rotation to illustrate the above:

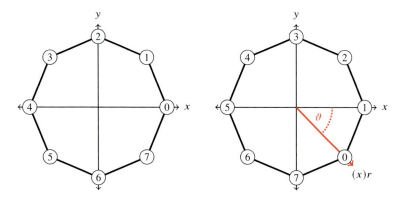

In the above we have explicitly noted that the identity matrix is simply the trivial rotation $\mathrm{id}_2 = r^n$. Finally, it is not hard to see that every element $r^k \in C_n$ has an inverse element, namely $r^{n-k} \in C_n$. (Simply multiply the matrices together to check this!) Thus we conclude that the 2-dimensional rotations of an n-gon, C_n, do indeed form a matrix group.

Given $n \in \mathbb{Z}_{\geqslant 2}$, we consider the 3-dimensional symmetries of a regular n-gon (flipping over the piece of paper the n-gon is drawn on). Or equivalently, the 2-dimensional *rotational* and *reflection* symmetries of a regular n-gon. In order to do this, we first fix the "initial n-gon" by labelling the vertices $\{v_0, \ldots, v_{2(k-1)}\}$ clockwise from $(1, 0)$ as follows:

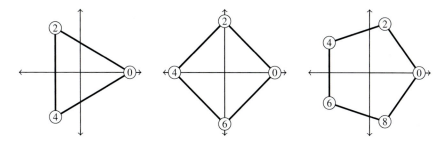

and we set e_{2k+1} to be the edge connecting v_{2k} and v_{2k+2} so that $\{e_1, e_3, \ldots, e_{2k-1}\}$ is a labelling of all the edges of the n-gon.

With this labelling in place we are able to define the reflection f_{2k} to be the line through the origin and the vertex labelled by $2k$. We define f_{2k+1} to be the reflection through the line that passes through the origin and the midpoint of e_{2k+1}th edge. We have that $f_k = f_{n+k}$ for $0 \leqslant k < n$ and these provide a complete list of n distinct flips. (As any flip must fix the origin and an edge or a vertex.) For example, we now picture the reflections f_2 and f_1 in the above examples in red and blue respectively, as follows

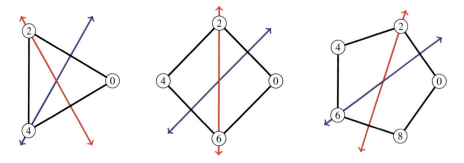

For $0 \leq k < 2n$, we have that

$$f_k = \begin{bmatrix} \cos(k\vartheta) & \sin(k\vartheta) \\ \sin(k\vartheta) & -\cos(k\vartheta) \end{bmatrix}$$

where ϑ is equal to π/n radians. We notice that this gives us a total of $2n$ distinct symmetries

$$D_{2n} = \{\mathrm{id}_2, r, r^2, \ldots, r^{n-1}\} \cup \{f_0, f_1, \ldots, f_{n-1}\}.$$

The simplest of these reflections is

$$f_0 = \begin{bmatrix} 1 & 0 \\ 0 & -1 \end{bmatrix}$$

and, in fact, we can rewrite each flip as follows

$$r^k f_0 = \begin{bmatrix} \cos(k\vartheta) & -\sin(k\vartheta) \\ \sin(k\vartheta) & \cos(k\vartheta) \end{bmatrix} \begin{bmatrix} 1 & 0 \\ 0 & -1 \end{bmatrix} = \begin{bmatrix} \cos(k\vartheta) & \sin(k\vartheta) \\ \sin(k\vartheta) & -\cos(k\vartheta) \end{bmatrix} = f_k$$

for $0 \leq k < n$. We now try multiplying them the other way around and obtain

$$f_0 r^k = \begin{bmatrix} 1 & 0 \\ 0 & -1 \end{bmatrix} \begin{bmatrix} \cos(k\vartheta) & -\sin(k\vartheta) \\ \sin(k\vartheta) & \cos(k\vartheta) \end{bmatrix}$$

$$= \begin{bmatrix} \cos(k\vartheta) & -\sin(k\vartheta) \\ -\sin(k\vartheta) & -\cos(k\vartheta) \end{bmatrix}$$

$$= \begin{bmatrix} \cos(-k\vartheta) & \sin(-k\vartheta) \\ \sin(-k\vartheta) & -\cos(-k\vartheta) \end{bmatrix}$$

$$= f_{n-k}$$

for $0 \leq k < n$ where the third and forth equality follow by basic trigonometry (namely, $\sin(\vartheta) = -\sin(-\vartheta)$ and $\cos(\vartheta) = \cos(-\vartheta)$ and both periodic functions with repetition on an interval of width 2π). We record this pair of rules as follows

1.3 Abstract groups and the Sudoku property

$$r^k f_0 = f_k = f_0 r^{n-k} \tag{1.2.10}$$

We now have enough information in order to understand the product of any two elements of the group (and hence check the group axioms for D_{2n}). We have already seen (in (1.2.9)) that the composition of two rotations

$$r^i \times r^j = r^{i+j \pmod{n}}$$

is a rotation; we can now use (1.2.10) to deduce that the composition of two flips

$$f_j \times f_i = (r^j f_0)(r^i f_0) = r^j (f_0 r^i) f_0 = r^j (r^{n-i} f_0) f_0 = (r^j r^{n-i})(f_0 f_0) = r^{n-i+j}$$

is a rotation; and similarly that the composition of a rotation and a flip

$$r^j * f_i = r^j (r^i f_0) = (r^j r^i) f_0 = r^{i+j \pmod{n}} f_0 = f_{i+j \pmod{n}}$$

is a flip; and finally that the composition of a flip and a rotation

$$f_j * r^i = (r^j f_0) r^i = r^j (f_0 r^i) = r^j (r^{n-i} f_0) = r^{n-i+j} f_0 = f_{n-i+j}$$

is again a flip. Thus the composition of any two elements of D_{2n} is again an element of D_{2n}; hence the closure axiom for D_{2n} holds. We have already seen that any rotation r^k has inverse r^{n-k}. Each flip is its own inverse,

$$f_k^2 = \begin{bmatrix} \cos(k\vartheta) & \sin(k\vartheta) \\ \sin(k\vartheta) & -\cos(k\vartheta) \end{bmatrix} \begin{bmatrix} \cos(k\vartheta) & \sin(k\vartheta) \\ \sin(k\vartheta) & -\cos(k\vartheta) \end{bmatrix} = \begin{bmatrix} 1 & 0 \\ 0 & 1 \end{bmatrix}$$

using only the fact that $\sin^2(k\vartheta) + \cos^2(k\vartheta) = 1$. Thus we conclude that the 3-dimensional rotations of an n-gon, D_{2n}, do indeed form a matrix group.

1.3 Abstract groups and the Sudoku property

Now that we have played around with concrete matrix groups, we are ready to axiomatise this idea by defining an abstract group. We then prove our first theorem: that the multiplication tables of abstract groups satisfy a *Sudouku-like* property.

Definition 1.3.1. Let G be a set and let $* : (G \times G) \to G$. We say that $(G, *)$ is a group if it satisfies the axioms:

(G1) there exists an element $\mathrm{id}_G \in G$ such that $\mathrm{id}_G * g = g = g * \mathrm{id}_G$;
(G2) $a, b \in G$ implies $a * b \in G$;
(G3) for each $a \in G$ there exists an element $a^{-1} \in G$ such that

$$a * a^{-1} = \mathrm{id}_G = a^{-1} * a.$$

(G4) we have that $*$ is associative, that is:
$$a * (b * c) = (a * b) * c$$
for $a, b, c \in G$.

Notice the addition of a final "bracketing" axiom which automatically holds for matrix groups (and in this way every matrix group is an example of a group, as one would expect).

Definition 1.3.2. We define the *order* of a group, G, to be the number of elements in the underlying set, and denote this by $|G|$.

Example 1.3.3. Consider the set $\mathbb{Q} \setminus \{0\}$ of all rational numbers except zero. Let $* = \times$ be the usual multiplication of rational numbers. We have that $1 \in \mathbb{Q} \setminus \{0\}$ is the identity element ($1 \times \frac{a}{b} = \frac{a}{b} = \frac{a}{b} \times 1$ for all $\frac{a}{b} \in \mathbb{Q} \setminus \{0\}$). We have that $\frac{a}{b} \times \frac{c}{d} = \frac{ac}{bd} \in \mathbb{Q} \setminus \{0\}$ and so the closure axiom is satisfied. Any element $\frac{a}{b} \in \mathbb{Q} \setminus \{0\}$ has multiplicative inverse $\frac{b}{a} \in \mathbb{Q} \setminus \{0\}$. Associativity can easily be checked by hand. Thus $(\mathbb{Q} \setminus \{0\}, \times)$ is a group. One can repeat this example with the real and complex numbers (with zero removed). We note that all of these groups are of infinite order.

Exercise 1.3.4. Does the set of non-zero rational numbers, $\mathbb{Q} \setminus \{0\}$, under the operation of *division* form a group?

We now consider the *Sudouku-like* property for the multiplication table of a group. To illustrate the idea, we refer to the (1.1.2), in which the reader can readily check that each $g \in D_6$ appears precisely once in each row and each column of the matrix.

Theorem 1.3.5. *If $(G, *)$ is a finite group then in its multiplication table every element occurs precisely once in each row and in each column. In particular, the group G has a unique identity element, and each $g \in G$ has a unique inverse element.*

Proof. We prove this theorem by building up a series of properties for groups, starting with the cancellation rules:

Claim 1: Let $x, y, z \in G$. Then

- if $x * y = x * z$, this implies that $y = z$;
- if $y * x = z * x$, this implies that $y = z$.

Proof of Claim 1. We consider the first case, as the the second case is similar. By (G3), we know that there is an element $x^{-1} \in G$ such that $x * x^{-1} = x^{-1} * x = \text{id}$. By assumption, we have that
$$x * y = x * z$$
and multiplying both sides of the equation by x^{-1} on the left, we obtain
$$x^{-1} * (x * y) = x^{-1} * (x * z)$$

1.3 Abstract groups and the Sudoku property

using the associativity property (G4), we can re-bracket this as follows

$$(x^{-1} * x) * y = (x^{-1} * x) * z,$$

now we can use the inverse axiom (G3) in order to deduce that

$$\text{id}_G * y = \text{id}_G * z$$

and finally, using the identity axiom (G1) we obtain that $y = z$ as required. Thus Claim 1 holds.

Claim 2: Given $a, b \in G$, any equation of the form

$$a * x = b$$

has a unique solution, namely: $x = a^{-1} * b \in G$.

Proof of Claim 2. It is clear that $x = a^{-1} * b$ is a solution of this equation since

$$a * (a^{-1} * b) = (a * a^{-1}) * b = \text{id}_G * b = b$$

by applying the associativity, inverse, and identity axioms in turn. Now suppose that we had another solution $x' \in G$, then

$$a * x' = b = a * x.$$

By Claim 1, we have that $x' = x = a^{-1} * b$, as required. Thus Claim 2 holds.

We are now ready to prove two particular instances of the Sudoku property: firstly, that the group G has a *unique* identity element, and secondly that each $g \in G$ has a *unique* inverse element. For uniqueness of the identity, we note that by Claim 2, the equation $a * x = a$ has a unique solution $x = a^{-1} * a = \text{id}_G$. So there is only one element such that $a * x = a$, namely id_G. Now, for the uniqueness of inverses we note that by Claim 2 the equation $a * x = \text{id}_G$ has a unique solution $x = a^{-1}$, and so a has a unique inverse.

Now we are now ready to prove the Sudoku property in full. Let $G = \{g_1, g_2, \ldots g_n\}$. Then the multiplication table is given by

*	g_1	g_2	$\cdots$	g_n
g_1	$g_1 * g_1$	$g_1 * g_2$	$\cdots$	$g_1 * g_n$
g_2	$g_2 * g_1$	$g_2 * g_2$	$\cdots$	$g_2 * g_n$
$\vdots$				
g_i	$g_i * g_1$	$g_i * g_2$	$\cdots$	$g_i * g_n$
$\vdots$				
g_n	$g_n * g_1$	$g_n * g_2$	$\cdots$	$g_n * g_n$

If $g_i * g_s = g_i * g_t$ then using Claim 1 we get $g_s = g_t$. So the elements in each row are all distinct and as there are n of them, each element of G appears precisely once. One can argue similarly for $g_s * g_i = g_t * g_i$ and hence deduce the uniqueness statement for columns. □

1.4 Presenting groups

It is too much work to write down a full multiplication table every time we wish to discuss a given group. We wish, instead, to come up with a more succinct "presentation" of any group of interest. Firstly, we need a compact way of listing all the elements of a group; this is provided by the notion of generators.

Definition 1.4.1. Let G be a finite group and let $S \subset G$ be a subset of elements of G. We say that S **generates** G if every element of G can be written as a product of elements from S, together with their inverses.

Secondly, we need a succinct way of encoding the multiplication table of the group, without writing the entire ($|G| \times |G|$)-table; this is provided by the notion of relations.

Definition 1.4.2. Let G be a finite group and let $S \subset G$ be a set of generators of G. For $s_{i_1}, \ldots, s_{i_p}, s_{j_1}, \ldots, s_{j_q} \in S$, we define a **relation** to be any equality $s_{i_1} s_{i_2} \ldots s_{i_p} = s_{j_1} s_{j_2} \ldots s_{j_q}$ which holds in G. We say that a set of relations, R, is **complete** if one can deduce the entire multiplication table of G using only (repeated applications of) the set of relations R.

Definition 1.4.3. Given a group G, we define a **presentation** of G to be a set of generators, S, together with a complete set of relations, R. In which case, we write $G = \langle S \mid R \rangle$.

For cyclic and dihedral groups we have already encoded the multiplication tables quite succinctly. This was an implicit construction of a group presentation, which we now revisit.

Example 1.4.4. Recall $C_n = \{\text{id}, r, \ldots, r^{n-1}\}$ is the group of 2-dimensional symmetries of an n-gon. The multiplication rule for these rotations is easily summarised as follows
$$r^i \times r^j = r^{i+j \bmod n}$$
We refer to C_n as the **cyclic group of order** n. Rather than write down the full multiplication table every time we discuss this group, we instead list the group elements and their product rule as succinctly as possible. The element r is the generator of the group (as all group elements are powers r^k for some $0 \leq k < n$) and the rule $r^n = \text{id}$ is the sole relation for the group. We record this in the following presentation:
$$C_n = \langle r \mid r^n = \text{id} \rangle.$$

1.5 Isomorphisms

Example 1.4.5. The rotation, r, together with the flip, f_0, generate D_{2n}. To see this, simply note that all elements

$$\{\mathrm{id}_2, r, r^2, \ldots, r^{n-1}\} \cup \{f_0, f_1, \ldots, f_{n-1}\}.$$

of D_{2n} are of the form $r^i f_0^j$ for $0 \leqslant i < n$ and $0 \leqslant j \leqslant 1$ using the fact that $f_k = r^k f_0$. The rules $r^n = \mathrm{id}$ and $f_0^2 = e$ are obvious. Thus it remains to understand the products of rotations and the flip, which we have done in (1.2.10). We record this in the following presentation:

$$D_{2n} = \langle r, f_0 \mid r^n = \mathrm{id}, r^k f_0 = f_0 r^{n-k} \text{ for } 0 \leqslant k < n \rangle.$$

Exercise 1.4.6. Let $G = \langle a, b, c, \ldots z \mid$ English words that are homonyms$\rangle$, that is the group generated by all letters of the alphabet, with two words considered to be the same if their pronunciation is identical (for example ate=eight). What is the order of G?

1.5 Isomorphisms

We have already encountered the idea that two groups of the same size are not necessarily 'the same group'. We have seen that while both the symmetry groups of the triangular prism and the hexagonal based pyramid both have size 6, they are very different groups. This is because the 'size' of the group is merely the size of the underlying set. A group has more structure than a set, and we need to take this structure into account. This need is addressed in the following definition, which provides us with the necessary notion of "sameness" for groups.

Definition 1.5.1. Let $(G, *_G)$ and $(H, *_H)$ be groups. We say that G and H are isomorphic if there exists a bijection $\phi : G \to H$ such that

$$\phi(g_1 *_G g_2) = \phi(g_1) *_H \phi(g_2)$$

for all $g_1, g_2 \in G$. In this case, we call the map ϕ an **isomorphism** between G and H and we write $G \cong H$.

Notice that the conditions of the isomorphism are that it preserves *all the defining structure of the group* (the size of the set and the rule for multiplying elements together). Thus, in what follows, we will always regard isomorphic groups as "the same group" even if, on a superficial level, they look different.

Example 1.5.2. Consider the (matrix) groups $U_2(\mathbb{R})$ and $\mathbb{R}^\times$ and the map $\phi : U_2(\mathbb{R}) \to \mathbb{R}^\times$ given by

$$\phi \begin{bmatrix} 1 & x \\ 0 & 1 \end{bmatrix} = e^x$$

Then it is easy to see that ϕ is a bijection, with inverse map given by

$$\phi^{-1}(y) = \begin{bmatrix} 1 & \ln(y) \\ 0 & 1 \end{bmatrix}.$$

Moreover we have

$$\phi\left(\begin{bmatrix} 1 & a \\ 0 & 1 \end{bmatrix}\begin{bmatrix} 1 & b \\ 0 & 1 \end{bmatrix}\right) = \phi\begin{bmatrix} 1 & a+b \\ 0 & 1 \end{bmatrix} = e^{a+b} = e^a \times e^b = \phi\begin{bmatrix} 1 & a \\ 0 & 1 \end{bmatrix}\phi\begin{bmatrix} 1 & b \\ 0 & 1 \end{bmatrix}$$

and so the map ϕ is indeed an isomorphism. Thus $U_2(\mathbb{R})$ and $\mathbb{R}^\times$ are regarded as being *the same*, even though they look quite different.

Definition 1.5.3. Let G be a finite group and let $g \in G$. The **order of** g is the smallest positive integer r such that $g^r = \text{id}_G$.

Example 1.5.4. Consider the group D_6. The elements r and r^2 are both of order 3. The elements f_0, rf_0, and $f_0 r^2$ all have order 2.

Exercise 1.5.5. How many elements of order 2 are there in D_{2n}?

Exercise 1.5.6. For any two elements $g, h \in G$, show that $|g| = |hgh^{-1}|$.

(The order of an element g in a group G should not be confused with the order of the group, $|G|$, defined earlier.)

Proposition 1.5.7. *Let $(G, *_G)$ and $(H, *_H)$ be groups. If $\phi : G \to H$ is an isomorphism then we have*
(i) $\phi(\text{id}_G) = \text{id}_H$.
(ii) $\phi(g^{-1}) = (\phi(g))^{-1}$ for all $g \in G$.
(iii) $\phi(g^r) = (\phi(g))^r$ for all $g \in G$.
(iv) the order of g is equal to the order of $\phi(g)$ for any $g \in G$.

Proof. (i) As $\phi(g) = \phi(\text{id}_G *_G g) = \phi(\text{id}_G) *_H \phi(g)$, and the identity element of H is unique, we must have that $\phi(\text{id}_G) = \text{id}_H$.
(ii) As
$$\text{id}_H = \phi(\text{id}_G) = \phi(g^{-1} *_G g) = \phi(g^{-1}) *_H \phi(g),$$
and $\phi(g) \in H$ has a unique inverse, we must have that $\phi(g^{-1}) = (\phi(g))^{-1}$.
(iii) We have that
$$\phi(g^r) = \phi(g *_G g *_G \ldots *_G g) = \phi(g) *_H \phi(g) *_H \ldots *_H \phi(g) = \phi(g)^r.$$

(iv) Let r be the order of g. Using (i) and (iii) we have that $(\phi(g))^r = \text{id}_H$. Now suppose, for a contradiction, that $(\phi(g))^s = \text{id}_H$ for some $s < r$. Now as ϕ is a bijection it has an inverse ϕ^{-1}. So we get
$$g^s = (\phi^{-1} \circ \phi)(g^s) = \phi^{-1}(\phi(g^s)) = \phi^{-1}((\phi(g))^s) = \phi^{-1}(\text{id}_H) = \text{id}_G.$$

But this contradicts our assumption that r is the smallest integer satisfying $g^r = \text{id}_G$. □

1.6 Subgroups and direct products of groups

We now consider the idea of how groups can "sit inside" one another. This idea is formalised via the definition of a *sub*group, as follows:

Definition 1.6.1. A subset H of a group G is called a **subgroup** of G if H is itself a group with the multiplication inherited from G. In this case we write $H \leqslant G$.

Remark 1. Any group G will always have the subgroups $\{\mathrm{id}_G\} \leqslant G$ and $G \leqslant G$. We refer to a subgroup $H \leqslant G$ as a **proper subgroup** if $H \neq \{\mathrm{id}_G\}$ or G.

Example 1.6.2. Clearly the set of elements $C_\infty = \{\cos(\vartheta) + i\sin(\vartheta) \mid 0 \leqslant \vartheta < 2\pi\}$ is a subset of the set of all non-zero complex numbers, $\mathbb{C} \setminus \{0\}$. We have already checked in Example 1.2.6 that C_∞ forms a group under the multiplication of complex numbers. Therefore $C_\infty \leqslant \mathbb{C} \setminus \{0\}$ is a subgroup. One can picture C_∞ as the subgroup of elements on the circle of radius 1 within the complex plane.

Example 1.6.3. Let $n \geqslant 2$ be any integer. Consider the subset of all elements

$$\{\cos(2k\pi/n) + i\sin(2k\pi/n) \mid 0 \leqslant k < n\} \leqslant C_\infty.$$

One can check that this set of elements forms a subgroup using elementary trigonometry. In fact, this set of elements forms a subgroup isomorphic to the cyclic group C_n. This isomorphism is given by the map

$$\cos(2k\pi/n) + i\sin(2k\pi/n) \mapsto \begin{bmatrix} \cos(2k\pi/n) & -\sin(2k\pi/n) \\ \sin(2k\pi/n) & \cos(2k\pi/n) \end{bmatrix}.$$

This isomorphism identifies the vertex of a regular n-gon at point $p = \cos(2k\pi/n) + i\sin(2k\pi/n) \in \mathbb{C} \setminus \{0\}$ with the rotation around $(0,0)$ which takes the vertex $p \in \mathbb{C} \setminus \{0\}$ to the point $(1,0) \in \mathbb{C} \setminus \{0\}$. For example, under this isomorphism the groups C_3, C_4, and C_5 consist of the points

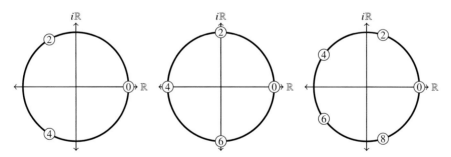

on the unit circle, where the vertex with label $2k$ is the point $p = \cos(2k\pi/n) + i\sin(2k\pi/n) \in \mathbb{C}$.

Example 1.6.4. The following are examples of (chains of) subgroups

$$\{1, -1\} \leqslant \mathbb{R} \setminus \{0\} \leqslant \mathbb{C} \setminus \{0\} \qquad C_n \leqslant D_{2n} \leqslant \mathrm{GL}_2(\mathbb{R}).$$

Exercise 1.6.5. Check whether of not the following are subgroups:

- Does $U_2(\mathbb{Z})$ the subset of all (2×2)-upper triangular matrices with integer coefficients form a subgroup of $U_2(\mathbb{R})$?
- Does $SL_2(\mathbb{Z})$ the subset of all (2×2)-matrices with integer coefficients and determinant 1 form a subgroup of $SL_2(\mathbb{R})$?

Proposition 1.6.6. *Let $(G, *)$ be a group and let H be a subset of G. Then H is a subgroup of G if and only if the following conditions are satisfied.*
(S1) If $a \in H$ and $b \in H$, then $a * b \in H$.
(S2) If $a \in H$, then $a^{-1} \in H$.

Proof. Associativity is automatically satisfied as it holds in G. If $a, a^{-1} \in H \neq \emptyset$, then $a * a^{-1} = \mathrm{id}_H \in H$ by (S1). □

It is natural to ask: Can we find all subgroups of a given group?

Example 1.6.7. Can we find all the subgroups of $D_6 = \{\mathrm{id}, r, r^2, f, fr, fr^2\}$? Every subgroup must contain the identity element and in particular $\{\mathrm{id}\}$ itself forms a subgroup.

The pair of elements $\{\mathrm{id}, f\}$ forms a subgroup and in fact, the same is true for the other flips: $\{\mathrm{id}, fr\}$, and $\{\mathrm{id}, fr^2\}$ form subgroups of D_6. All these subgroups are isomorphic to C_2. One can check this simply by constructing the multiplication tables directly.

Now, the set $\{\mathrm{id}, r\}$ cannot form a subgroup as $r^{-1} = r^2 \notin \{\mathrm{id}, r\}$; however we can add this element and hence obtain $H = \{\mathrm{id}, r, r^2\}$. This now does form a subgroup and $H \cong C_3$. One can similarly see that $\{\mathrm{id}, r^2\}$ is not a subgroup of D_6 and that, in fact, H is the smallest subgroup containing the rotation r^2.

Thus we have listed all the smallest subgroups of D_6 (those determined by a single non-identity element). One can now consider the smallest subgroups of D_6 containing an arbitrary pair of (non-identity) elements. The smallest subgroup containing such a pair is either H (if the pair is r and r^2) or is the whole of D_6. For an example of the latter case, let $f, r^2 \in H$, which implies that $(r^2)^{-1} = r \in H$ and so $r, f \in H$; but we know that G is generated by r and f.

Exercise 1.6.8. For subgroups $H_1, H_2 \leqslant G$, prove that $H_1 \cap H_2 = \{h \mid h \in H_1 \text{ and } h \in H_2\}$ is a subgroup of G.

The procedure of Example 1.6.7 illustrates an important idea: we can consider subgroups *generated* by a given list of elements of G.

Definition 1.6.9. Let $X \subseteq G$ be a non-empty subset of a group G. We define the subgroup generated by X (denoted $\langle X \rangle$) to be the subgroup whose elements are given by all possible products of elements of X and their inverses. That is, the set of all elements of the form

$$x_1^{m_1} * x_2^{m_2} * \ldots * x_n^{m_n}$$

where $x_i \in X$ (not necessarily distinct), $m_i \in \mathbb{Z}$ and $n \in \mathbb{Z}_{\geqslant 1}$.

1.6 Subgroups and direct products of groups

We have specifically defined $\langle X \rangle$ so that it is the smallest possible group containing the elements $X \subseteq G$ which also obeys the closure and inverses axioms.

Example 1.6.10. If $X = \{x\}$ contains only one element, then $\langle X \rangle = \langle x \rangle = \{x^i \mid i \in \mathbb{Z}\}$. If x has order n then $\langle x \rangle \cong C_n$ the cyclic group of order n. And if x has infinite order then $\langle x \rangle \cong C_\infty$.

Exercise 1.6.11. Determine the subgroup $\langle r^2, fr^2 \rangle \leqslant D_{2n}$ for n odd and even respectively.

Definition 1.6.12. Let $(G, *_G)$ and $(H, *_H)$ be two groups. Then we can form a new group $(G \times H, *)$ called the **direct product** of G and H by setting

$$G \times H = \{(g, h) \mid g \in G, h \in H\}$$

and

$$(g_1, h_1) * (g_2, h_2) = (g_1 *_G g_2, h_1 *_H h_2).$$

Example 1.6.13. We consider the group of all 3-dimensional symmetries of a proper rectangle (that is, a rectangle which is not a square). There are two flips we can perform, which are illustrated as follows:

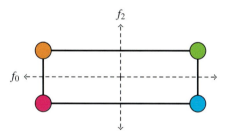

and we observe that, by performing neither, one, or both of these flips we obtain the following set of rectangles

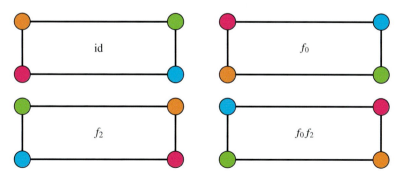

where the final rectangle could also have been obtained from the first one by rotation through π radians (recall that the composition of two reflections is a rotation!). We let

$$\text{Rect} = \{\text{id}, f_0, f_2, f_0 f_2\}$$

be the group of symmetries of the rectangle; since these reflections are merely through the x and y-axes, their corresponding matrices are very easy to construct and are equal to

$$\begin{bmatrix} 1 & 0 \\ 0 & 1 \end{bmatrix} \begin{bmatrix} 1 & 0 \\ 0 & -1 \end{bmatrix} \begin{bmatrix} -1 & 0 \\ 0 & 1 \end{bmatrix} \begin{bmatrix} -1 & 0 \\ 0 & -1 \end{bmatrix}$$

respectively. Clearly Rect is not isomorphic to D_8 (these groups are not of the same size!). However, we notice that D_8 has a subgroup consisting of the elements $\{\text{id}, f_0, f_2, f_0, f_2\}$ (or if you prefer, the subgroup generated by f_0 and f_2) and it is not difficult to check that Rect is isomorphic to this subgroup (in particular, the matrices corresponding to these reflections are all equal to one another!).

We now observe that the subgroups of Rect given by $\{\text{id}, f_0\}$ and $\{\text{id}, f_2\}$ are both isomorphic to C_2 and that, moreover, Rect is isomorphic to the direct product

$$\{\text{id}, f_0\} \times \{\text{id}, f_2\} \to \text{Rect}$$

via the map

$$(\text{id}, \text{id}) \mapsto \text{id} \quad (\text{id}, f_2) \mapsto f_2 \quad (f_0, \text{id}) \mapsto f_0 \quad (f_0, f_2) \mapsto f_0 f_2$$

and so the group of symmetries of the rectangle is isomorphic to the direct product $C_2 \times C_2$.

In other words, the symmetries of the rectangle do not provide a particularly interesting or "new" group. They can be reduced to products of smaller groups we have already seen.

Remark 2. Notice that we have not defined a dihedral group D_4 (as there is no such thing as a 2-sided polygon). One can think of a 2-gon as an infinite straight line, which one can in turn think of as an infinitely long and skinny rectangle. In this way, Rect provides the symmetry group D_4. We will use the latter notation in what follows.

Chapter 2
Coxeter groups and the 15-puzzle

In this chapter, we introduce the symmetric and alternating groups and use these ideas to solve a famous $1,000 question of Samuel Loyd. This gives us a taste of how group theory can be used to solve "real world" problems. We go on to consider more general groups of a similar flavour, namely the Coxeter groups.

2.1 The symmetric group

Definition 2.1.1. A permutation matrix of degree n is an $(n \times n)$-matrix that has exactly one entry equal to 1 in each row and each column, and zeros elsewhere. The set of all permutation matrices of degree n is called the **symmetric group of degree** n and is denoted $\mathfrak{S}_n$.

To any permutation matrix of degree n we associate a diagram: we first draw a rectangular frame with the numbers $1, \ldots, n$ in descending order along the left and right edges of the rectangle. The nodes of the left edge of the rectangle are associated to the rows of the matrix and the nodes of the right edge of the rectangle are associated to the columns of the matrix. Associated to each entry 1 in a given row and column, we draw a strand connecting these nodes. For example

$$\begin{bmatrix} 0 & 1 & 0 & 0 & 0 \\ 0 & 0 & 0 & 0 & 1 \\ 1 & 0 & 0 & 0 & 0 \\ 0 & 0 & 1 & 0 & 0 \\ 0 & 0 & 0 & 1 & 0 \end{bmatrix} \mapsto$$

where we see, for example, that the non-zero entry in the first row is in the second column and so we draw an arrow $1 \longrightarrow 2$ in the diagram. This provides us a graphical calculus for understanding the symmetric group. The multiplication of matrices can be encoded by concatenation of these diagrams. For example,

$$\begin{bmatrix} 0 & 1 & 0 \\ 0 & 0 & 1 \\ 1 & 0 & 0 \end{bmatrix} \begin{bmatrix} 0 & 1 & 0 \\ 1 & 0 & 0 \\ 0 & 0 & 1 \end{bmatrix} = \begin{bmatrix} 1 & 0 & 0 \\ 0 & 0 & 1 \\ 0 & 1 & 0 \end{bmatrix}$$

or equivalently, in diagrammatic form

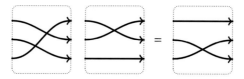

where here "composition" of strand diagrams is simply given by horizontally concatenating the diagrams. We do not care about the double-crossings of strands, we merely care about the starting point (on the lefthand-side of the diagram) and the ending point (on the righthand-side of the diagram) of each strand.

Example 2.1.2. There are 6 elements of $\mathfrak{S}_3$ which can be pictured as follows:

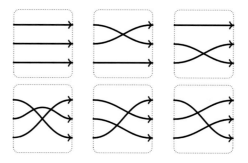

Exercise 2.1.3. Compute the products of all strand diagram in Example 2.1.2 and hence write down the (6×6)-multiplication matrix for the group $\mathfrak{S}_3$.

More generally, we have the following theorem:

Proposition 2.1.4. *The order of $\mathfrak{S}_n$ is $n!$.*

Proof. The number of ways to pair the numbers from left-to-right is given by

$$n(n-1)(n-2)\ldots 1 = n!$$

To see this, we note that there are n choices for where to send the first strand (starting at 1 on the lefthand-side of the frame); then there are only $n-1$ choices as to where to send the second strand (as we cannot send the second strand to *the same place as the first strand*)... Continuing in this manner, the result follows. □

Clearly the product of any two of these strand diagrams is again a strand diagram and so the group multiplication satisfies the closure axiom. Now, the inverse of a

2.1 The symmetric group

permutation is given by flipping a diagram through its y-axis, for example the pair of diagrams

are mutual inverses of one another. To see this, simply concatenate the diagrams and notice that every strand starts and finishes at the same point.

We now introduce one final notation for permutations. This notation is not as powerful as the diagrammatic notation, but it is much more compact. Consider the following permutation in $g \in \mathfrak{S}_5$,

$$g = \begin{bmatrix} 0 & 0 & 1 & 0 & 0 \\ 0 & 0 & 0 & 0 & 1 \\ 0 & 1 & 0 & 0 & 0 \\ 1 & 0 & 0 & 0 & 0 \\ 0 & 0 & 0 & 1 & 0 \end{bmatrix} \qquad (2.1.5)$$

We will write g in cycle notation as

$$g = (1, 3, 2, 5, 4).$$

Here a bracketed sequence means that each number is mapped to the next one within the bracketed sequence, and the last number in the bracketed sequence is mapped to the first number in the bracketed sequence. Of course, this is a "cyclic" way of writing down a sequence (the first number in the sequence is irrelevant, as we always go around in a circle back to the starting point). One can picture this cycle quite literally as a circle

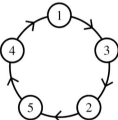

and, since it does not matter which of the 5 vertices we choose to start from when we walk around the circle, this gives us 5 distinct ways to record this circle in *cycle notation* as follows:

$$g = (1, 3, 2, 5, 4) = (3, 2, 5, 4, 1) = (2, 5, 4, 1, 3) = (5, 4, 1, 3, 2) = (4, 1, 3, 2, 5)$$

all of which correspond to the same permutation matrix, g from (2.1.5). Of course, not all permutation matrices can be written down as a single cycle; for example,

$$h = \begin{bmatrix} 0 & 1 & 0 & 0 & 0 \\ 1 & 0 & 0 & 0 & 0 \\ 0 & 0 & 0 & 0 & 1 \\ 0 & 0 & 1 & 0 & 0 \\ 0 & 0 & 0 & 1 & 0 \end{bmatrix} \qquad (2.1.6)$$

We will write h in cycle notation as follows

$$h = (1,2)(3,5,4) = (1,2)(5,4,3) = (2,1)(5,4,3) = (2,1)(3,5,4) = \ldots$$

We now introduce a little bit more language around our cycle notation. A cycle of length n, $(x_1, x_2, \ldots, x_n)$ is called an n-cycle. A 2-cycle is often called a transposition. The permutation h depicted in (2.1.6) is a product of a single transposition and a single three-cycle: the transposition maps $1 \to 2$ and $2 \to 1$; the three-cycle maps $3 \to 5$, and $5 \to 4$, and $4 \to 3$.

We have already seen that it is easy to calculate the inverse of a strand diagram (we simply reflect the diagram through its y-axis). It is similarly easy to calculate the inverse of a permutation depicted in cycle notation; we simply flip the cycle from left to right. For example $((1,3,4)(2,5))^{-1} = (5,2)(4,3,1)$.

The product of two elements, written in cycle notation, is easy to compute. For example, with $g = (1,3,2,5,4)$ and $h = (1,2)(3,5,4)$ as above, we compute the product, gh, by calculating where each number "goes" as read from left-to-right. Let's start with seeing where 1 goes in the composition gh, as follows

$$(1,3,2,5,4) \quad (1,2)(3,5,4)$$

here $1 \to 3$ in the 5-cycle, this 3 is then ignored by the 2-cycle, and then $3 \to 5$ in the 3-cycle; thus, in the overall product, we have that $1 \to 5$. Having calculated that $1 \to 5$, we now wish to see where 5 goes in the composition gh:

$$(1,3,2,5,4) \quad (1,2)(3,5,4)$$

here $5 \to 4$ in the 5-cycle, this 4 is then ignored by the 2-cycle, and then $4 \to 3$ in the 3-cycle; thus, in the overall product, we have that $5 \to 3$. Carrying on in this manner, we have that

$$gh = (1,3,2,5,4)(1,2)(3,5,4) = (1,5,3)(2,4).$$

Exercise 2.1.7. Calculate the product gh with $g = (1,3,2,5,4)$ and $h = (1,2)(3,5,4)$ using strand diagrams and cycle notation simultaneously.

Exercise 2.1.8. Show that the group $\mathfrak{S}_n$ is abelian if and only if $n = 1$ or 2.

2.1 The symmetric group

Exercise 2.1.9. What is the order of the subgroup $\langle (1,2)(3,4), (1,3)(2,4) \rangle \leqslant \mathfrak{S}_4$? Does this remind you of a group you have already seen?

Example 2.1.10. The multiplication table for $\mathfrak{S}_3$ can easily be computed by composing all the permutations. It is given by

	id	$(1,2,3)$	$(1,3,2)$	$(2,3)$	$(1,2)$	$(1,3)$
id	id	$(1,2,3)$	$(1,3,2)$	$(2,3)$	$(1,2)$	$(1,3)$
$(1,2,3)$	$(1,2,3)$	$(1,3,2)$	id	$(1,3)$	$(2,3)$	$(1,2)$
$(1,3,2)$	$(1,3,2)$	id	$(1,2,3)$	$(1,2)$	$(1,3)$	$(2,3)$
$(2,3)$	$(2,3)$	$(1,3)$	$(1,2)$	id	$(1,2,3)$	$(1,3,2)$
$(1,2)$	$(1,2)$	$(2,3)$	$(1,3)$	$(1,3,2)$	id	$(1,2,3)$
$(1,3)$	$(1,3)$	$(1,2)$	$(2,3)$	$(1,2,3)$	$(1,3,2)$	id

We say that two cycles $(x_1, \ldots, x_k)$ and $(y_1, \ldots, y_\ell)$ are disjoint if they have no numbers in common, that is $\{x_1, \ldots, x_k\} \cap \{y_1, \ldots, y_\ell\} = \emptyset$.

Proposition 2.1.11. *Every permutation can be written as a product of (not necessarily disjoint) transpositions. In other words, the set of all transpositions $\{(i, j) \mid 1 \leqslant i < j \leqslant n\}$ generates $\mathfrak{S}_n$.*

Proof. Clearly every permutation can be written as a product of cycles. Therefore it is enough to show that every cycle can be written as a product of transpositions. Now we have

$$(x_1, x_2, \ldots, x_n) = (x_1, x_n)(x_2, x_n)(x_3, x_n)(x_4, x_n) \ldots (x_{n-2}, x_n)(x_{n-1}, x_n)$$

as required. $\square$

Now we zoom in a little further and see that we need only "adjacent" transpositions in our set of generators:

Proposition 2.1.12. *Every permutation can be written as a product of the elements $s_i = (i, i+1)$ for $1 \leqslant i < n$. In other words, the set $\{s_i \mid 1 \leqslant i < n\}$ generates $\mathfrak{S}_n$.*

Proof. By Proposition 2.1.11, it will suffice to show that any transposition (i, j) for $1 \leqslant i < j \leqslant n$ can be written as a product of the s_i elements. We have that

$$(i, j) = (j-1, j) \ldots (i+1, i+2)(i, i+1)(i+1, i+2) \ldots (j-1, j)$$

and the result follows. $\square$

Exercise 2.1.13. Prove that $\mathfrak{S}_n$ is generated by the pair of elements $(1, 2)$ and $(1, 2, 3, \ldots, n)$.

Example 2.1.14. One of the advantages of the strand diagrams is that they make very clear the manner in which one can write a permutation, π, as a product of the s_i for $1 \leqslant i < n$. Simply draw the diagram of π and enumerate the crossing from left to right, each crossing involves two adjacent strands and we can stretch the diagram out as follows:

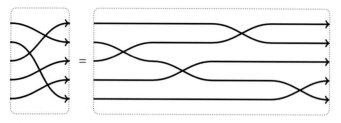

or, equivalently, in cycle notation

$$(1, 2, 5, 4, 3) = (2, 3)(3, 4)(1, 2)(4, 5).$$

Notation. We have introduced the strand diagrams so that strands go from left-to-right; this was in order to match-up the diagram concatenation with matrix multiplication. In the future, we will often rotate these diagrams anti-clockwise by $\pi/2$ so that the group multiplication is now given by vertical concatenation. This convention uses less space on the page.

2.2 The length function and the alternating group

The first context in which we encounter the notion of parity is in integers under addition. We instinctively know that

$$\text{odd} + \text{odd} = \text{even} \quad \text{odd} + \text{even} = \text{odd} \quad \text{even} + \text{even} = \text{even}.$$

Is there a generalisation of this notion to the symmetric group?

Psuedo-definition 2.2.1. Define the length of $\sigma \in \mathfrak{S}_n$ to be the minimal number of crossings in the strand diagram of $\sigma \in \mathfrak{S}_n$.

Note that we can draw a given permutation in lots of different ways, for example

of which our favourite depiction is the leftmost one. We think of it as being the diagram in which all the strands are pulled taut (and so there are no "extra" crossings, in which a pair of strands cross each other 2, 3, 4,... times). Notice that the total number of crossings in each case is 2, 4, 4. Even though this number is not well-defined, notice that the parity is! Rewriting this in terms of cycle-notation, we have

2.2 The length function and the alternating group

that

$$(1, 2, 3) = (2, 3)(1, 2) = (2, 3)(1, 2)(1, 2)(1, 2) = (1, 2)(2, 3)(1, 2)(2, 3)$$

Notice that the total number of transpositions involved in each case is 2, 4, 4 (and that this preserves our notion of parity). Before we prove this, we first require a rigorous definition of length.

Definition 2.2.2. Define the length of $\sigma \in \mathfrak{S}_n$ as follows:

$$\ell(\sigma) = |\{(i, j) \mid 1 \leq i < j \leq n \text{ and } (i)\sigma > (j)\sigma\}|.$$

Proposition 2.2.3. *Given $\sigma \in \mathfrak{S}_n$ and $s_i = (i, i+1) \in \mathfrak{S}_n$ we have that*

$$\ell(\sigma s_i) = \ell(s_i \sigma) = \ell(\sigma) \pm 1.$$

Proof. The proof is best visualised in terms of strand diagrams, which we assume have been drawn in our favourite manner, so that there are no double-crossings. We provide both the diagrammatic argument and an equivalent (bracketed) non-diagrammatic argument. There are two possible cases to consider. In the first case, the strands beginning at i and $i+1$ in the strand diagram for σ do cross (equivalently, $(i)\sigma > (i+1)\sigma$). In this case the effect of composing with s_i on the left is to untangle them (equivalently, $(i)s_i\sigma < (i+1)s_i\sigma$). No other crossings are affected (equivalently, $(j)s_i\sigma = (j)\sigma$ for $j \neq i, i+1$) and therefore $\ell(s_i\sigma) = \ell(\sigma) - 1$. The only other possibility is that strands beginning at i and $i+1$ in the strand diagram for σ do not cross (that is, $(i)\sigma < (i+1)\sigma$). In this case the effect of composing with s_i on the left is to introduce a crossing between them (equivalently, $(i)s_i\sigma > (i+1)s_i\sigma$). No other crossings are affected (equivalently, $(j)s_i\sigma = (j)\sigma$ for $j \neq i, i+1$) and so $\ell(s_i\sigma) = \ell(\sigma) + 1$. □

Corollary 2.2.4. *We have that $\ell(\sigma\tau) \equiv \ell(\sigma) + \ell(\tau)$ modulo 2.*

Proof. Suppose $\sigma = s_{i_1} s_{i_2} \ldots s_{i_l}$ and $\tau = s_{j_1} s_{j_2} \ldots s_{j_k}$. We have that

$$\sigma\tau = s_{i_1} s_{i_2} \ldots s_{i_l} s_{j_1} s_{j_2} \ldots s_{j_k}$$

and the result follows by repeated applications of Proposition 2.2.3. □

Definition 2.2.5. We say that a permutation $\sigma \in \mathfrak{S}_n$ is **even** or **odd** if $\ell(\sigma)$ is even or odd, respectively.

Proposition 2.2.6. *We let A_n be the set of all even permutations of degree n. Then A_n together with composition of permutations is a group. We call this the* **alternating group of degree** n.

Proof. First observe that if $\sigma, \tau \in A_n$ then $\ell(\sigma) \equiv 0$ and $\ell(\tau) \equiv 0$ modulo 2 and so $\ell(\sigma\tau) \equiv 0$ modulo 2. Therefore $\sigma, \tau \in A_n$ implies $\sigma\tau \in A_n$. If $\sigma \in A_n$, say $\sigma = s_{i_1} s_{i_2} \ldots s_{i_{2m}}$, then $\sigma^{-1} = s_{i_{2m}} \ldots s_{i_2} s_{i_1} \in A_n$. Therefore $\sigma \in A_n$ implies $\sigma^{-1} \in A_n$. The result follows from Proposition 1.6.6. □

Proposition 2.2.7. *The alternating group is generated by the 3-cycles of the form $(a, a+1, a+2)$ and $(a, a+2, a+1)$ for $1 \leq a \leq n-2$.*

Proof. Any element of A_n can be written as a product of an even number of transpositions of the form $s_i = (i, i+1)$ for $1 \leq i < n$. So given any

$$\sigma = s_{i_1} s_{i_2} \ldots s_{i_{2k-1}} s_{i_{2k}} = (s_{i_1} s_{i_2}) \ldots (s_{i_{2k-1}} s_{i_{2k}})$$

we wish to write any of the factors as a product of 3-cycles. That is, we wish to write any product of 2 distinct transpositions, $s_i s_j$, as a product of (possibly many) 3 cycles of the form $(a, a+1, a+2)$ and $(a, a+2, a+1)$ for $1 \leq a \leq n-2$. We assume that $i < j$. We have that

$$s_i s_j = s_i(s_{i+1} s_{i+1}) \ldots (s_{j-1} s_{j-1}) s_j$$
$$= (s_i s_{i+1})(s_{i+1} s_{i+2}) \ldots (s_{j-1} s_j)$$
$$= (i, i+1, i+2)(i+1, i+2, i+3) \ldots (j-1, j, j+1).$$

Similarly, we can write any $s_i s_j$ with $i > j$ as a product of elements of the form $(a+2, a+1, a)$ simply by taking inverses of the above. $\square$

Proposition 2.2.8. *Let $n \geq 2$. Then $|A_n| = n!/2$.*

Proof. We must calculate how many even permutations there are in $\mathfrak{S}_n$. We will prove that there are exactly the same number of even and odd permutations by defining a one-to-one correspondence between them. Let f be the map that sends a permutation σ to $\sigma(1,2)$ (i.e., it composes σ with the transposition $(1,2)$). If σ is even, $f(\sigma)$ has one more transposition so is odd. Similarly, odd permutations are sent to even permutations by f. The map f is its own inverse as $\sigma(1,2)(1,2) = \sigma$. So this map is a bijection and we have a one-to-one correspondence between the set of odd permutations in $\mathfrak{S}_n$ and the set of even permutations in $\mathfrak{S}_n$. So exactly half of all permutations are even and half are odd, and $|A_n| = n!/2$. $\square$

2.3 The 15-puzzle

We will now discuss a famous example of a "real life" problem which can be solved using ideas from symmetric and alternating groups. In the 1870s, Sam Loyd offered a $1,000 prize for anyone who could solve his "14-15 puzzle". This puzzle consists of 15 numbered tiles in a (4×4)-grid. We refer to the empty space as being "occupied by an empty tile". A "legal move" consists of "sliding the empty tile" that is, exchanging the empty tile with one of its horizontal or vertical neighbours. Sam then asked "can you obtain the latter puzzle in Figure 2.1 from the former puzzle using only legal moves?".

2.3 The 15-puzzle

1	2	3	4
5	6	7	8
9	10	11	12
13	14	15	

1	2	3	4
5	6	7	8
9	10	11	12
13	15	14	

Fig. 2.1: Can you use legal moves to obtain the latter puzzle from the former?

Theorem 2.3.1. *Any puzzle can be obtained from one and only one of the two puzzles in Figure 2.1. Moreover, of the total of* 16! *puzzles, half can be obtained from the former and half from the latter.*

We now turn to proving the theorem. Consider an arbitrary arrangement of the 15-puzzle, we define the snake reading word to be given by reading the entries of the puzzle, starting from the top-rightmost corner and snaking down the grid first from right-to-left and then left-to-right draw as depicted in Figure 2.2.

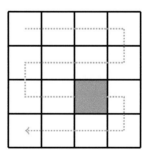

Fig. 2.2: The ordering for the snake reading word of a 15-puzzle.

Example 2.3.2. The reading words of the puzzles in Figure 2.3 are given by

$$\text{read}(P) = \text{read}(Q) = [1, 2, 3, 4, 8, 7, 6, 5, 9, 10, 11, 12, 15, 14, 13];$$
$$\text{read}(R) = [1, 2, 3, 4, 7, 6, 5, 9, 10, 11, 8, 12, 15, 14, 13].$$

Notice that Q is obtained from P by moving the empty tile (pictured as a grey square) *along the snake*. More generally, notice that if we start with any given puzzle and move the gray square along the snake, we *always* preserve the reading word of the puzzle. Given two puzzles, P and Q, we obtain a **puzzle-permutation**, $w_Q^P \in \mathfrak{S}_{15}$, by placing the reading word of P above the reading word of Q and matching-up the numbers with strands. For the puzzles in Figure 2.3, we obtain

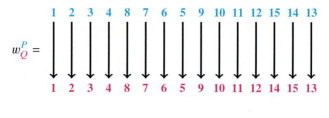

and

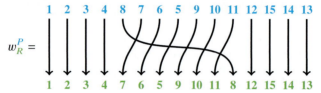

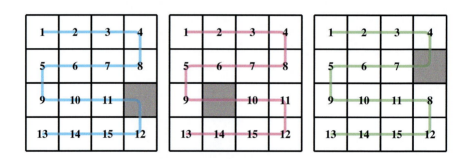

Fig. 2.3: Three examples of puzzles, P, Q and R.

We define an equivalence relation on the set of puzzles: two puzzles are equivalent if we can obtain one from the other by moving the empty tile along the snaking path. Each equivalence class is called a **configuration**, and contains 16 puzzles, one for each tile the empty-tile can occupy. There are 16! puzzles, and 15! configurations. Thus the puzzle-permutations are all elements of $\mathfrak{S}_{15}$.

We say that a puzzle-permutation w_Q^P is **legal** if P and Q differ by a legal puzzle move, that is by sliding the empty tile up/down or left/right. We let $G_{\text{puzz}} \leqslant \mathfrak{S}_{15}$ be the subgroup generated by the set of legal puzzle-permutations. We now return to Sam Loyd's favourite two puzzles, which we re-picture with their reading words in Figure 2.4.

2.3 The 15-puzzle

A \$1,000 group theoretic question. *Now, Sam Loyd's question can be recast as asking whether or not the puzzle-permutation w_T^S depicted below*

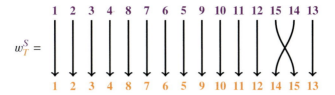

belongs to the group generated by all legal puzzle-permutations. In other words, is $w_T^S \in G_{\text{puzz}}$? Having recast Sam Loyd's question in group theoretic terms, we now dedicate our efforts to understanding G_{puzz}.

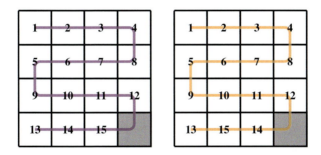

Fig. 2.4: The reading words for Sam Loyd's pair of puzzles, S and T. Their reading words are $\text{read}(S) = [1, 2, 3, 4, 8, 7, 6, 5, 9, 10, 11, 12, 15, 14, 13]$ and $\text{read}(T) = [1, 2, 3, 4, 8, 7, 6, 5, 9, 10, 11, 12, 14, 15, 13]$.

We label the tiles in the grid by the pair $[r, c]$ for $1 \leqslant r, c \leqslant 4$ the row and column indices (ascending upwards and left-to-right). We choose to label the entries of the initial puzzle, P, so that the subscripts increase along the snake, see Figure 2.5.

Left to right. Let P be a puzzle and suppose that Q is obtained from P by moving the empty-tile from left to right or vice versa, i.e. moving the empty-tile $[r, c] \to [r, c+1]$ or $[r, c] \to [r, c-1]$. Clearly, moving the empty-tile along the snake ordering leaves the reading word unchanged, i.e. $\text{read}(P) = \text{read}(Q)$. Therefore, moving the box from *left-and-right* induces the identity permutation $w_Q^P = \text{id}_{\mathfrak{S}_{15}}$.

Up and down. Let P be a puzzle and suppose that Q is obtained from P by moving the empty tile up or down, i.e. moving the empty tile $[r, c] \to [r + 1, c]$ or $[r, c] \to [r - 1, c]$ to obtain a new puzzle, Q. There are 12 such possible moves in total (up to swapping P and Q). If the empty-tile is moving along the snake (in other words, if we move $[1, 4] \to [2, 4]$, or $[2, 1] \to [3, 1]$, or $[3, 4] \to [4, 4]$) then the reading word of P is equal to that of Q and so $w_Q^P = \text{id}_{\mathfrak{S}_{15}}$. We now consider the remaining 9 cases. We take P to be an arbitrary puzzle and consider the effect of moving the empty-tile $[3, 1] \to [4, 1]$ to obtain a second puzzle Q. Such a pair P and Q are depicted in Figure 2.5.

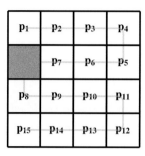 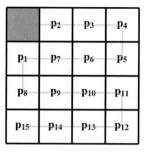

Fig. 2.5: The initial puzzle U and the puzzle V obtained from the move $[3, 1] \to [4, 1]$. We have that $w_V^U = (1, 7, 6, 5, 4, 3, 2)$.

For a general puzzle, we only record the subscripts in the entries, for ease of notation. Thus, for $[3, 1] \to [4, 1]$ (pictured in Figure 2.5) we obtain

$$w_V^U = \quad = (7, 6, 5, 4, 3, 2, 1)$$

and when we consider the inverse move $[4, 1] \to [3, 1]$ we obtain

$$w_U^V = \quad = (1, 2, 3, 4, 5, 6, 7).$$

and we note that the latter permutation is the inverse of the former.

There are eight further ways to move the empty-tile upwards in this puzzle (in addition to those moves already considered). Treating the moves $[3, 2] \to [4, 2]$, $[3, 3] \to [4, 3]$, $[2, 2] \to [3, 2]$, $[2, 3] \to [3, 3]$, and $[2, 4] \to [3, 4]$, in a similar fashion, we obtain the permutations

$$(2, 6, 5, 4, 3) \quad (3, 5, 4) \quad (7, 9, 8) \quad (6, 10, 9, 8, 7) \quad (5, 11, 10, 9, 8, 7, 6)$$

and treating the moves $[1, 3] \to [2, 3]$, $[1, 2] \to [2, 2]$, and $[1, 1] \to [2, 1]$, again in a similar fashion we obtain

2.3 The 15-puzzle

$$(11, 13, 12) \quad (10, 14, 13, 12, 11) \quad (9, 15, 14, 13, 12, 11, 10).$$

Pushing the empty box downward we obtain the inverses of these permutations (just as we did above for $w_V^U = (1, 7, 6, 5, 4, 3, 2) = (1, 2, 3, 4, 5, 6, 7)^{-1} = (w_U^V)^{-1}$ for U, V pictured in Figure 2.5 above).

Sam Loyd's prize: Notice that all the cycles we have obtained are of even length! Therefore any product of these cycles will also be of even length (because even + even = even, or more precisely Corollary 2.2.4). In particular, the transposition (14, 15) is not an element of the group generated by all possible puzzle moves (as it is of odd length, equal to 1). Thus we can deduce that Sam Loyd's puzzle was a con and he would never have to pay out the $1,000.

We can now go beyond Sam Loyd's prize question. We know that the subgroup $G_{\text{puzz}} \leqslant \mathfrak{S}_{15}$ consists only of even permutations, thus $G_{\text{puzz}} \leqslant A_{15}$. We claim that, in fact, G_{puzz} consists of *all* even permutations and so $G_{\text{puzz}} = A_{15}$.

Beyond Sam Loyd's prize: We claim that we can obtain any cycle of the form $(a, a+1, a+2)$ or $(a+2, a+1, a)$ as a product of the above legal puzzle permutations. To see this note that

$$(7, 6, 5, 4, 3, 2, 1)(3, 4, 5)(1, 2, 3, 4, 5, 6, 7) = (2, 3, 4)$$

and more generally, for $0 \leqslant n < 7$, we have

$$(7, 6, 5, 4, 3, 2, 1)^n (3, 4, 5)(1, 2, 3, 4, 5, 6, 7)^n = (3 - n, 4 - n, 5 - n)$$

where the numbers are read modulo 7 and so, in this fashion, we obtain

$$(3, 4, 5), \quad (2, 3, 4), \quad (1, 2, 3), \quad (7, 1, 2), \quad (6, 7, 1), \quad (5, 6, 7), \quad (4, 5, 6)$$

respectively. We can repeat this with the cycle (5, 4, 3) and hence obtain the inverses of each of these 7 elements. We then repeat all of the above with

$$(11, 10, 9, 8, 7, 6, 5)^n (7, 8, 9)(5, 6, 7, 8, 9, 10, 11)^n$$
$$(15, 14, 13, 12, 11, 10, 9)^n (11, 12, 13)(9, 10, 11, 12, 13, 14, 15)^n$$

to obtain all 3-cycles of the form $(a, a + 1, a + 2)$ for $1 \leqslant a \leqslant 5$ (similarly for the inverses). By Proposition 2.2.7, these cycles generate the group A_{15}. Finally, to finish the proof, we notice that

- all legal permutations can be obtained by compositions of the moves we have considered
- therefore these moves generate the group $G_{\text{puzz}} = A_{15} < \mathfrak{S}_{15}$.
- The subgroup $G_{\text{puzz}} = A_{15} < \mathfrak{S}_{15}$ consists of half of the elements of $\mathfrak{S}_{15}$.

Thus any puzzle configuration can be obtained from one and only one of the two configurations in Figure 2.1. Moreover, of the total of 16! puzzles, half can be obtained from the former and half from the latter. Theorem 2.3.1 follows.

2.4 The hyper-octahedral groups

We now consider the permutations satisfying a certain symmetry property. We take the element of $\mathfrak{S}_{2n}$ and we draw a vertical line down the centre of the diagram; we say that an element is **symmetric** if it is fixed by reflection through this this line. This is best illustrated via an example, as follows:

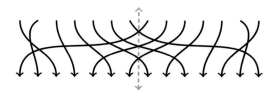

Clearly vertical concatenation of such symmetric diagrams preserves the underlying symmetry. We recall that the inverse map is given by flipping the diagram through the horizontal axis, and this clearly preserves the property of being a symmetric diagram. Therefore, for a given $n \in \mathbb{N}$, the set of symmetric diagrams $H_n \leqslant \mathfrak{S}_{2n}$ forms a subgroup, which we call the **hyper-octahedral group**. This group is generated by the $(n-1)$ "symmetrized" or "doubled-up" transpositions of the form

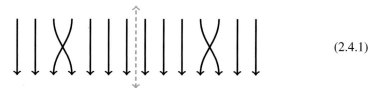

(2.4.1)

along with the one additional "wall-crossing" transposition

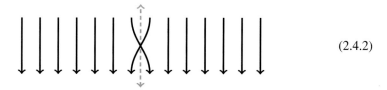

(2.4.2)

We label the strands from left-to-right by

$$-n, \ 1-n, \ \ldots, \ -2, \ -1, \ 1, \ 2, \ \ldots, \ n-1, \ n$$

and in this way we can label the $(n-1)$ symmetrized transpositions in a similar fashion to the classical symmetric group. For example, the element in (2.4.1) can be labelled by s_4. We can label the "wall-crossing" transposition (swapping 1 and -1, pictured in (2.4.2)) by s_0. In this manner, H_n is generated by s_0 together with $s_1, \ldots, s_{n-1}$. This also makes it clear that we have a proper subgroup $\mathfrak{S}_n < H_n$ generated by $s_1, \ldots, s_{n-1}$. With this labelling, we can also see that $s_0 s_1 s_0 s_1 = s_1 s_0 s_1 s_0$ and this element can be pictured as follows

2.5 Groups of decorated strand diagrams

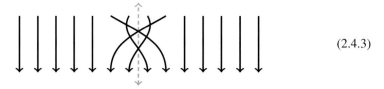

(2.4.3)

and we note that s_0 commutes with the remaining generators $s_2, s_3, \ldots, s_{n-1}$.

Exercise 2.4.4. Construct an isomorphism $\varphi : D_8 \to H_2$.

Exercise 2.4.5. Calculate the order of H_n for $n \geq 2$.

2.5 Groups of decorated strand diagrams

We now define a way of constructing "new groups from old". This is given by a procedure which takes as inputs an arbitrary group, G, and the symmetric group, $\mathfrak{S}_n$, and outputs a "wreath product" of these groups.

Definition 2.5.1. Let $(G, *_G)$ be a group. We define $G \wr \mathfrak{S}_n$ to be the wreath product group whose elements are given by the n-strand permutations with at n-tuple of decorations: given by attaching an element of G to each strand. The group multiplication of two decorated-strand diagrams is given by: first concatenation the diagrams to obtain a new strand diagram; then perform the $*_G$-operation strand-wise to obtain a new n-tuple of decorations. This is illustrated below.

For concreteness let $n = 3$, and take $g_1, g_2, g_3, h_1, h_2, h_3 \in G$ for some arbitrary group G. We have the following product

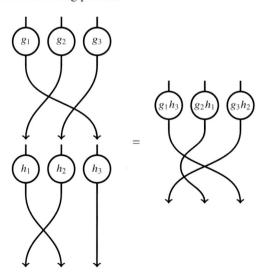

where the strands are redrawn as usual, with labels given by the products of the constituent labels. Now it is clear that there are $|G|^n n!$ such diagrams and that their product will always be such a diagram (although keeping track of the labels is a bit tedious). With this notation in place, it becomes clear that the inverse of an element can be given by flipping the diagram through the vertical axis and inverting each strand-label. So, for example

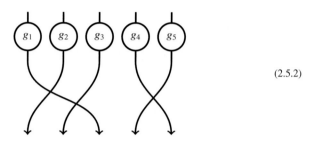

(2.5.2)

has inverse given by the decorated strand diagram

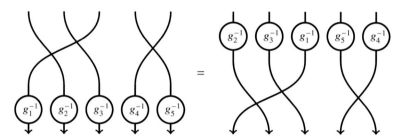

where the lefthand-diagram is the "naive flip" of the diagram in (2.5.2) and the righthand-diagram is given by moving the labels along their corresponding strands (both the left and righthand diagrams are equal, we merely prefer the uniformity of putting "all labels at the top"). Should one so wish, one can think of each strand as being "coloured" by an element $g \in G$. Finally, we note that the identity is given by placing $\mathrm{id}_G \in G$ label on every strand of the identity permutation diagram.

Exercise 2.5.3. Let C_2 be the cyclic group of order 2 and H_n be the hyperoctahedral group. Construct an isomorphism $\varphi : H_n \to C_2 \wr \mathfrak{S}_n$.

Exercise 2.5.4. Construct a subgroup of $C_3 \wr \mathfrak{S}_4$ of order 972.

Now, we say that a matrix is monomial if it has precisely one non-zero entry in each row and each column. Notice that the $(n \times n)$-permutation matrices are precisely the $(n \times n)$-monomial matrices whose non-zero entries are *all equal to 1*, and recall that these matrices form a group, $\mathfrak{S}_n$.

Exercise 2.5.5. Let G_n be the set of monomial $(n \times n)$-matrices whose non-zero entries are all ± 1. We define

$$W_n = \{x \in G_n, | \text{ the matrix } x \text{ has an even number of entries equal to } -1\}.$$

For example, the following matrices

$$\begin{bmatrix} 1 & 0 & 0 \\ 0 & 0 & -1 \\ 0 & -1 & 0 \end{bmatrix}, \begin{bmatrix} 0 & -1 & 0 \\ 0 & 0 & -1 \\ 1 & 0 & 0 \end{bmatrix}, \begin{bmatrix} 0 & 0 & 1 \\ 1 & 0 & 0 \\ 0 & 1 & 0 \end{bmatrix}$$

are all elements of W_3. Verify that $W_n \leqslant G_n$ is a subgroup. Calculate the order of this group, $|W_n|$. Find all elements $w \in W_3$ of order 6.

Exercise 2.5.6. Let $G = D_8 \wr \mathfrak{S}_2$. Write down a set of generators and relations of the group G. Let $H \leqslant \mathrm{GL}_4(\mathbb{C})$ be the subgroup generated by the following five matrices

$$\begin{bmatrix} 0 & 1 & 0 & 0 \\ -1 & 0 & 0 & 0 \\ 0 & 0 & 1 & 0 \\ 0 & 0 & 0 & 1 \end{bmatrix}, \begin{bmatrix} 1 & 0 & 0 & 0 \\ 0 & -1 & 0 & 0 \\ 0 & 0 & 1 & 0 \\ 0 & 0 & 0 & 1 \end{bmatrix}, \begin{bmatrix} 0 & 0 & 0 & 0 \\ 0 & 0 & 0 & 0 \\ 0 & 0 & 0 & 1 \\ 0 & 0 & -1 & 0 \end{bmatrix}, \begin{bmatrix} 1 & 0 & 0 & 0 \\ 0 & 1 & 0 & 0 \\ 0 & 0 & 1 & 0 \\ 0 & 0 & 0 & -1 \end{bmatrix}, \begin{bmatrix} 0 & 0 & 1 & 0 \\ 0 & 0 & 0 & 1 \\ 1 & 0 & 0 & 0 \\ 0 & 1 & 0 & 0 \end{bmatrix}.$$

Construct an isomorphism from G to H.

2.6 Coxeter groups

All the groups we have considered thus far in this chapter fit into the common framework of *Coxeter groups*.

Definition 2.6.1. We say that a group W is a Coxeter group if it is generated by some Coxeter generators $s_i \in W$ indexed by elements of a finite set $i \in S_W$, subject to relations solely of the form $s_i^2 = 1$ and $(s_i s_j)^{m(i,j)} = 1$ for $i, j \in S_W$ and some $m_{i,j} \geqslant 2$. We refer to such a presentation as a Coxeter presentation of W.

Proposition 2.6.2. *The symmetric group, $\mathfrak{S}_n$, admits the Coxeter presentation*

$$\langle s_1, \ldots, s_{n-1} \mid s_i^2 = 1, (s_i s_{i+1})^3 = 1, (s_i s_j)^2 = 1 \text{ for } |i - j| > 1 \rangle.$$

Proof. We define $G_n = \langle s_1, \ldots, s_{n-1} \mid s_i^2 = 1, (s_i s_{i+1})^3 = 1, (s_i s_j)^2 = 1 \text{ for } |i-j| > 1 \rangle$ and we claim that the map $\phi : G_n \to \mathfrak{S}_n$ given by $\phi : s_i \mapsto (i, i+1)$ is an isomorphism. It is immediate that ϕ is surjective and that

$$\phi(g_1 *_{G_n} g_2) = \phi(g_1) *_{\mathfrak{S}_n} \phi(g_2)$$

for all $g_1, g_2 \in G$. The three relations in the presentation follow from the fact that each 2-cycle has order 2, each 3-cycle has order 3, and non-adjacent transpositions commute (respectively). It remains to check that the map is injective (or equivalently, that $G_n = n!$). To see this, simply note that every element of G_n is equal to some $\pi \in G_{n-1} \leqslant G_n$ or is of the form $s_i \ldots s_{n-1} \pi$ for some $1 \leqslant i < n$ and $\pi \in G_{n-1}$ and so the result follows by induction. $\square$

We summarise the information from the Coxeter presentation in a corresponding Coxeter graph, as follows. We draw a node for each generator s_i for some $i \in S_W$ and we draw a total of $(m_{i,j} - 2)$ edges between each pair of nodes $i, j \in S_W$. In this way (by subtracting 2 from each $m_{i,j}$) we can suppress the vast number of commuting relations from cluttering up our graph. For example, the Coxeter graph of $\mathfrak{S}_n$ is depicted in Figure 2.6.

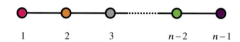

Fig. 2.6: The Coxeter graph of the symmetric group $\mathfrak{S}_n$.

Proposition 2.6.3. *The hyper-octahedral groups are Coxeter groups with graphs pictured in Figure 2.7.*

The Coxeter graph of H_n is obtained from that of $\mathfrak{S}_n$ be adding a single node s_0 corresponding to the "wall-crossing" transposition. While we do not provide a proof of Proposition 2.6.3, we remark that the proof for the hyper-octahedral groups is very similar to that of the symmetric groups; the only interesting relation is that between s_0 and s_1. We checked this relation in (2.4.3).

Fig. 2.7: The Coxeter graph of the hyper-octahedral group H_n

Exercise 2.6.4. Let W_n be the group defined in Exercise 2.5.5. Show that the group W_4 is generated by the matrices

$$\begin{bmatrix} -1 & 0 & 0 & 0 \\ 0 & -1 & 0 & 0 \\ 0 & 0 & 1 & 0 \\ 0 & 0 & 0 & 1 \end{bmatrix} \begin{bmatrix} 0 & 1 & 0 & 0 \\ 1 & 0 & 0 & 0 \\ 0 & 0 & 1 & 0 \\ 0 & 0 & 0 & 1 \end{bmatrix} \begin{bmatrix} 1 & 0 & 0 & 0 \\ 0 & 0 & 1 & 0 \\ 0 & 1 & 0 & 0 \\ 0 & 0 & 0 & 1 \end{bmatrix} \begin{bmatrix} 1 & 0 & 0 & 0 \\ 0 & 1 & 0 & 0 \\ 0 & 0 & 0 & 1 \\ 0 & 0 & 1 & 0 \end{bmatrix}$$

and provide a complete list of relations for W_4. Hence show that W_4 is a Coxeter group and construct its Coxeter graph. Generalise the above to W_n for $n > 4$.

Exercise 2.6.5. Construct the Coxeter graph of the dihedral group, D_{2n} for $n \in \mathbb{N}$.

Remark 3. The Weyl groups are Coxeter groups of particular importance as they arise as reflection groups of Euclidean spaces.

Chapter 3
Composition series

We now delve a little deeper into the structure theory of groups. We will study groups by *dividing* them into smaller pieces (a normal subgroup and a quotient group) and studying maps which pass this structural information from one group to another (homomorphisms). This provides us with a notion of a "simple group" (one that cannot be divided!) and a "composition series" (whereby we factorise a group into its simple constituents).

3.1 Cosets and Lagrange's theorem

The main result of this section will be Lagrange's theorem, which states that if $H \leqslant G$ and G is a finite group, then $|H|$ divides $|G|$. The idea of the proof (which will occupy most of this section) is to partition G in such a way that each part has the same size as H. So if we have k parts then $|G| = k|H|$ and hence we have that $|H|$ divides $|G|$.

Definition 3.1.1. Let $H \leqslant G$ and let $g \in G$. The left coset gH of H is defined by

$$gH = \{x \in G \mid x = gh \text{ for some } h \in H\}.$$

The right coset Hg of H is defined by

$$Hg = \{x \in G \mid x = hg \text{ for some } h \in H\}.$$

Example 3.1.2. We begin by constructing the left cosets of $H = \{\text{id}, f\} \leqslant D_6 = G$. We first circle the elements of $H = \{\text{id}, f\}$. We then pick any uncircled element, say $r \in G$ and circle the elements $rH = r\{\text{id}, f\} = \{r, rf\}$. We then pick any uncircled element, say $r^2 \in G$ and circle the elements $r^2 H = r^2\{\text{id}, f\} = \{r^2, r^2 f\}$. We have now run out of uncircled elements and so we have finished constructing the left cosets (these are depicted in the leftmost diagram of Figure 3.1).

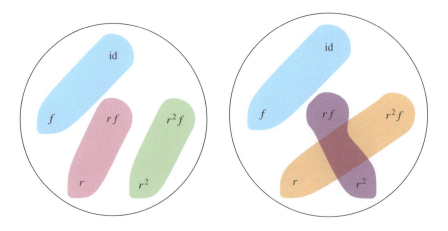

Fig. 3.1: The left and right cosets of $H = \{\mathrm{id}, f\} \leqslant D_6 = G$ respectively.

We now construct the right cosets of $H = \{\mathrm{id}, f\} \leqslant D_6 = G$. We again circle the elements of $H = \{\mathrm{id}, f\}$. We then pick any uncircled element, say $r \in H$ and circle the elements $Hr = \{\mathrm{id}, f\}r = \{r, fr\} = \{r, r^2f\}$. We then pick any uncircled element, say $r^2 \in G$ and circle the elements $Hr^2 = \{\mathrm{id}, f\}r^2 = \{r^2, fr^2\} = \{r^2, rf\}$. We have now run out of uncircled elements and so we have finished constructing the right cosets (these are depicted in the rightmost diagram of Figure 3.1).

Importantly, we notice that the left and right cosets of $H = \{\mathrm{id}, f\} \leqslant D_6 = G$ do not coincide.

Example 3.1.3. We now construct the left (and right) cosets of $H = \{\mathrm{id}, r^2\} \leqslant D_8 = G$. We first circle the elements of $H = \{\mathrm{id}, r^2\}$. We then pick any uncircled element, say $r \in G$ and circle the elements $rH = r\{\mathrm{id}, r^2\} = \{r, rr^2\} = \{r, r^3\}$. We then pick any uncircled element, say $f \in G$ and circle the elements $fH = f\{\mathrm{id}, r^2\} = \{f, fr^2\} = \{f, r^2f\}$. We then pick any uncircled element, say $rf \in G$ and $rfH = rf\{\mathrm{id}, r^2\} = \{rf, rfr^2\} = \{rf, rr^2f\} = \{rf, r^3f\}$. We have now finished constructing the left cosets.

One can now construct the right cosets similarly. We remark that the generators r and f both commute with both elements of $\{\mathrm{id}, r^2\}$ and so the right cosets are precisely the same sets of elements as the left cosets. These cosets are depicted in Figure 3.2.

Notice that in both Examples 3.1.2 and 3.1.3 we have choices of coset "representative". For example if $H = \{\mathrm{id}, r^2\} \leqslant D_8$, we have that

$$r\{\mathrm{id}, r^2\} = \{r * \mathrm{id}, r * r^2\} = \{r, r^3\} = \{r^3 * \mathrm{id}, r^3 * r^2\} = r^3\{\mathrm{id}, r^2\}$$

3.1 Cosets and Lagrange's theorem

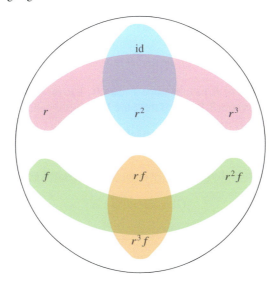

Fig. 3.2: The left and right cosets of $H = \{\text{id}, r^2\} \leq D_8 = G$. As these cosets coincide, we only require one picture (unlike in Figure 3.1).

and so rH and $r^3 H$ are the same and r and r^3 are both representatives of the same coset. More generally, each left/right coset gH has precisely $|H|$ distinct left/right coset representatives.

Lemma 3.1.4. *If $H \leq G$ is a finite subgroup and $g \in G$ then $|gH| = |Hg| = |H|$.*

Proof. Say $H = \{h_1, h_2, \ldots, h_m\}$. Then gH has elements $gh_1, gh_2, \ldots, gh_m$. Note that these are all distinct as if we had $gh_i = gh_j$ then by Theorem 1.3.5 we would have that $h_i = h_j$. Thus we have $|gH| = |H|$. One can argue similarly for $|Hg|$. □

Theorem 3.1.5. *Let H be a subgroup of G and let $g_1, g_2 \in G$. Then the left cosets $g_1 H$ and $g_2 H$ are either identical or they have no element in common, that is, either $g_1 H = g_2 H$ or $g_1 H \cap g_2 H = \emptyset$.*

Proof. We will assume that $g_1 H$ and $g_2 H$ do have a common element, and we will prove that they are identical. Suppose $x \in g_1 H \cap g_2 H$, then we have that

$$x = g_1 h_1 = g_2 h_2 \tag{3.1.6}$$

for some $h_1, h_2 \in H$. We need to show that $g_1 H = g_2 H$. First we show that $g_1 H \subseteq g_2 H$. Let $y \in g_1 H$ then $y = g_1 h$ for some $h \in H$. By (3.1.6), we have that

$$y = g_1 h = (g_2 h_2 h_1^{-1})h = g_2(h_2 h_1^{-1} h) \in g_2 H.$$

and hence $g_1 H \subseteq g_2 H$. Similarly we can show that $g_2 H \subseteq g_1 H$, hence we have $g_1 H = g_2 H$. □

This brings us to our first substantial theorem on general group-theoretic structures, Lagrange's theorem:

Theorem 3.1.7. *If G is a finite group and H is a subgroup of G then $|H|$ divides $|G|$.*

Proof. Each left coset of H has the same size as H. The set of distinct cosets form a partition of G by Theorem 3.1.5. So if there are k distinct cosets then $|G| = k|H|$. □

Definition 3.1.8. The number of distinct left (or right) cosets of H in G is called the index of H in G, denoted by $[G : H]$.

Using this definition, Lagrange's theorem can be restated as

$$|G| = |H| \times [G : H]$$

where $[G : H] \in \mathbb{Z}_{>0}$.

Corollary 3.1.9. *Suppose that G is a group with $|G| = p$ a prime number. Then G is isomorphic to the cyclic group C_p. Moreover, its only subgroups are $\{\mathrm{id}\}$ and G itself.*

Proof. Since $p > 1$, G has an element $g \neq \mathrm{id}$. So $|\langle g \rangle| > 1$ and by Lagrange's theorem it must divide p. So we have $|\langle g \rangle| = p$ and hence $\langle g \rangle = G$. Thus $G = \{\mathrm{id}, g, g^2, \ldots, g^{p-1}\} \cong C_p$. Now using Lagrange's theorem again we see that any subgroup of G must have order 1 or p, and hence must be either $\{\mathrm{id}\}$ or G itself. □

We have seen that, generally speaking, left cosets are not equal to right cosets. In the next chapter we consider the subgroups for which left and right cosets *do* coincide.

Example 3.1.10. We have that $A_n \leqslant \mathfrak{S}_n$ is a subgroup of index 2, for all $n \geqslant 2$.

Exercise 3.1.11. Label the vertices of a regular n-gon as follows

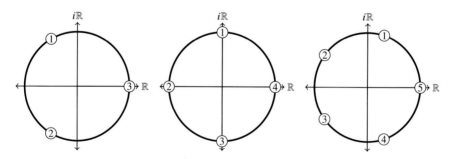

Write down the effect of applying a rotation or flip $r^j f^k \in D_{2n}$ to the set of vertices (as a permutation of the set $\{1, \ldots, n\}$). Using this, construct an injective group homomorphism $\varphi : D_{2n} \to \mathfrak{S}_n$ and hence identify D_{2n} as a subgroup of $\mathfrak{S}_n$. Calculate the index of D_{2n} as a subgroup of $\mathfrak{S}_n$.

Exercise 3.1.12. Calculate the index of $\mathfrak{S}_n \leqslant H_n$ and of $H_n \leqslant \mathfrak{S}_{2n}$.

3.2 Simple groups and normal subgroups

We have seen that the order of a subgroup divides that of the ambient group, thus we have a way of "dividing" the *size* of a group, but this *does not* give us a way of dividing the *group theoretic* structure. In order to remedy this, we introduce the idea of a *normal* subgroup, as follows:

Definition 3.2.1. A subgroup H of a group G is called a normal subgroup if its left and right cosets coincide, that is, if $gH = Hg$ for all $g \in G$. In this case we write $H \trianglelefteq G$.

Example 3.2.2. We have already seen in Examples 3.1.2 and 3.1.3 that $\{\mathrm{id}, f\} \leq D_6$ is not a normal subgroup, whereas $\{\mathrm{id}, r^2\} \leq D_8$ is a normal subgroup.

Exercise 3.2.3. Is the subgroup $\langle r \rangle \leq D_6$ a normal subgroup?

Example 3.2.4. For $n = pq$, the group $C_n = \langle g \mid g^n = \mathrm{id}\rangle$ has a subgroup

$$C_p = \{g^q, g^{2q}, \ldots, g^{pq-q}, g^{pq}\} = \{g^q, g^{2q}, \ldots, g^{(p-1)q}, \mathrm{id}_G\}.$$

To see that $C_p \leq C_n$ is a normal subgroup, we simply note that

$$\begin{aligned} gC_p &= g\{g^q, g^{2q}, \ldots, g^{(p-1)q}, \mathrm{id}_G\} = \{g * g^q, g * g^{2q}, \ldots, g * g^{(p-1)q}, g * \mathrm{id}_G\} \\ &= \{g^{q+1}, g^{2q+1}, \ldots, g^{(p-1)q+1}, g\} \\ &= \{g^q * g, g^{2q} * g, \ldots, g^{(p-1)q} * g, \mathrm{id}_G * g\} \\ &= \{g^q, g^{2q}, \ldots, g^{(p-1)q}, \mathrm{id}_G\}g = C_p g. \end{aligned}$$

Exercise 3.2.5. Let G be an abelian group and $H \leq G$ an arbitrary subgroup. Prove that H is a normal subgroup of G.

Example 3.2.6. Assume $H \leq G$ and $[G : H] = 2$. In which case $H \leq G$ is a normal subgroup. To see this, simply note that $\mathrm{id} * H = H * \mathrm{id}$ and that for any $g \notin H$, we have that gH has trivial intersection with H (and is of order $|H|$) and therefore consists of every element in $G \setminus H$. The same argument holds for Hg and so $gH = \{g \in G \mid g \notin H\} = Hg$. In particular, $A_n \leq \mathfrak{S}_n$ and $C_n \leq D_{2n}$ for $n \geq 2$ are both infinite families of normal subgroups.

Exercise 3.2.7. Show that $\mathfrak{S}_4$ has a normal subgroup isomorphic to Rect.

Exercise 3.2.8. Find all normal subgroups of $\mathfrak{S}_4$.

Example 3.2.9. We have that $\mathsf{Rect} = \{\mathrm{id}, r^2, f, r^2 f\} \leq D_8$ is a subgroup of index 2 and so by Example 3.2.6 we deduce that $\mathsf{Rect} \triangleleft D_8$. We note that Rect is an abelian group and so by Exercise 3.2.5 we deduce that $H = \{\mathrm{id}, f\} \triangleleft \mathsf{Rect}$. However, H is not a normal subgroup of D_8 as we can see by checking that $rH = \{r, rf\} \neq \{r, r^3 f\} = Hr$. Thus we conclude that the property of being a normal subgroup is not transitive.

We want to try to define a group structure on the set of (left, say) cosets by defining $(g_1 H)(g_2 H) = (g_1 g_2) H$. Does this make sense? Or does the product depend on the given choice of coset representatives? In fact, this idea only makes sense for normal subgroups...

Theorem 3.2.10. *Let H be a normal subgroup in G. Then coset multiplication given by $(g_1H)(g_2H) = (g_1g_2)H$ for all $g_1, g_2 \in G$ is well-defined.*

We emphasise that this iw *only* well-defined when H is a normal subgroup in G. It does not work for subgroups which are not normal.

Proof of Theorem 3.2.10. Choose $g_1' \in g_1H$ and $g_2' \in g_2H$. Then we have $g_1' = g_1h_1$ and $g_2' = g_2h_2$ for some $h_1, h_2 \in H$. So we get

$$(g_1'H)(g_2'H) = (g_1'g_2')H = (g_1h_1g_2h_2)H.$$

We have to show that $(g_1h_1g_2h_2)H = (g_1g_2)H$. Now $h_1g_2 \in Hg_2 = g_2H$, so we have $h_1g_2 = g_2h_1'$ for some $h_1' \in H$. Thus we get

$$(g_1h_1g_2h_2)H = (g_1g_2h_1'h_2)H = (g_1g_2)H$$

as required. □

Corollary 3.2.11. *Let H be a normal subgroup of a group G. Then the set of cosets of H in G forms a group with the group multiplication given by*

$$(g_1H)(g_2H) = (g_1g_2)H$$

for all $g_1, g_2 \in G$. We refer to this group as the **quotient group** *of G by H, and denote it by G/H.*

Proof. We have that $\mathrm{id}_{G/H} = H$ as $(gH)H = gH = H(gH)$ and so the identity axiom holds. The closure axiom holds because we have defined the product of two left cosets of H to be a left coset of H (this multiplication is well-defined by Theorem 3.2.10). We can define inverses $(gH)^{-1} := (g^{-1})H$, and then one has that $(gH)(g^{-1}H) = (gg^{-1})H = H$ and so the inverse axiom holds. Finally, one can check that

$$(g_1H)((g_2H)(g_3H)) = (g_1H)((g_2g_3)H) = (g_1(g_2g_3))H$$

and

$$((g_1H)(g_2H))(g_3H) = ((g_1g_2)H)(g_3H) = ((g_1g_2)g_3)H$$

coincide using associativity in G. □

Example 3.2.12. We have that $A_n \triangleleft \mathfrak{S}_n$ and $\mathfrak{S}_n/A_n \cong C_2$.

Exercise 3.2.13. Let C_{24} denote the cyclic group of order 24. Construct normal subgroups of C_{24} isomorphic to C_3 and C_4; and calculate the resulting quotient groups.

Exercise 3.2.14. Verify that $\{\mathrm{id}, (1,2)(3,4), (1,3)(2,4), (1,4)(2,3)\} \leqslant A_4$ is a normal subgroup. Calculate the resulting quotient group.

3.2 Simple groups and normal subgroups

Exercise 3.2.15. Find all the normal subgroups of D_8 and D_{10}. Now generalise to calculate all normal subgroups of D_{2n}; calculate all the resulting quotient groups.

Thus we have achieved our goal: we have found a way to *divide* a group G by another group H (a normal subgroup) and hence obtain a *quotient* group G/H. Thus we can think of groups as being built-up from smaller groups, this leads us to the question: what are the smallest groups from which all other groups are built?

Definition 3.2.16. We say that a finite group G is simple if it has no normal subgroups.

Finite simple groups provide the "atoms" from which we build all other finite groups. Before providing our favourite examples of simple groups, we first prove a useful result. Given $g, h \in G$ we refer to the element ghg^{-1} as the conjugate of $h \in G$ by $g \in G$.

Proposition 3.2.17. *Let H be a subgroup of a group G. Then H is a normal subgroup if and only if we have $ghg^{-1} \in H$ for all $h \in H, g \in G$.*

Proof. Suppose that $gH = Hg$ for all $g \in G$. Then we have $gh = h'g$ for some $h' \in H$ and hence we get $ghg^{-1} = h' \in H$ as required.

Now suppose $ghg^{-1} \in H$ for all $h \in H$ and for all $g \in G$. Then we have that $ghg^{-1} = h'$ for some $h' \in H$. So we have $gh = h'g \in Hg$. This shows that $gH \subseteq Hg$. On the other hand, replacing g by g^{-1} we also have $g^{-1}hg \in H$, that is $g^{-1}hg = h''$ for some $h'' \in H$. So we get $hg = gh'' \in gH$ and hence $Hg \subseteq gH$ as required. □

We are now ready to construct our first interesting infinite family of simple groups.

Theorem 3.2.18. *The alternating groups A_n for $n \neq 4$ are simple.*

Proof. For $n = 3$ the result is trivial, we suppose $n > 4$. We first prove two claims, which will provide the backbone of the proof.

Claim 1: For any fixed choice of $1 \leq a < b \leq n$, the group A_n is generated by the set of $(n-2)$ elements: (a, b, k) for $k \in \{1, \ldots, n\} \setminus \{a, b\}$.

Proof of Claim 1: Without loss of generality we fix $a = 1$ and $b = 2$. By Proposition 2.2.7 it suffices to write any 3-cycle as a product of 3-cycles $(1, 2, k)$ for $3 \leq k \leq n$ and their inverses. We have that $(1, 2, k)^{-1} = (1, k, 2)$ and therefore we can obtain any $(1, k, \ell)$ as

$$(1, k, \ell) = (1, k, 2)(1, 2, \ell)$$

for any $1 < k, \ell \leq n$. As in the previous step, we can obtain any

$$(2, k, \ell) = (1, 2, k)(1, \ell, 2)$$

for any $1 < k, \ell \leq n$. Finally, we can obtain any

$$(k, \ell, m) = (1, 2, k)(1, \ell, 2)(1, 2, m)(1, k, 2)$$

for any $2 < k, \ell, m \leq n$ as required.

Claim 2: Let $H \triangleleft A_n$ be a normal subgroup containing a 3-cycle $(a, b, c) \in H$. Then $H = A_n$.

Proof of Claim 2: We can assume that $a = 1, b = 2, c = 3$ so that $(1, 2, 3) \in H$ (and therefore $(1, 3, 2) \in H$). Conjugating by $(1, 2)(3, k) \in \mathfrak{S}_n$ for $3 \leq k \leq n$, we have that
$$(1, 2, k) = (1, 2)(3, k)(1, 3, 2)(1, 2)(3, k) \in H$$
by Proposition 3.2.17 and the normality of the subgroup $(1, 3, 2) \in H \triangleleft A_n$. Therefore $(1, 2, k) \in H$ for all $k \geq 3$ and $H = A_n$ by Claim 1.

Armed with our above claims, we now consider $H \triangleleft A_n$ an arbitrary normal subgroup. We need only show that H contains a 3-cycle to deduce our result.

(i) First suppose that $g \in H$ has disjoint cycle decomposition $g = (1, 2, \ldots, r)\sigma$ with $r \geq 4$. Conjugating $g \in H$ by $(1, 2, 3) \in \mathfrak{S}_n$ and then multiplying by $g^{-1} \in H$, we have that
$$\begin{aligned}((1, 3, 2)g(1, 2, 3))g^{-1} &= ((1, 3, 2)(1, 2, \ldots, r)\sigma(1, 2, 3))\sigma^{-1}(r, \ldots, 2, 1) \\ &= ((1, 3, 2)(1, 2, \ldots, r)(1, 2, 3))(r, \ldots, 2, 1) \\ &= (2, 3, 1, 4, 5, \ldots, r)(r, \ldots, 3, 2, 1) \\ &= (1, 3, r) \in H\end{aligned}$$
and so $H = A_n$ by Claim 2. (The second equality follows as σ and $(1, 2, 3)$ are disjoint.)

(ii) Now suppose that $g \in H$ has disjoint cycle decomposition $g = (1, 2, 3)(4, 5, 6)\sigma$. We have that
$$((1, 4, 2)g(1, 2, 4))g^{-1} = (1, 4, 2, 6, 3) \in H$$
and so $H = A_n$ by (i) above, because $(1, 4, 2, 6, 3)$ has length greater than 4.

(iii) Now suppose that $g \in H$ has disjoint cycle decomposition $g = (1, 2, 3)\sigma$ with σ a product of disjoint transpositions. Similarly to above, we have that $g^2 = (1, 3, 2) \in H$ and so $H = A_n$ by Claim 2.

(iv) By (i) to (iii) it only remains to consider the case that any non-identity $g \in H$ has disjoint cycle decomposition $g = (1, 2)(3, 4)\sigma \in H$ with σ a product of disjoint transpositions. We have that
$$((1, 3, 2)g(1, 2, 3))g = (1, 3)(2, 4) \in H.$$
Conjugating $(1, 3)(2, 4) \in H$ by $(1, 5, 3) \in \mathfrak{S}_n$ and multiplying by $(1, 3)(2, 4) \in H$, we obtain
$$((1, 5, 3)(1, 3)(2, 4)(1, 3, 5))(1, 3)(2, 4) = (1, 3, 5) \in H$$
and so $H = A_n$ by Claim 2.

This completes the proof. □

3.3 Homomorphisms

We now define maps which respect group-theoretic structures. These maps will allow us to transport information from one group to another.

Definition 3.3.1. Let G, H be groups. A homomorphism from G to H is a map $\phi : G \longrightarrow H$ satisfying
$$\phi(g_1 *_G g_2) = \phi(g_1) *_H \phi(g_2)$$
for all $g_1, g_2 \in G$.

Note that an isomorphism is a bijective homomorphism, but homomorphisms need not be injective or surjective in general.

Example 3.3.2. Consider the group D_8 of symmetries of the square. We label the vertices anticlockwise by $1, 2, 3$ and 4 (starting from the top left vertex) as follows

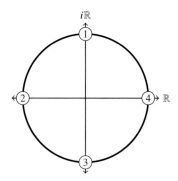

The clockwise rotation through $90°$ can be identified with the permutation of the vertex labels $(1, 4, 3, 2) \in \mathfrak{S}_4$. Identify the flip with an element of $\mathfrak{S}_4$ and hence define a homomorphism $\vartheta : D_8 \to \mathfrak{S}_4$. Prove that your map is an injective homomorphism.

Proposition 3.3.3. *Let $\phi : G \to H$ be a homomorphism. Then we have the following properties.*
(i) $\phi(\mathrm{id}_G) = \mathrm{id}_H$.
(ii) $\phi(g^{-1}) = \phi(g)^{-1}$ for all $g \in G$.
(iii) $\phi(g^r) = \phi(g)^r$ for all $g \in G$.

Proof. Follow the proof of Proposition 1.5.7. □

Definition 3.3.4. Let $\phi : G \to H$ be a homomorphism. The kernel of ϕ is defined by
$$\ker(\phi) = \{g \in G \mid \phi(g) = \mathrm{id}_H\}.$$
The image of ϕ is defined by
$$\mathrm{Im}(\phi) = \{\phi(g) \mid g \in G\}.$$

We note that ϕ is injective if and only if $\ker(\phi) = \{\mathrm{id}_G\}$, and ϕ is surjective if and only if $\mathrm{Im}\,\phi = H$.

Theorem 3.3.5. *Let $\phi : G \to H$ be a homomorphism. Then we have*
(i) $\ker(\phi) \trianglelefteq G$, and
(ii) $\mathrm{Im}(\phi) \leqslant H$.

Proof. We first verify (i). We first show that $K = \ker(\phi)$ is a subgroup of G. If $g_1, g_2 \in G$ then $\phi(g_1 g_2) = \phi(g_1)\phi(g_2) = \mathrm{id}_H \mathrm{id}_H = \mathrm{id}_H$, then $g_1 g_2 \in K$ and hence (S1) holds. If $g \in K$ then, using Proposition 3.3.3(ii), we have $\phi(g^{-1}) = \phi(g)^{-1} = \mathrm{id}_H^{-1} = \mathrm{id}_H$ and so $g^{-1} \in K$, and hence (S2) holds.

Next we show that $K \trianglelefteq G$. Let $g \in G$ and let $k \in K$ then we have

$$\begin{aligned}\phi(gkg^{-1}) &= \phi(g)\phi(k)\phi(g^{-1}) \\ &= \phi(g)\phi(k)\phi(g)^{-1} \\ &= \phi(g)\mathrm{id}_H \phi(g)^{-1} \\ &= \phi(g)\phi(g)^{-1} \\ &= \mathrm{id}_H.\end{aligned}$$

where the first equality follows as ϕ is a homomorphism, the second from Proposition 3.3.3(ii), the third as $k \in K$. Thus we have $gkg^{-1} \in K$ (and therefore K is normal by Proposition 3.2.17).

We now verify (ii). We need to check that $\mathrm{Im}(\phi)$ satisfies (S1) and (S2). Let $h_1, h_2 \in H$, that is we have $h_1 = \phi(g_1)$ and $h_2 = \phi(g_2)$ for some $g_1, g_2 \in G$; then we get $h_1 h_2 = \phi(g_1)\phi(g_2) = \phi(g_1 g_2) \in \mathrm{Im}(\phi)$ and so (S1) holds. Let $h \in \mathrm{Im}(\phi)$, that is $h = \phi(g)$ for some $g \in G$; then we have $h^{-1} = \phi(g)^{-1} = \phi(g^{-1}) \in \mathrm{Im}(\phi)$ and so (S2) holds. $\square$

Exercise 3.3.6. Construct a surjective homomorphism $\varphi : H_n \to \mathfrak{S}_n$ and calculate $\ker(\varphi)$.

Exercise 3.3.7. Let G be the group of monomial (3×3)-matrices with entries from the group $D_8 = \langle r, f_0 \mid f_0^2 = r^4 = \mathrm{id}, r^k f_0 = f_0 r^{4-k} \rangle$ as constructed in Exercise 2.5.6. Construct a surjective homomorphism $: G \to \mathfrak{S}_3$ and calculate its kernel.

3.4 The Jordan–Hölder Theorem

We now put together the ideas from this section into an overall picture, in which homomorphisms and normal subgroups form complementary pieces. We then prove the Jordan–Hölder theorem, which states that any finite group can be dissected into simple pieces in a well-defined manner. This serves as a prototype for the study of all mathematical objects: identify a notion of simplicity for a given mathematical

3.4 The Jordan–Hölder Theorem

object, and then break down complicated examples of these objects into their simple pieces, and do so *in a well-defined* (Jordan–Hölder-esque) manner. The Jordan–Hölder theorem is the central abstract result of Part 1 of this book.

Theorem 3.4.1 (First isomorphism theorem). *Let $\phi : G \to H$ be a homomorphism with $\ker(\phi) = K$ then the map*

$$\mu : G/K \longrightarrow \mathrm{Im}(\phi) \qquad : gK \mapsto \phi(g)$$

is well-defined and is an isomorphism.

Proof. First we need to show that μ is well defined. Let $g' \in gK$, we want to show that $\phi(g') = \phi(g)$. If $g' \in gK$ then $g' = gk$ for some $k \in K$ and so

$$\phi(g') = \phi(gk) = \phi(g)\phi(k) = \phi(g)\mathrm{id}_H = \phi(g)$$

as $k \in K$. Next we show that μ is a homomorphism. On the one hand we have

$$\mu((g_1 K)(g_2 K)) = \mu((g_1 g_2)K) = \phi(g_1 g_2).$$

And on the other hand we have

$$\mu(g_1 K)\mu(g_2 K) = \phi(g_1)\phi(g_2).$$

But these coincide as ϕ is a homomorphism.

Now we show that μ is injective. Suppose that $\mu(gK) = \mathrm{id}_H$. As $\mu(gK) = \phi(g) = \mathrm{id}_H$ we have that $g \in K$. But this implies that $gK = K$. So we have that the kernel of μ is given by $\{K = \mathrm{id}_{G/K}\}$ and hence μ is injective.

Finally we show that μ is surjective. Let $h \in \mathrm{Im}(\phi)$, then we have $h = \phi(g)$ for some $g \in G$. So we get $\mu(gK) = \phi(g) = h$ as required. $\square$

Definition 3.4.2. For $\ell \geq 1$, we fix ξ a primitive ℓth root of unity. We set $G(\ell, 1, n) \leq \mathrm{GL}_n(\mathbb{C})$ to be the subgroup of monomial matrices with non-zero entries of the form ξ^d for $0 \leq d < \ell$. Then $G(\ell, 1, n)$ forms a group under matrix multiplication.

Example 3.4.3. Let ξ be a primitive 3rd root of unity, that is $\ell = 3$. We have that

$$\begin{bmatrix} 0 & 0 & 1 & 0 & 0 \\ \xi & 0 & 0 & 0 & 0 \\ 0 & 0 & 0 & 1 & 0 \\ 0 & 0 & 0 & 0 & \xi^2 \\ 0 & 1 & 0 & 0 & 0 \end{bmatrix}$$

is an element of $G(3, 1, 5)$ and the determinant of this matrix is equal to $\xi \times 1 \times 1 \times 1 \times \xi^2 = 1$.

Example 3.4.4. We have a surjective group homomorphism $\varphi : G(\ell, 1, n) \to \mathfrak{S}_n$ given by replacing each monomial entry ξ^i with the number 1. The identity of the symmetric group is the unique diagonal matrix; therefore the kernel of the map φ is the set of all diagonal matrices, which we denote D. We have that $\psi : C_\ell \times C_\ell \times \cdots \times C_\ell \to D$ is an isomorphism, where

$$\psi(\xi^{i_1}, \xi^{i_2}, \ldots, \xi^{i_\ell}) = \begin{bmatrix} \xi^{i_1} & 0 & 0 & \cdots & 0 \\ 0 & \xi^{i_2} & 0 & \cdots & 0 \\ 0 & 0 & \xi^{i_3} & \cdots & 0 \\ \vdots & \vdots & \vdots & \ddots & 0 \\ 0 & 0 & 0 & 0 & \xi^{i_n} \end{bmatrix}$$

Thus we have a sequence of homomorphisms

$$0 \to C_\ell \times C_\ell \times \cdots \times C_\ell \xrightarrow{\psi} G(\ell, 1, n) \xrightarrow{\varphi} \mathfrak{S}_n \to 0$$

with $\ker(\varphi) \cong \operatorname{Im}(\psi)$.

Theorem 3.4.5 (Second isomorphism theorem). *Let G be a group with $H \leqslant G$, and $N \triangleleft G$. Then the following hold:*

(i) *The product subgroup $HN = \{hn \mid h \in H, n \in N\}$ is a subgroup of G*
(ii) *The subgroup $N \leqslant HN$ is a normal subgroup of HN.*
(iii) *The subgroup $H \cap N$ is a normal subgroup of H.*
(iv) *The quotient groups HN/N and $H/H \cap N$ are isomorphic.*

Proof. For part (i) we check the closure axiom as follows, given $h_1 n_1, h_2 n_2 \in HN$ we have that

$$(h_1 n_1)(h_2 n_2) = h_1(h_2 h_2^{-1}) n_1 h_2 n_2 = (h_1 h_2)(h_2^{-1} n_1 h_2) n_2 = (h_1 h_2)(n_1' n_2) \in HN$$

where $N \ni n_1' := h_2^{-1} n_1 h_2$ by Proposition 3.2.17 and $h_1 h_2 \in H$ and $n_1' n_2 \in N$ by the closure axioms for H and N. The inverse axiom for NH can be checked similarly and the identity axiom is trivial. For part (ii) we recall that $N \triangleleft G$ by assumption and so $gN = Ng$ for all $g \in HN \leqslant G$, as required. For part (iii) we note that if $x \in H \cap N$, then $hxh^{-1} \in N$ for all $h \in H \leqslant G$ (as $x \in N$ and $N \triangleleft G$) and $hxh^{-1} \in H$ for all $h \in H$ (as $x \in H$), as required.

Finally, we consider part (iv). We define a surjective map $\phi : H \to HN/N$ by the rule $\phi(h) = hN$. For $x, y \in H$, we have that

$$\phi(xy) = xyN = (xN)(yN) = \phi(x)\phi(y)$$

by part (ii) and so ϕ is a homomorphism. We have that

$$\ker(\phi) = \{h \in H \mid hN = N\} = \{h \in H \mid h \in N\} = H \cap N$$

and so the result follows by Theorem 3.4.1. $\square$

3.4 The Jordan–Hölder Theorem

Theorem 3.4.6 (The correspondence theorem). *Let G be a group and let $N \triangleleft G$ be a normal subgroup. Then every subgroup of G/N is of the form H/N for some $N \leqslant H \leqslant G$. Conversely, if $N \leqslant H \leqslant G$, then $H/N \leqslant G/N$. Thus we have a bijection between subgroups of G/N and subgroups of G containing N; this bijection preserves the property of being a normal subgroup.*

Proof. Let $N \leqslant H \leqslant G$, it is easy to verify that $\alpha(H) := \{hN \mid h \in H\}$ is a subgroup of G/N. Conversely, let $H' \leqslant G/N$, we claim that $\beta(H') := \{g \in G \mid gN \in H'\}$ is a subgroup of G. We check the closure axiom for $\beta(H')$. For $x, y \in \beta(H')$ we have that $xN, yN \in H'$ (by the definition of β) and therefore $(xN)(yN) = (xy)N \in H'$ (by the definition of the group G/N) and this implies $xy \in \beta(H')$. The inverse axiom is similar and the identity axiom is trivial.

In order to prove the claimed bijection, we must check that $\alpha \circ \beta$ and $\beta \circ \alpha$ are the identity maps. We have that

$$\beta \circ \alpha(H) = \beta(H/N) = \{g \in G \mid gN \in H/N\} = H$$

by the definitions of α, β, and a coset of H/N, respectively. We have that

$$\alpha \circ \beta(H') = \alpha\{g \in G \mid gN \in H'\} = \{gN \in H'\} = H'$$

by the definitions of β, α, and H', respectively. Thus we have the required bijection.

It remains to check that the bijective maps α and β preserve the property of being a normal subgroup. For $N \leqslant H \triangleleft G$ we have that

$$(gN)(hN)(gN)^{-1} = ghg^{-1}N = h'N \in H/N$$

for $h' = ghg^{-1} \in H$ by assumption. Conversely, if $H' \triangleleft G/N$, it is easy to check that $\beta(H') = \{g \in G \mid gN \in H'\}$ is a normal subgroup of G. $\square$

We say that a series of subgroups $G = G_0 > G_1 > \cdots > G_k = 1$ is a **composition series** of G if $G_{i+1} \triangleleft G_i$ and each successive quotient G_i/G_{i+1} is a simple group. We refer to these successive quotients as the **composition factors** of the series; and we refer to the number of these factors, k, as the **composition length**. The rest of this chapter is dedicated to understanding (finite) groups in terms of their composition factors; our first port-of-call is to prove that this notion makes sense.

Proposition 3.4.7. *Finite groups have composition series.*

Proof. We proceed by induction on $|G|$. If G is simple, then $G > \{\mathrm{id}_G\}$ is a composition series and we are done. Otherwise, G has a normal subgroup and we can suppose that $N \triangleleft G$ is a maximal normal subgroup (that is, if $N < M < G$, then M is not normal). By the correspondence theorem G/N is a simple group; by our inductive assumption $|N| < |G|$ and so N has a composition series. The result follows by induction. $\square$

Example 3.4.8. The group $C_6 = \{\mathrm{id}, r, r^2, r^3, r^4, r^5, r^6\}$ has precisely two composition series:

$$C_6 > \{\mathrm{id}, r^2, r^4\} > \{\mathrm{id}\} \quad \text{and} \quad C_6 > \{\mathrm{id}, r^3\} > \{\mathrm{id}\}.$$

The resulting composition factors are C_2 and C_3 (regardless of which of the two composition series we take, but note that the order in which they appear as factors *does change*).

Example 3.4.9. The group $\mathfrak{S}_n$ for $n \neq 4$ has composition series $\mathfrak{S}_n > A_n > \{\mathrm{id}\}$. The resulting composition factors are C_2 and A_n.

Exercise 3.4.10. Calculate a composition series of $\mathfrak{S}_4$.

We are now ready to prove that these composition series are, in some sense, canonical. Thus finite groups have well-defined sets of composition factors, and we can hope to piece-together our understanding of these simple factor groups in order to learn something about the ambient (non-simple) group.

Theorem 3.4.11 (The Jordan–Hölder Theorem). *Let G be a finite group. Then any two composition series of G have the same composition length and the same composition factors, up to permutation and isomorphism.*

Proof. We proceed by induction on $|G|$. Let $G = G_0 > G_1 > \cdots > G_r = \{\mathrm{id}_G\}$ and $H = H_0 > H_1 > \cdots > H_s = \{\mathrm{id}_G\}$ be two composition series for G. We refine our induction on $r \geq 1$ with the $r = 1$ case being trivial. Now let $r > 1$, if $G_1 = H_1$ then this group has a unique composition series (up to permutation of factors) of length $r - 1 = s - 1$ and so the result follows by induction.

Now suppose that $G_1 \neq H_1$. We have that $G_1 \not\leq H_1$ (since G/G_1 is simple) and therefore $H_1 < G_1 H_1$ and thus $G_1 H_1 = G$ (since G/H_1 is simple). For the remainder of the proof we set $K = G_1 \cap H_1 \trianglelefteq G$. By Theorem 3.4.5(*iv*), we have that

$$G/G_1 \cong H_1/K \text{ and } G/H_1 \cong G_1/K \tag{3.4.12}$$

are both simple. Now, the group K has a composition series $K = K_0 > K_1 > \cdots > K_t = \{\mathrm{id}_G\}$ by induction on $|K| < |G|$. We now have two composition series of G_1, namely

$$G_1 > \cdots > G_r = \{\mathrm{id}_G\} \quad \text{and} \quad G_1 > K > K_1 > \cdots > K_t = \{\mathrm{id}_G\}$$

of lengths $r - 1$ and $t + 1$ respectively. By induction, $r - 1 = t + 1$ and they have the same composition factors, up to permutation and isomorphism. Similarly, we now have two composition series of H_1, namely

$$H_1 > \cdots > H_r = \{\mathrm{id}_G\} \quad \text{and} \quad H_1 > K > K_1 > \cdots > K_t = \{\mathrm{id}_G\}$$

of lengths $s - 1$ and $t + 1$ respectively. By induction, $s - 1 = t + 1$ and they have the same composition factors, up to permutation and isomorphism. Putting this together with the isomorphisms of (3.4.12) we obtain two composition series of G, namely

3.4 The Jordan–Hölder Theorem

$$G > G_1 > K > K_1 > \cdots > K_t = \{\mathrm{id}_G\}$$
$$\text{and} \quad G > H_1 > K > K_1 > \cdots > K_t = \{\mathrm{id}_G\}$$

of length $r = t + 2 = s$ with the same composition factors, up to permutation and isomorphism. The result follows. □

Further Reading. The notion of uniqueness (up to permutation and isomorphism) of composition series is quite a universal idea in "finitistic" algebra (for example modules over Artinian rings [CR81] and finite dimensional Hopf algebras [Nat15]). One of the reasons we have introduced this idea here, in the group theoretic context, is so that the reader can warm up to the idea before we embark on representation theory of k-algebras. Later in this book we will not prove the Jordan–Hölder theorem for finite-dimensional k-algebras and instead we will just say that it is similar to the group-theoretic proof.

Chapter 4
Platonic and Archimedean solids and special orthogonal groups

In this section, we consider the (infinite) group $SO_3(\mathbb{R}) \leq GL_3(\mathbb{R})$ of the symmetries of the sphere; that is, the group consisting of all rotations about the origin in 3-dimensional Euclidean space, $\mathbb{R}^3$. Every non-trivial rotation is determined by its axis of rotation and its angle of rotation (depicted in Figure 4.1).

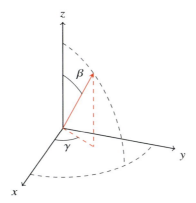

Fig. 4.1: Expressing an arbitrary axis of rotation in terms of a $(\frac{\pi}{2} - \beta)$ rotation in the xz-plane followed by a γ-rotation through the xy-plane (applied to the $(1, 0, 0) \in \mathbb{R}^3$).

The axis of rotation is simply a vector through the origin, in the case that this vector is the x-, y-, or z-axis of $\mathbb{R}^3$, then the matrix is of the familiar form

$$\begin{bmatrix} 1 & 0 & 0 \\ 0 & \cos\alpha & -\sin\alpha \\ 0 & \sin\alpha & \cos\alpha \end{bmatrix} \begin{bmatrix} \cos\beta & 0 & -\sin\beta \\ 0 & 1 & 0 \\ \sin\beta & 0 & \cos\beta \end{bmatrix} \begin{bmatrix} \cos\gamma & -\sin\gamma & 0 \\ \sin\gamma & \cos\gamma & 0 \\ 0 & 0 & 1 \end{bmatrix}$$

respectively for $0 \leq \alpha, \beta, \gamma < 2\pi$; an arbitrary rotation can be obtained by composing such elementary rotations (in other words, $SO_3(\mathbb{R})$ is generated by these elementary rotation matrices). In fact, it is not difficult to see that we only require two of these three planar rotations in order to generate the whole of $SO_3(\mathbb{R})$. In this section, we study the finite subgroups of $SO_3(\mathbb{R})$ and the manner in which they arise as the symmetry groups of the Platonic solids.

We start the story at the end, rather than the beginning, by first proving that there are precisely 5 Platonic solids. We then prove the orbit-stabiliser theorem and use this to determine the symmetry groups of these Platonic solids. We then prove that the symmetry groups of these Platonic solids realise all possible finite subgroups of $SO_3(\mathbb{R})$. Finally, we use this classification theorem in order to understand the symmetry groups of more general (Archimedean) solids.

4.1 The platonic solids

A convex regular polyhedron in $\mathbb{R}^3$ is called a Platonic solid. Here, a polyhedron is a region bounded by planes in $\mathbb{R}^3$. It has two-dimensional faces which meet in one-dimensional edges, which meet in vertices. We say that a polyhedron is regular if all its faces, edges and vertices are equal. That is, all the faces meet at the same angle and the same number of edges meet at the same angles at each vertex. In particular, this implies that all the faces are the same regular polygon.

Theorem 4.1.1. *There are* 5 *convex regular solids:*

- *the tetrahedron (which has 4 triangular faces)*
- *the cube (which has 6 square faces)*
- *the octahedron (which has 8 triangular faces)*
- *the dodecahedron (which has 12 pentagonal faces)*
- *the icosahedron (which has 20 triangular faces)*

Sketch of proof. Consider an arbitrary vertex in our convex regular solid. At least $p \geq 3$ faces must meet at this vertex (by our assumption that the shape is 3-dimensional and that the faces are flat). Recall our assumption that these p distinct faces are all identical regular q-sided polygons for some $q \geq 3$. We remark that each internal angle of the a regular q-sided polygon is $2\pi/q$.

At each vertex of the solid, the sum of the angles formed by the faces meeting there must be strictly less than 2π radians. To see this, note that if the sum was equal to 2π radians, then the faces which meet at the vertex would all lie in the same flat plane (and so this would not be a vertex at all). Therefore, if there are 3 faces meeting at a given vertex, then each face must have internal angle strictly less than $2\pi/3$; thus we deduce that each face is a triangle, square or pentagon. Similarly, if there are 4 or 5 faces meeting at a given vertex, then each face must have internal angle strictly less than $2\pi/4$ or $2\pi/5$; therefore each face is a triangle. Finally, if there are 6 faces meeting at a given vertex, then each face must have internal angle strictly less than $2\pi/6$ and there are no polygons satisfying this property.

4.2 Group actions 57

So, in summary, we know that there are only five possibilities for the pair of integers (p, q) — namely, $(3, 3), (3, 4), (3, 5), (4, 3)$ and $(5, 3)$. To see that there is at least one Platonic solid corresponding to a pair (p, q), all you need to do is construct it out of regular q-sided polygons, with p of them meeting at every vertex. On the other hand, to see that there is at most one Platonic solid corresponding to a pair (p, q) is a more difficult matter, but should seem believable. This is because if you start to glue together regular q-sided polygons, with p of them meeting at every vertex, then you do not have any choice in what the resulting shape will look like. □

We will revisit the ideas of this proof (classifying Platonic solids) in a more rigorous fashion (but using some of the same ideas) later on in this section in the proof of Theorem 4.6.2 (when we classify finite subgroups of $SO_3(\mathbb{R})$).

4.2 Group actions

We wish to make (abstract) groups more concrete by allowing them to "act" as the symmetries of some other object, for example a set (this set could have interesting structure attached to it, for example if might be the set of faces of a Platonic solid or the set of edges of a polygon).

Definition 4.2.1. We let $X = \{x_1, \ldots, x_n\}$ be an ordered set, we define the symmetry group of X to be the group $\text{Sym}(X) \cong \mathfrak{S}_n$ which permutes the elements of the set (by swapping the subscripts $1, \ldots, n$). A group G acts on a set X if there is a homomorphism
$$\pi : G \to \text{Sym}(X).$$

We usually write π_g instead of $\pi(g)$ for $g \in G$ in order to simplify notation. If the map π is injective then we say that G acts faithfully on X (that is, different group elements give different permutations on X).

Theorem 4.2.2. *Let G be a group acting on a set X. Define a relation $\sim$ on X by setting $x_1 \sim x_2$ if and only if there exists $g \in G$ such that $(x_1)\pi_g = x_2$. Then $\sim$ is an equivalence relation on X.*

Proof. We check that $\sim$ is reflexive, symmetric, and transitive.

Reflexive: We need to check that $x \sim x$ for all $x \in X$. This is true as $x = (x)\pi_{\text{id}}$ (as π is a homomorphism).

Symmetric: We need to check that if $x \sim y$ then $y \sim x$. Suppose that $x \sim y$, that is $\pi_g(x) = y$ for some $g \in G$, then we have
$$(y)\pi_{g^{-1}} = ((x)\pi_g)\pi_{g^{-1}} = (x)\pi_{gg^{-1}} = (x)\pi_{\text{id}} = x.$$

So we have $y \sim x$ as required.

Transitive: We need to check that if $x \sim y$ and $y \sim z$ then $x \sim z$. Suppose that $x \sim y$ and $y \sim z$, that is $(x)\pi_g = y$ and $(y)\pi_h = z$ for some $g, h \in G$. Then we have

$$(x)\pi_{gh} = ((x)\pi_g)\pi_h = (y)\pi_h = z.$$

So we have $x \sim z$ as required. □

Definition 4.2.3. The equivalence classes of $\sim$ are called the G-orbits on X. For each $x \in X$ we denote the G-orbit containing x by $\mathrm{Orb}_G(x)$.

Example 4.2.4. Consider the group D_8. Then the action of G on the set of vertices of the square has only one orbit.

Example 4.2.5. Consider the set of necklaces formed from 2 black and 2 white beads on a seamless strand. There are 6 such necklaces. Now consider the action of D_8 on these 6 necklaces. This action has 2 orbits. Namely, one orbit consists of the necklaces whose adjacent beads have the same colouring:

and the other orbit consists of the two necklaces whose adjacent beads have the opposite colouring:

Exercise 4.2.6. Consider the set of necklaces formed from 3 black and 2 white beads on a seamless strand. How many such necklaces are there? Now consider the action of D_{10} on these necklaces: identify the D_{10}-orbits and calculate the size of each D_{10}-orbit.

Exercise 4.2.7. Consider the set of necklaces formed from 3 black and 3 white beads on a seamless strand. How many such necklaces are there? Now consider the action of D_{12} on these necklaces: identify the D_{12}-orbits and calculate the size of each such orbit.

Example 4.2.8. The symmetric group $\mathfrak{S}_n$ acts faithfully on the set $\{1, \ldots, n\}$ by permuting these numbers amongst themselves.

Example 4.2.9. Consider the action of $G = \{\mathrm{id}, (1, 2), (3, 4), (1, 2)(3, 4)\}$ on the set $X = \{1, 2, 3, 4\}$. This action has 2 orbits, namely $\{1, 2\}$ and $\{3, 4\}$.

Exercise 4.2.10. Consider the action of $G = \langle (1, 2), (2, 3), (4, 5) \rangle$ on the set $X = \{1, 2, 3, 4, 5\}$. How many G-orbits are there?

4.2 Group actions

Example 4.2.11. Arrange the numbers $\{1, \ldots, 15\}$ on tiles in a 15-puzzle, then we have already seen that the alternating group A_{15} acts faithfully on the set of all legal puzzle configurations by permuting the tiles according to the puzzle rules.

Definition 4.2.12. Let G be a group acting on a set X and let $x \in X$. We define the stabiliser of x in G, denoted by G_x, by $G_x = \{g \in G \mid (x)\pi_g = x\}$.

Exercise 4.2.13. Verify that G_x is a subgroup of G.

Theorem 4.2.14 (Orbit-Stabiliser Theorem). *Let G be a group acting on a set X and let $x \in X$ then we have*

$$|G| = |\mathrm{Orb}_G(x)| \times |G_x|.$$

Proof. By definition, $|\mathrm{Orb}_G(x)|$ = number of distinct elements $(x)\pi_g$, for $g \in G$. So we ask, when do we have $(x)\pi_g = (x)\pi_h$? Precisely when $((x)\pi_g)(\pi_h)^{-1} = x$, that is $(x)(\pi_g \pi_{h^{-1}}) = (x)\pi_{gh^{-1}} = x$. This is equivalent to saying that $gh^{-1} \in G_x$, that is $g \in G_x h$, or equivalently, $G_x g = G_x h$. We have shown that $(x)\pi_g = (x)\pi_h$ precisely when g and h are in the same right coset of G_x in G. This means that there is a one-to-one correspondence between the elements of $\mathrm{Orb}_G(x)$ and the right cosets of G_x in G. By Lagrange's theorem we have

$$|\mathrm{Orb}_G(x)| = [G : G_x] = \frac{|G|}{|G_x|}$$

as required. □

For the rest of this chapter we will assume that G is a *finite* group acting on a *finite* set X. To simplify notation, we will write $(x)g$ instead of $(x)\pi_g$ for $g \in G$, $x \in X$ providing that the action of the group G on the set X is understood from the context.

Definition 4.2.15. We set $\mathrm{Fix}_g(X) = |\{x \in X \mid (x)g = x\}|$, we sometimes write $\mathrm{Fix}(g) := \mathrm{Fix}_g(X)$ when X is clear from context.

Theorem 4.2.16 (Burnside's counting theorem). *Let G be a finite group acting on a finite set X then the number of G-orbits on X is given by*

$$\frac{1}{|G|} \sum_{g \in G} \mathrm{Fix}_g(X).$$

To prove this theorem we will need the following lemma.

Lemma 4.2.17. *If $x, y \in X$ are in the same G-orbit then we have $|G_x| = |G_y|$.*

Proof. As x and y are in the same G-orbit, there exists $g \in G$ such that $(x)g = y$ and so $(y)g^{-1} = x$. Now if $a \in G_x$ then we have

$$(y)(g^{-1}ag) = (x)ag = (x)g = y,$$

so $g^{-1}ag \in G_y$. So we get a map

$$\vartheta : G_x \to G_y : a \mapsto gag^{-1}.$$

We need to check that ϑ is injective and surjective. If $g^{-1}ag = g^{-1}bg$ then $a = b$ (using the cancellation rules). Now let $c \in G_y$, then it's easy to check that $gcg^{-1} \in G_x$ and $\vartheta(gcg^{-1}) = c$ as required. □

Exercise 4.2.18. Check that the map ϑ given in the above proof is a group homomorphism (and thus an isomorphism).

Proof of Theorem 4.2.16. Consider the number

$$n = |\{(g,x) \mid g \in G, x \in X \text{ and } (x)g = x\}|.$$

We will calculate this number in two different ways. First observe that the number of x's appearing with a given g is precisely Fix(g). So we have

$$n = \sum_{g \in G} \text{Fix}(g). \tag{4.2.19}$$

On the other hand, the number of g's appearing with a given x is precisely $|G_x|$. So we have $n = \sum_{x \in X} |G_x|$. But using the previous lemma, if x and y are in the same orbit then we have $|G_x| = |G_y|$, so we get

$$\sum_{y \in \text{Orb}_G(x)} |G_y| = \sum_{y \in \text{Orb}_G(x)} |G_x| = |\text{Orb}_G(x)| \times |G_x| = |G|,$$

using the Orbit-Stabiliser Theorem. So this gives

$$n = \sum_{x \in X} |G_x| = (\text{number of orbits}) \times |G|. \tag{4.2.20}$$

Combining (4.2.19) and (4.2.20) we get that

$$\text{number of orbits} = \frac{1}{|G|} \sum_{g \in G} \text{Fix}(g).$$

The result follows. □

Example 4.2.21. Suppose that a manufacturer wants to make ID cards from plastic squares, marked with a 3 × 3 grid on both sides and punched with three holes, for example

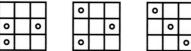

Let X be the set of all $\binom{9}{3} = 84$ possible ID cards formed in this fashion. We will now use Burnside's counting theorem to calculate the total number of D_8-orbits on

4.2 Group actions

X, or if you prefer, the number of distinct ID cards *up to symmetry*. In order to do this, we must calculate the $|\text{Fix}_g(X)|$ for $g \in G$.

Clearly, the identity fixes every possible ID card (as it does nothing!) and so $|\text{Fix}_{\text{id}}(X)| = 84$. On the other hand, the rotations r and r^3 (in 90° and −90° respectively) will not fix any ID card ($|\text{Fix}_r(X)| = 0 = |\text{Fix}_{r^3}(X)|$); to see this, consider the effect of putting a hole in any non-central square — rotating through 90° we realise that we need to put at least four holes in the card in order that it be fixed by the rotation (and we are only allowed to make 3).

Now, we consider the rotation r^2 which fixes the following ID cards

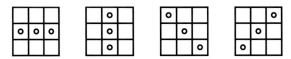

which can be seen as follows: we first make a hole in any non-central box (this forces us to make one additional hole in the box obtained by rotation through r^2); the final hole we make must then go in the central box.

Now we consider the flip through the x-axis, which fixes the following ID cards

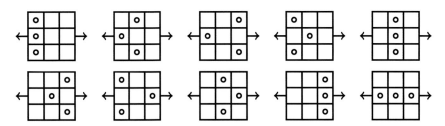

where the first 3^2 diagrams come from choosing one hole (in 3 possible positions) along the line of symmetry and then picking a pair of holes on opposite sides of the line of symmetry (in 3 possible positions). The final diagram has 3 holes all along the line of symmetry. The flip through the y-axis is similar.

Now we consider the northwest-to-southeasterly flip, which fixes the following ID cards

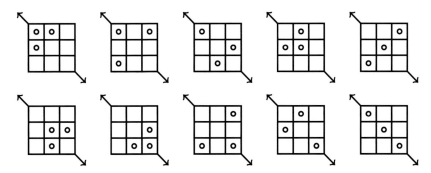

where we obtain this list in a similar fashion to the previous case. Therefore by Burnside's counting theorem we have a total of

$$16 = \tfrac{1}{8}(84 + 0 + 0 + 4 + 10 + 10 + 10 + 10)$$

distinct ID cards, up to symmetry.

One might think that it would be easier to simply write down the distinct ID cards up to symmetry directly. However, this becomes more and more difficult as we increase the size of our problem.

Example 4.2.22. Suppose that a manufacturer wants to make ID cards from plastic squares, marked with a 4 × 4 grid on both sides and punched with four holes. How many distinct ID cards can be made *up to symmetry*? First try listing them directly, then try using Burnside's counting theorem.

Example 4.2.23. Repeat the previous questions with a 6 × 6 grid on both sides, punched with six holes.

4.3 The symmetry group of the tetrahedron

We begin by constructing as many symmetries of the tetrahedron as we can possibly think up (spoiler alert: there are 12 of them). We then use the orbit-stabiliser theorem to deduce that we have found them all. This is an exhaustive approach and so we do not go into this level of detail with the other Platonic solids, but we do encourage the reader to give this a go.

Proposition 4.3.1. *The symmetry group of the tetrahedron is isomorphic to A_4.*

Proof. We first fix a labelling for the identity tetrahedron by placing the numbers 1, 2, 3, and 4 on its vertices as in the leftmost diagram in Figure 4.2. The symmetry group of the tetrahedron permutes these vertices and so is a subgroup of $\mathfrak{S}_4$.

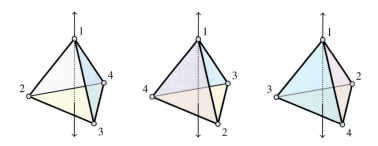

Fig. 4.2: The identity tetrahedron and the rotations (2, 3, 4) and (2, 4, 3) respectively.

4.3 The symmetry group of the tetrahedron

We first consider the rotations which fix a given vertex. For the 1st vertex, these are depicted in Figure 4.2 (and again in Figure 4.3 in flat pack form); these rotations are id, $(2, 3, 4)$, and $(2, 4, 3)$.

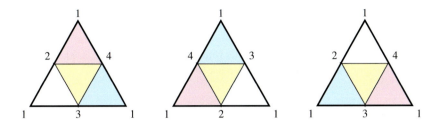

Fig. 4.3: The flat pack versions of the tetrahedra depicted in Figure 4.2 (the identity tetrahedron and the rotations $(2, 3, 4)$ and $(2, 4, 3)$ respectively).

For the 2nd and 3rd vertices, the rotated (non-identity) tetrahedra are depicted in Figure 4.4 and the permutations of the vertices are $(1, 4, 3)$ and $(1, 3, 4)$ for the former and $(1, 2, 4)$ and $(1, 4, 2)$ for the latter. The rotations which fix the 4th vertex are not pictured, but the non-identity rotations of That fix 4 are the permutations $(1, 2, 3)$ and $(1, 3, 2)$.

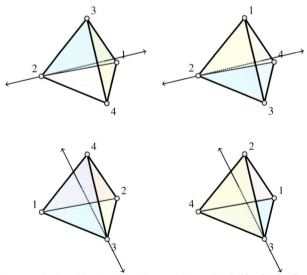

Fig. 4.4: The tetrahedra labelled by the rotations $(1, 4, 3)$, $(1, 3, 4)$, $(1, 2, 4)$ and $(1, 4, 2)$ respectively.

In addition to the above 9 rotations fixing a given vertex, there are 3 more rotations which we have yet to picture. The axes of rotation for these reflections bisect a pair of opposite edges. The first axis bisects the edges connecting vertices 1 to 4 and 2 to 3; this permutation of the vertices is equal to $(1,4)(2,3)$; this rotation is equal to the product $(1,4)(2,3) = (1,3,4)(1,3,2)$ and so is a composite of 2 rotations which we have already considered. Arguing in a similar fashion, one obtains two further rotations. The first of which is through the line bisecting the edges 1 to 2 and 3 to 4; this rotation is equal to $(1,2)(3,4)$. The second of which is through the line bisecting the edges 2 to 3 and 1 to 4; this rotation is equal to $(1,4)(2,3)$. These are pictured in Figure 4.5 (and again in flat pack form in Figure 4.6).

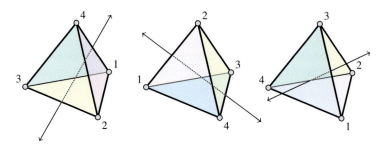

Fig. 4.5: The tetrahedra labelled by the rotations $(1,4)(2,3)$, $(1,2)(3,4)$, and $(1,3)(2,4)$.

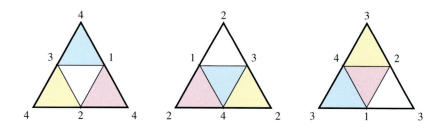

Fig. 4.6: The flat pack versions of the tetrahedra depicted in Figure 4.5 (the rotations $(1,4)(2,3)$, $(1,2)(3,4)$, and $(1,3)(2,4)$ respectively).

Thus we have found 12 distinct symmetries of the tetrahedron. To verify that these are all the symmetries, we consider the action of G on the vertices $\{1,2,3,4\}$ and use the orbit-stabiliser theorem. The stabiliser of a given vertex is the cyclic group of order 3 (which permutes the 3 triangles meeting at said vertex) and the orbit of this vertex has order 4 (as there are 4 vertices in total and the action is transitive). By the orbit-stabiliser theorem, the order of the group of symmetries is $4 \times 3 = 12 = |A_4|$ and so the result follows. □

4.4 The symmetry group of the cube

Exercise 4.3.2. Consider the action of G on the faces (respectively edges) in place of the vertices. Calculate $|G|$ in both cases using the orbit-stabiliser theorem.

Exercise 4.3.3. We now colour two faces of the tetrahedron black and two of them white. We let G be the group of all rotational symmetries of the cube which preserve this colouring of the vertices. Use the orbit-stabiliser theorem to calculate the order of G. Find $|G|$ rotations that preserve the colouring and hence determine the group G.

4.4 The symmetry group of the cube

We begin by using the orbit-stabiliser theorem to deduce the size of the symmetry group, G, of the cube. The cube has 6 square faces and it is easy to see that these faces form a single G-orbit; each face is a square and so the stabiliser of any given face, F, consists of the rotations (through $\pi/2$, π, $3\pi/2$, and 2π) around the axis connecting the midpoint of F with the midpoint of the face opposite F. Thus

$$|G| = |\text{Orb}_G(F)| \times |\text{Stab}_G(F)| = 6 \times 4 = 24.$$

We wish to consider the action of G on some aspect of the cube. The cube has 8 vertices, 12 edges, and 6 faces. The resulting symmetric groups are very large indeed (even the smallest, $\mathfrak{S}_6$ is 30 times bigger than G). So instead we consider the diagonals connecting opposite vertices of the cube. There are 8 vertices and therefore 4 such diagonals (an encouraging sign, since $\mathfrak{S}_4 = 4! = 24 = |G|$). We label the diagonals by the numbers 1, 2, 3, 4 and, for ease of notation, we place these labels on the vertices at both ends. Two of these diagonals (and the labelling of the vertices) can be seen as follows:

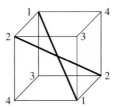

Proposition 4.4.1. *The symmetry group of the cube is isomorphic to* $\mathfrak{S}_4$.

Proof. Given a fixed pair $1 \leq i < j \leq 4$, we consider the axis of rotation which fixes the pair of edges connecting the ith and jth vertices (as in the second rotation depicted in Figure 4.7, for $\{i, j\} = \{1, 4\}$). This rotation swaps the ith and jth diagonals (i.e. the pairs of ith and jth vertices) but fixes the other two diagonals. In this manner we obtain all the transpositions $(i, j) \in \mathfrak{S}_4$ for $1 \leq i < j \leq 4$. Together these transpositions generate $\mathfrak{S}_4$ and the result follows (since $|G| = |\mathfrak{S}_4| = 24$). □

4 Platonic and Archimedean solids and special orthogonal groups

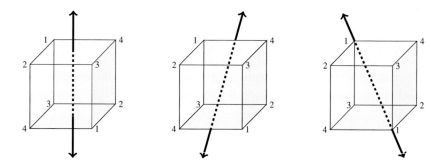

Fig. 4.7: Rotations of the cube corresponding to elements $(1,2,3,4)$, $(1,4)$, and $(2,3,4)$ of $\mathfrak{S}_4$ respectively.

Exercise 4.4.2. Consider the action of G on the faces (respectively edges) in place of the vertices. Calculate $|G|$ in both cases using the orbit-stabiliser theorem.

Exercise 4.4.3. We now colour the vertices of the cube black or white as shown in the picture below. We let G be the group of all rotational symmetries of the cube which preserve this colouring of the vertices.

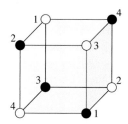

Use the orbit-stabiliser theorem to calculate the order $n = |G|$. Find 8 rotations of order 3 that belong to $G \leqslant \mathfrak{S}_4$. Hence determine G.

Example 4.4.4. In how many ways can the six faces of a cube be marked with the numbers $1, \ldots, 6$ to form a die? Initially there are $6! = 720$ possibilities. However, if one marked cube can be obtained from another by rotating the cube then these two markings will give the same die.

Let $G \cong \mathfrak{S}_4$ be the group of all rotational symmetries of the cube and let X be the set of all marked cubes (we have $|X| = 720$). Now G acts on X, that is G permutes the markings of the cube. Two markings give the same die if there are in the same G-orbits. In other words, the number of different dice is exactly the number of G-orbits on X. This is given by Burnside's counting theorem.

We have $\mathrm{Fix}(\mathrm{id}) = 720$ as id fixes every marked die. And for every $g \neq \mathrm{id}$ we have $\mathrm{Fix}(g) = 0$ as none of the rotations fixes any marked die (as the label on each face is a distinct number from the set $1, 2, \ldots, 6$). Thus the number of (distinguishable)

4.5 The symmetry group of the dodecahedron

dice is given by

$$\frac{1}{|G|} \sum_{g \in G} \text{Fix}(g) = \frac{1}{24}(720 + 0 + \cdots + 0) = 30.$$

Remark 4. The symmetry group of the octahedron is the same as that of the cube. To see this, note that if we choose the centres of the six square faces of a cube, these are the vertices of an octahedron. We say that the octahedron is the dual of the cube.

4.5 The symmetry group of the dodecahedron

We begin by using the orbit-stabiliser theorem to deduce the size of the symmetry group, G, of the dodecahedron. The dodecahedron has 12 pentagonal faces and it is easy to see that these faces form a single G-orbit; each face is a pentagon and so the stabiliser of any given face, F, consists of the rotations (through $2\pi/5$, $4\pi/5$, $6\pi/5$, $8\pi/5$, and 2π) around the axis connecting the midpoint of F with the midpoint of the face opposite F. Thus

$$|G| = |\text{Orb}_G(F)| \times |\text{Stab}_G(F)| = 12 \times 5 = 60.$$

The plan of the dodecahedron is given in Figure 4.8.

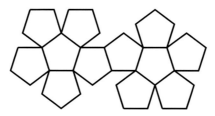

Fig. 4.8: The plan of the dodecahedron

We wish to consider the action of G on some aspect of the dodecahedron. The dodecahedron has 20 vertices, 30 edges, and 12 faces. The resulting symmetric groups are very large indeed (even the smallest, $\mathfrak{S}_{12}$ is $7,983,360$ times bigger than G). Instead, we consider the 5 diagonals of any pentagonal face of the dodecahedron. We notice that each vertex of the dodecahedron is incident to 3 pentagons and that we can pick a triple of diagonals (one from each pentagon meeting said vertex) so that they form the vertex of a cube (in other words the three diagonals are at angle $\pi/2$ to one another). In fact, this allows us to construct 5 different cubes within the dodecahedron. The plans of these cubes (within the plan of the dodecahedron) are given in Figure 4.9.

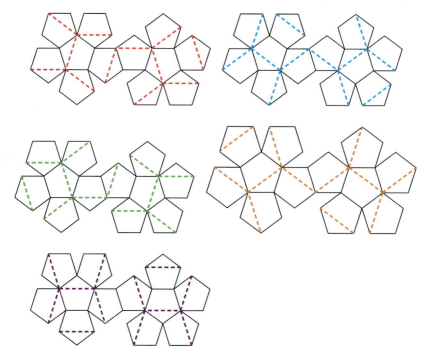

Fig. 4.9: The five cubes within the dodecahedron

Proposition 4.5.1. *The symmetry group of the dodecahedron is isomorphic to A_5.*

Proof. Take any pair of opposite vertices, x and x' say, of the dodecahedron and consider the rotation through this axis. We wish to understand the effect of this rotation on the 5 cubes. We notice that any vertex of a pentagonal face has exactly 3 diagonals which do not meet this vertex — in other words there are 3 cubes in the dodecahedron which do not meet the vertex x (and similarly for x'). In fact, one can check that the 3 cubes which do not meet x are precisely the same 3 cubes which do not meet x' (by checking the plans in Figure 4.9). With a little more thought, one can see that the rotation through x and x' permutes these 3 cubes amongst themselves and fixes the remaining 2 cubes (although it does act by non-trivially rotating the vertices of each given cube amongst themselves, we do not need to consider that here). By labelling the 5 cubes by the numbers $1, 2, 3, 4, 5$ and considering each rotation in turn, we hence obtain all 3-cycles of 5 integers. These 3-cycles together generate the group A_5, which has order 60, and so the result follows. □

Exercise 4.5.2. Suppose that you require a die with the numbers $1, 2, \ldots, 12$ each appearing once (in order to play Dungeons and Dragons, say). How many rotationally distinct dice can you make by decorating a dodecahedron?

Remark 5. The symmetry group of the icosahedron is the same as that of the dodecahedron as they are dual to one another (in a similar fashion to the cube and octahedron).

Exercise 4.5.3. Place four balls on the vertices of dodecahedron as follows:

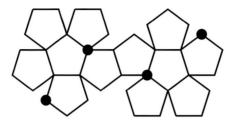

what is the resulting symmetry group which preserves this decorated shape?

4.6 The classification of 3-dimensional rotational symmetry groups

We now consider the group of all symmetries of 2-dimensional real space, $O_2(\mathbb{R})$, and the subgroup of all rotational symmetries of 2-dimensional real space, $SO_2(\mathbb{R})$. The latter is the subgroup of the former consisting of the rotations in 2-dimensions. The former *also* consists of the mirror flips which we have already considered in the construction of the dihedral groups (where we flip-over the piece of paper on which we have drawn $\mathbb{R}^2$). Indeed, we will see that the finite subgroups of $SO_2(\mathbb{R})$ are precisely the cyclic groups and that the finite subgroups of $O_2(\mathbb{R})$ are precisely the cyclic and dihedral groups.

The general form of a 2-dimensional rotation through angle ϑ, or a general 2-dimensional flip through the line through angle ϑ, is given by

$$\begin{bmatrix} \cos \vartheta & \sin \vartheta \\ -\sin \vartheta & \cos \vartheta \end{bmatrix} \quad \begin{bmatrix} -\cos \vartheta & \sin \vartheta \\ \sin \vartheta & \cos \vartheta \end{bmatrix}$$

respectively for $0 \leqslant \vartheta < 2\pi$. Thus

$$O_2(\mathbb{R}) = \left\langle \begin{bmatrix} \cos \vartheta & \sin \vartheta \\ -\sin \vartheta & \cos \vartheta \end{bmatrix}, \begin{bmatrix} -1 & 0 \\ 0 & 1 \end{bmatrix} \mid 0 \leqslant \vartheta < 2\pi \right\rangle$$

and $SO_2(\mathbb{R})$ is the subgroup consisting solely of the rotations.

One can picture a mirror flip in 2-dimensions instead as a rotation of the plane in 3-dimensions as in and therefore every finite subgroup of $O_2(\mathbb{R})$ will appear as a subgroup of $SO_3(\mathbb{R})$. This is why we consider these here.

Theorem 4.6.1. *Let $G \leqslant O_2(\mathbb{R})$ be a finite subgroup. Then G is isomorphic to precisely one of the cyclic groups: C_n for $n \geqslant 1$ or the dihedral groups D_{2n} for $n \geqslant 2$.*

Proof. Let G be a finite (non-trivial) subgroup of $O_2(\mathbb{R})$. Suppose first that $G \leqslant SO_2(\mathbb{R})$. Then every element of G is a rotation in the plane. Write r_ϑ for the rotation anticlockwise by ϑ (where $0 \leqslant \vartheta < 2\pi$) around the origin $(0,0)$. Choose $r_\phi \in G$ with $\phi > 0$ as small as possible (which we can do, by our assumption that G is finite and non-trivial). We claim that every other rotation in G is of the form

$$r_{m\phi} = (r_\phi)^m$$

for some $m \geqslant 1$. Let $r_\vartheta \in G$, then $\vartheta = m\phi + \psi$ where $0 \leqslant \psi < \phi$ and $m \in \mathbb{Z}_{\geqslant 0}$. Now

$$r_\vartheta = r_{m\phi + \psi} = (r_\phi)^m r_\psi$$

So $r_\psi = (r_\phi)^{-m} r_\vartheta \in G$ and $0 \leqslant \psi < \phi$. However, ϕ is the smallest non-zero angle (by assumption) we have that $\psi = 0$ and $\vartheta = m\phi$ as required. Therefore, G is generated by r_ϕ and so G is cyclic.

We now suppose that G contains a reflection f (in particular $f^2 = e$). We set $H = G \cap SO_2(\mathbb{R})$. Then H is a subgroup of $SO_2(\mathbb{R})$ and, by the above, H is cyclic. So we have

$$H = \{\mathrm{id}, r, r^2, \ldots, r^{n-1}\}$$

for some $n \geqslant 1$. Take any reflection $f' \in G$. Then $f'f$ is a rotation, so $f'f = r^i$ for some i (by the trigonometric double angle formulae). Thus we get $f' = r^i f^{-1} = r^i f$. This shows that

$$G = \{\mathrm{id}, r, r^2, \ldots, r^{n-1}, f, rf, r^2 f, \ldots, r^{n-1} f\}$$

and satisfies

$$r^n = \mathrm{id}, \quad f^2 = \mathrm{id}, \quad fr = r^{n-1} f$$

Hence we get that $G \cong D_{2n}$. $\square$

We now show that, in addition to the cyclic and dihedral groups, the symmetry groups of the Platonic solids provide an exhaustive list of the finite rotation groups of $\mathbb{R}^3$.

Theorem 4.6.2. *Let G be a finite subgroup of $SO_3(\mathbb{R})$. Then G is isomorphic to precisely one of the following groups:*

- C_n, $(n \geqslant 1)$: *rotational symmetry group of an n-pyramid*
- D_{2n}, $(n \geqslant 2)$: *rotational symmetry group of an n-prism*
- A_4: *rotational symmetry group of a regular tetrahedron*
- $\mathfrak{S}_4$: *rotational symmetry group of a cube (or a regular octahedron)*
- A_5: *rotational symmetry group of a regular dodecahedron (or a regular icosahedron).*

4.6 The classification of 3-dimensional rotational symmetry groups

Proof. Let G be a finite subgroup of $SO_3(\mathbb{R})$. Each element of G (other than id) represents a rotation in $\mathbb{R}^3$ around an axis passing through the origin. Take the unit sphere centered at the origin $(0, 0, 0)$. Then each rotation gives two poles on the unit sphere which are the intersection of the axis of rotation with the unit sphere. Let X denote the set of all poles of all the elements in $G \setminus \{\text{id}\}$. We claim that G acts on the set X. To see this, let $g \in G$ and let $x \in X$. Say that x is a pole for $h \in G$ (i.e. $(x)h = x$). Then we have

$$((x)g)(g^{-1}hg) = (x)(gg^{-1})hg = (x)hg = (x)g.$$

So we have that $(x)g$ is a pole for $g^{-1}hg$ and so $(x)g \in X$. Now the idea of the proof is to apply Burnside Counting theorem to the action of G on X and show that X has to be a particularly 'nice' configuration of points on the sphere.

Let N be the number of orbits of G in X. Choose a representative from each orbit $x_1, x_2, \ldots, x_N$. Now the identity id fixes every pole and each $g \neq I$ fixes exactly two poles. So using Burnside Counting theorem we get

$$N = \frac{1}{|G|}(|X| + (|G| - 1)2) = \frac{1}{|G|}\left(2(|G| - 1) + \sum_{i=1}^{N} |\text{Orb}_G(x_i)|\right).$$

Rearranging and using the Orbit-Stabilizer theorem we get

$$2\left(1 - \frac{1}{|G|}\right) = N - \frac{1}{|G|}\sum_{i=1}^{N} |\text{Orb}_G(x_i)|$$

$$= N - \sum_{i=1}^{N} \frac{|\text{Orb}_G(x_i)|}{|G|}$$

$$= N - \sum_{i=1}^{N} \frac{1}{|G_{x_i}|}$$

$$= \sum_{i=1}^{N} \left(1 - \frac{1}{|G_{x_i}|}\right).$$

Now assuming that $G \neq \{\text{id}\}$ we have

$$1 \leq 2\left(1 - \frac{1}{|G|}\right) < 2.$$

And each $|G_{x_i}| \geq 2$ as it contains at least id and one rotation; so we have that

$$\frac{1}{2} \leq 1 - \frac{1}{|G_{x_i}|} < 1$$

for $1 \leq i \leq N$. This implies that $2 \leq N < 4$ and hence $N = 2$ or 3.

For $N = 2$ then we have that $|\mathrm{Orb}_G(x_1)| + |\mathrm{Orb}_G(x_2)| = 2$, each orbit contains one pole, and we have two poles in total. Thus each rotation has the same axis. The plane passing through the origin and perpendicular to this axis is preserved by G. So G is isomorphic to a subgroup of $SO_2(\mathbb{R})$. Using the previous Theorem we see that $G \cong C_n$ for some n.

For $N = 3$ the situation is more complicated. We set $x = x_1, y = x_2, z = x_3$. We have that
$$1 + \frac{2}{|G|} = \frac{1}{|G_x|} + \frac{1}{|G_y|} + \frac{1}{|G_z|} > 1.$$

So we have four possible cases:

(a) $\frac{1}{|G_x|} = \frac{1}{2}, \frac{1}{|G_y|} = \frac{1}{2}, \frac{1}{|G_z|} = \frac{1}{n}$ for $n \geq 2$.
(b) $\frac{1}{|G_x|} = \frac{1}{2}, \frac{1}{|G_y|} = \frac{1}{3}, \frac{1}{|G_z|} = \frac{1}{3}$.
(c) $\frac{1}{|G_x|} = \frac{1}{2}, \frac{1}{|G_y|} = \frac{1}{3}, \frac{1}{|G_z|} = \frac{1}{4}$.
(d) $\frac{1}{|G_x|} = \frac{1}{2}, \frac{1}{|G_y|} = \frac{1}{3}, \frac{1}{|G_z|} = \frac{1}{5}$.

We will consider each of these cases in turn.

Case (a): Suppose $|G_x| = |G_y| = |G_z| = 2$ then we get $|G| = 4$. Up to isomorphism, there only exist two groups of order 4, namely C_4 and D_4. If $|G_x| = |G_y| = 2$ and $|G_z| = n \geq 3$ then we get $|G| = 2n$. Consider G_z the subgroup of all rotations with axis passing through z and $-z$. This group is cyclic of order n, so
$$G_z = \{\mathrm{id}, g, g^2, \ldots, g^{n-1}\}$$
for some $g \in G$. We claim that $x, (x)g, (x)g^2, \ldots, (x)g^{n-1}$ are all distinct. To see this suppose that $(x)g^i = (x)g^j$ for some $i > j$. Then $(x)g^{i-j} = x$. But z and $-z$ are the only points fixed by G_z and $x \neq -z$ (as $|G_x| = 2$ and $|G_z| = |G_{-z}| = n \geq 3$). Now we have
$$|x - (x)g| = |(x)g - (x)g^2| = \cdots = |g^{n-1} - x|$$
and $|z - x| = |z - (x)g^i|$ for all $i = 1, 2, \ldots, n - 1$. This means that the points $x, (x)g, \ldots, (x)g^{n-1}$ all lie in the same plane and form a regular n-gon, P. Now we have that $|\mathrm{Orb}_G(x)| = |G|/|G_x| = n$, and so
$$\mathrm{Orb}_G(x) = \{x, (x)g, \ldots, (x)g^{n-1}\}.$$

Thus G maps P to P and we get a homomorphism
$$\phi : G \longrightarrow G'$$
where G' is the 3-dimensional rotational symmetries of P. Now every non-trivial rotation in G has only two fixed points in X and so does not fix P. This means that $\ker \phi = \{\mathrm{id}\}$. Now as $|G| = 2n = |G'| = |D_{2n}|$, we see that ϕ is an isomorphism and $G \cong G' \cong D_{2n}$.

4.6 The classification of 3-dimensional rotational symmetry groups

Case (b): Suppose $|G_x| = 2$, $|G_y| = |G_z| = 3$. Then we have that $|G| = 12$ and $|\text{Orb}_G(z)| = 4$. Let $u \in \text{Orb}_G(z)$ with $|z - u| < 2$ (this is always possible as all poles lie on the unit sphere and $|\text{Orb}_G(z)| > 2$). So $u \neq -z$. As $|G_z| = 3$ we have that $G_z \cong C_3$. Choose $g \in G_z$ with $\langle g \rangle = G_z$. Then $u, g(u), g^2(u)$ are all distinct (same argument as in Case (a)). As g preserves distances, they form an equilateral triangle and are all equidistant from z. Now the orbit $\text{Orb}_G(z) = \{z, u, g(u), g^2(u)\}$ is preserved under the action of G. For $h \in G_u$ we have $h(u) = u$ and h permutes $z, g(u), g^2(u)$. As h preserves distances we see that the distances from u to $z, g(u)$ and $g^2(u)$ are all equal. Hence we have that $\{z, u, g(u), g^2(u)\}$ form a regular tetrahedron T and we have a homomorphism

$$\phi : G \longrightarrow G'$$

where G' is the rotational symmetry group of T. No rotation (other than id) fixes T, so $\ker(\phi) = \{\text{id}\}$ and ϕ is one-to-one. Now as $|G| = |G'| = 12$ we have that ϕ is an isomorphism and $G \cong G' \cong A_4$.

Case (c): Suppose $|G_x| = 2$, $|G_y| = 3$ and $|G_z| = 4$. Here we get that $|G| = 24$ and $|\text{Orb}_G(z)| = 6$. Now choose $u \in \text{Orb}_G(z)$ with $u \neq z, -z$. As $G_z \cong C_4$ we have $G_z = \{\text{id}, g, g^2, g^3\}$ for some $g \in G$. We can show as before that $u, g(u), g^2(u), g^3(u)$ form a square equidistant from z. As $-z \notin \text{Orb}_G(x)$ or $\text{Orb}_G(y)$ (otherwise $|G_{-z}| = |G_x|$ or $|G_y|$) we have

$$\text{Orb}_G(z) = \{z, -z, u, (u)g, (u)g^2, (u)g^3\}.$$

Now, $-u \in \text{Orb}_G(z)$ (as $|G_{-u}| = |G_u| = |G_z|$) and $-u \neq z, -z$ (as $u \neq z, -z$). Also we have that $|(u)g - u| = |(u)g^3 - u| < 2$ as $u, (u)g, (u)g^2, (u)g^3$ form a square. Thus $-u = (u)g^2$. This shows that $z, -z, u, (u)g, (u)g^2, (u)g^3$ form the vertices of a regular octahedron. Let G' be the rotational symmetry group of this regular octahedron. Then we get a homomorphism

$$\phi : G \longrightarrow G'$$

with $\ker(\phi) = \{\text{id}\}$. So ϕ is one-to-one and as $|G| = |G'| = 24$ we have that ϕ is an isomorphism and we have $G \cong G' \cong S_4$.

Case (d): Suppose $|G_x| = 2$, $|G_y| = 3$ and $|G_z| = 5$. Then we get that $|G| = 60$ and $|\text{Orb}_G(z)| = 12$. As $G_z \cong C_5$ we can find $g \in G$ with $G_z = \{\text{id}, g, g^2, g^3, g^4\}$. It can be shown that we can pick $u \in \text{Orb}_G(z)$ with $u \neq z, -z$, and $v \in \text{Orb}_G(z)$ with $v \neq z, -z, u, (u)g, (u)g^2, (u)g^3, (u)g^4$. Moreover one can check that

$$\{z, -z, u, (u)g, (u)g^2, (u)g^3, (u)g^4, v, (v)g, (v)g^2, (v)g^3, (v)g^4\}$$

form a regular icosahedron. Using the same argument as before, if we let G' be the rotational symmetry group of this regular icosahedron, then we get a one-to-one homomorphism from G to G' and as $|G| = |G'| = 60$ this is in fact an isomorphism. Thus we get that $G \cong G' \cong A_5$. □

4.7 Archimedean solids and their symmetry groups

We have now seen that the symmetry group of any 3-dimensional polyhedron is a cyclic or dihedral group, or A_4, A_5, or $\mathfrak{S}_4$ and that these arise as the symmetry groups of regular polyhedra (pyramids, prisms, the tetrahedron, the cube/octahedron, and the icosahedron/dodecahedron). However, there are many more polyhedra that one might be interested in, and one might ask "what are their symmetry groups"?

In this section, we are interested in the "semi-regular" or "Archimedean" solids whose faces are given by two or more distinct types of regular polygons (as oppose to Platonic solids, whose faces were all the same polygon). The most famous of the Archimedean solids is the "football" (or "truncated icosahedron") formed from regular pentagons and hexagons (see Figure 4.11). Of course, we know the symmetry groups of the Archimedean solids are finite subgroups of $SO_3(\mathbb{R})$ and so must belong to our list of cyclic and dihedral groups, A_4, A_5, and $\mathfrak{S}_4$. We now highlight how Theorem 4.6.2 can be used to calculate the symmetry groups of these (possibly decorated) Archimedean solids.

Example 4.7.1. Let's determine the symmetry group, G, of the permutahedron (depicted in Figure 4.10) obtained by "cutting the 6 nibs" off of an octahedron (hence its other name: the truncated octahedron). This polyhedron is formed of 8 hexagons and 6 squares.

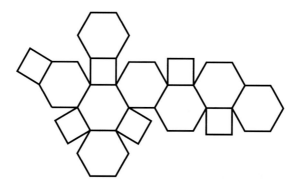

Fig. 4.10: The truncated octahedron, or "permutohedron".

Our first port-of-call is to determine $|G|$. Consider G acting on the set X consisting of all the hexagonal faces. There is precisely one orbit on X (as we can rotate any hexagonal face so that it lands on any other of the 8 hexagonal faces). The stabiliser of G acting on a given hexagonal face is not C_6 (as one might expect) but rather is $C_3 \leqslant C_6$; to see this, note that each hexagon $x \in X$ is adjacent to 3 squares and 3 hexagons (placed symmetrically around its edges) and the rotation through the face $x \in X$ must send hexagons to hexagons and square to squares. Therefore $|G| = 3 \times 8 = 24$.

4.7 Archimedean solids and their symmetry groups

Therefore G can only possibly be isomorphic to one of the three possible groups of size 24 from our list: C_{24}, D_{24}, or $\mathfrak{S}_4$. We must now determine which of these groups is isomorphic to G. We have already considered the symmetries of an arbitrary hexagonal face $x \in X$, so let's start here. We can pair-off the 8 hexagonal faces into 4 "opposite pairs" by placing a pole through the centre of such a pair of faces. We can rotate the permutahedron either 120° clockwise or 120° anti-clockwise through this pole and hence obtain two distinct rotations of order 3 belonging to G. Thus we have constructed $4 \times 2 = 8$ distinct rotations of order 3 as elements of G. Now we observe that the only elements of C_{24} of order 3 are $r^8, r^{16} \in C_{24}$; the only elements of D_{24} of order 3 are $r^4, r^8 \in C_{24}$; whereas $\mathfrak{S}_{24}$ does have 8 elements of order 3 (simply count the 3-cycles). Therefore $G \cong \mathfrak{S}_{24}$ by process of elimination.

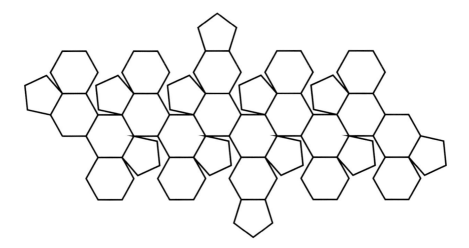

Fig. 4.11: The flat plan of a football, or "truncated icosahedron".

Exercise 4.7.2. Determine the symmetry groups of the truncated cube and its painted version, as depicted in Figure 4.12.

Exercise 4.7.3. Determine the symmetry group of the truncated icosahedron in Figure 4.11.

Exercise 4.7.4. Can you colour the hexagonal faces of the truncated icosahedron in such a way that the resulting symmetry group is A_4? What about $\mathfrak{S}_4$?

Exercise 4.7.5. Determine the symmetry groups of all thirteen Archimedean solids.

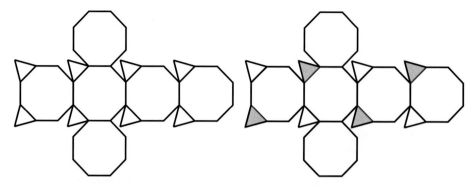

Fig. 4.12: The truncated cube and a painted version.

Part II
Algebras and representation theory

> "All mathematics is representation theory".
>
> Israel Gelfand

Representation theory is, in a nutshell, all about making abstract mathematical ideas more concrete. Israel Gelfand famously quipped that *"all mathematics is representation theory"*, but within the context of this book, we can take "representation theory" to mean the attempt to understand the sub-structures of algebraic objects, in a manner akin to how chemists break down molecules into atoms. More widely speaking, representation theory is a sort of unifying language that cuts across the natural sciences.

In some approaches to quantum gravity, particularly those emphasising the role of spacetime symmetries, it is hypothesised that these symmetries may be more fundamental than spacetime itself. Within these frameworks, analysing many-body systems often involves "multiplying together" representations of the underlying symmetry groups. Representation theory is used in chemistry to classify molecules based on their symmetry, which helps predict properties like polarity, chirality, and spectroscopic behaviour. The icosahedral symmetries considered in Chapter 4 are the starting point of a rich interplay between representation theory and mathematical virology [MST05].

Fermat's Last Theorem is one of the most famous statements in all of mathematics: the equation $a^n + b^n = c^n$ has no positive integer solutions for $n > 2$. This theorem was finally proven by Andrew Wiles [Wil95], in what has been heralded by John Conway as *"The proof of the twentieth century"*. Wiles' inspiration and strategy was based on Ribet's proof of the ε-conjecture [Rib90] (first proposed by Jean Pierre Serre) which makes properties of elliptic curves more concrete by rephrasing them in terms of corresponding properties of *Galois representations*. By proving that that certain *"modular representations"* could not exist, Wiles proved that no solution to $a^n + b^n = c^n$ could exist (for $n > 2$). This is part of a wider relationship between representation theory and number theory, known as the *Langland's programme* which has been described by Edward Frenkel as *"a kind of grand unified theory of mathematics"*.

What is a *"modular representation"* of an algebraic object? In a nutshell, it is a portrayal of the algebraic object in terms of concrete matrices whose entries are from the field $\mathbb{F}_p$ (consisting of the numbers $0, 1, \ldots, p-1$ with addition and multiplication modulo p). Studying representations over these finite fields leads to beautiful fractal-like behaviour, where the structure "repeats" in a manner related to the prime p. In Part 2 of this book we will encounter beautiful Sierpinski fractal behaviour when we construct the simple representations of binary Schur algebras in terms of colourings of Pascal's triangle (see Figure 6.5). Delving deeper into the structure of these modular representations, we can calculate their layers in terms of the p-adic evaluations of binomial coefficients, (see Section 6.11). Our hope with Part 2 of this book is that it will spark in readers a fascination for these elegant, fractal-like patterns which characterise modular representation theory. Generalising the analysis of Part 2 beyond the realms of the binary Schur algebras was the focus of Lusztig's conjecture.

Chapter 5
Non-invertible symmetry

The first part of this book dealt with "invertible symmetries". Every element of a group has an inverse and this, in some sense, is their defining feature. However, non-invertible symmetries occur in in physics, quantum mechanics, and across all of mathematics. For example, physicists wish to understand irreversible phase transitions — as we cool iron to below its Curie temperature, it transforms itself (irreversibly) into a permanent magnet.

This desire to understand non-invertible symmetries and transformations leads us to consider the more general constructs of $\Bbbk$-algebras. In concrete linear algebra-theoretic terms, we are widening our perspective from the set of invertible matrices $GL_n(\Bbbk)$ to the set of all (not necessarily invertible) matrices $Mat_n(\Bbbk)$. Whilst these algebraic objects are no longer constrained by the idea of invertibility, they admit stratifications whose "simple layers" do have invertible structures. The study of these invertible structures, and how they can be pieced together, is known as representation theory.

This chapter is comprised of a series of short sections, each of which introduces a new idea, or a new family of algebras. In the next chapter, we will systematically study the modular representation theory of these algebras.

5.1 Algebras

Let $\Bbbk$ be a field, for example the real or complex numbers. We let $\mathbb{F}_p$ be the field with p elements, that is the set of integers $\{0, 1, 2, \ldots, p-1\}$ equipped with the operations of p-modular addition and multiplication (for example $2 \times 3 \equiv 1 \equiv 4+2$ for $p = 5$). We let $Mat_n(\Bbbk)$ be the set of $(n \times n)$-matrices with entries from the field $\Bbbk$.

Definition 5.1.1. We define $A \subseteq Mat_n(\Bbbk)$ to be a **matrix $\Bbbk$-algebra** if for any $a, b \in A$ and $\alpha, \beta \in \Bbbk$, we have that $\alpha a + \beta b \in A$ and $ab \in A$. We say that the algebra A is **unital** if it contains the identity matrix, $id_n \in A$.

Example 5.1.2. Recall that the cyclic group C_2 consists of the two matrices

$$\text{id}_2 = \begin{bmatrix} 1 & 0 \\ 0 & 1 \end{bmatrix} \qquad r = \begin{bmatrix} 0 & 1 \\ 1 & 0 \end{bmatrix}.$$

The group algebra, $\mathbb{R}C_2$, consists of the (infinitely many) matrices of the form

$$\begin{bmatrix} \alpha & \beta \\ \beta & \alpha \end{bmatrix} = \alpha \text{id}_2 + \beta r$$

for $a, b \in \mathbb{R}$ together with the operations of matrix addition and multiplication, for example

$$(\alpha \text{id}_2 + \beta r) \times (\alpha' \text{id}_2 + \beta' r) = \begin{bmatrix} \alpha & \beta \\ \beta & \alpha \end{bmatrix} \begin{bmatrix} \alpha' & \beta' \\ \beta' & \alpha' \end{bmatrix}$$

$$= \begin{bmatrix} \alpha\alpha' + \beta\beta' & \alpha\beta' + \alpha'\beta \\ \alpha'\beta + \alpha\beta' & \alpha\alpha' + \beta\beta' \end{bmatrix}$$

$$= (\alpha\alpha' + \beta\beta')\text{id}_2 + (\alpha\beta' + \alpha'\beta)r.$$

Example 5.1.3. Let $G = \{g_1, \ldots, g_k\}$ be a finite matrix group, then we can take $\Bbbk G$ to be the set of all $\Bbbk$-linear combinations

$$\alpha_1 g_1 + \alpha_2 g_2 + \cdots + \alpha_k g_k.$$

Then $\Bbbk G$ is an example of a matrix $\Bbbk$-algebra. We refer to this as the **group algebra** of the group G over the field $\Bbbk$.

Example 5.1.4. We say that a matrix $M \in \text{Mat}_n(\Bbbk)$ is **generalised doubly stochastic** if all the row/columns sums are equal. For example, the matrix

$$\begin{bmatrix} -1 & -3 & 5 & -0.5 \\ 2 & 4 & -3 & -2.5 \\ -2 & 1 & -4 & 5.5 \\ 1.5 & -1.5 & 2.5 & -2 \end{bmatrix}$$

is generalised doubly stochastic, with common row/column sum equal to 0.5. We let $G_n(\Bbbk)$ be the set of all $(n \times n)$-generalised doubly stochastic matrices. Let $\alpha \in \Bbbk$ and let $A \in G_n(\Bbbk)$ with row/column sum equal to $a \in \Bbbk$; it is easy to check that the row/column sum of αA is equal to $\alpha a \in \Bbbk$. Similarly, it is not difficult to check that if $A, B \in G_n(\Bbbk)$ and these matrices have row/columns sums equal to a and b respectively, then $A + B \in G_n(\Bbbk)$ and has row/column-sum equal to $a + b$ and that $AB \in G_n(\Bbbk)$ and has row/column-sum ab. Thus $G_n(\Bbbk)$ forms a matrix $\Bbbk$-algebra.

The formal definition of a $\Bbbk$-algebra merely extracts the properties of matrix addition and multiplication as follows.

5.1 Algebras

Definition 5.1.5. Let A be a $\Bbbk$-linear vector space and let $* : A \times A \to A$. We say that A is a $\Bbbk$-algebra if is satisfies the axioms

- right distributivity $(x + y) * z = x * z + y * z$;
- left distributivity: $z * (x + y) = z * x + z * y$;
- scalar compatibility $(\alpha x) * (\beta y) = (\alpha \beta)(x * y)$;

for all $x, y, z \in A$ and $\alpha, \beta \in \Bbbk$. We say that the algebra is unital if there exists an identity element $\mathrm{id} \in A$ such that $x * \mathrm{id} = x = x * \mathrm{id}$ for all $x \in A$.

Let $(A, *)$ be a $\Bbbk$-algebra with basis (as a $\Bbbk$-vector space) given by $\{a_1, a_2, \ldots, a_n\}$. We define the corresponding multiplication table of A to be the $(n \times n)$-table with rows and columns indexed by $a_1, a_2, \ldots, a_n$ and whose ith row and jth column contains the product $a_i * a_j$ (see Example 5.1.6 for an example). While this table depends on the choice of basis, we note that any two bases of A differ only by a linear transformation; therefore the corresponding multiplication tables also differ only by linear transformation. Thus an algebra A is completely determined by any choice of basis, together with its corresponding multiplication table.

Example 5.1.6. The algebra of (3×3)-generalised doubly stochastic matrices $G_3(\Bbbk)$ has basis given by the five permutation matrices

$$\begin{bmatrix} 1 & 0 & 0 \\ 0 & 1 & 0 \\ 0 & 0 & 1 \end{bmatrix}, \begin{bmatrix} 0 & 1 & 0 \\ 1 & 0 & 0 \\ 0 & 0 & 1 \end{bmatrix}, \begin{bmatrix} 1 & 0 & 0 \\ 0 & 0 & 1 \\ 0 & 1 & 0 \end{bmatrix}, \begin{bmatrix} 0 & 1 & 0 \\ 0 & 0 & 1 \\ 1 & 0 & 0 \end{bmatrix}, \begin{bmatrix} 0 & 0 & 1 \\ 1 & 0 & 0 \\ 0 & 1 & 0 \end{bmatrix}$$

which we denote by $\mathrm{id}_3, g_1, g_2, g_3, g_4$ respectively. We have that

$$g_1 * g_3 = \mathrm{id}_3 - g_1 - g_2 + g_3 + g_4.$$

Exercise 5.1.7. Write down the multiplication table for $G_3(\Bbbk)$ with respect to the basis of Example 5.1.6.

Example 5.1.8. The set of all $(n \times n)$-matrices, $\mathrm{Mat}_n(\Bbbk)$ is itself a matrix algebra. We let E_{ij} denote the matrix whose only non-zero entry is a single 1 in the ith row and jth column. The algebra $\mathrm{Mat}_n(\Bbbk)$ is n^2-dimensional with basis $\{E_{i,j} \mid 1 \leq i \leq j \leq n\}$.

Definition 5.1.9. Let A be a (unital) $\Bbbk$-algebra and let $S \subset A$ be a subset of elements of A. We say that S generates A as a (unital) $\Bbbk$-algebra if every (non-identity) element of A can be written as a linear combination of products of elements from A. We define a relation to be any equality

$$\sum \alpha_{i_1,\ldots,i_p}(a_{i_1} * a_{i_2} * \cdots * a_{i_p}) = 0$$

for $\alpha_{i_1,\ldots,i_p} \in \Bbbk$ and $a_{i_1}, a_{i_2}, \ldots, a_{i_p} \in A$. We say that a set of relations, R, is complete if one can deduce the entire multiplication table of A using only (repeated applications of) the set of relations R. Given a $\Bbbk$-algebra, A, we define a presentation of A to be a set of generators, S, together with a complete set of relations, R.

Exercise 5.1.10. Continuing with Example 5.1.6, show that the algebra $G_3(\Bbbk)$ is generated by the elements g_1 and $g_2 \in G_3(\Bbbk)$. Provide a complete list of relations for these generators.

Exercise 5.1.11. Continuing with Example 5.1.8, show that the algebra $\mathrm{Mat}_n(\Bbbk)$ is generated by the elements $E_{i,i\pm 1} \in \mathrm{Mat}_n(\Bbbk)$. Provide a complete list of relations for these generators.

Besides presentations, many other group-theoretic notions have obvious counterparts for $\Bbbk$-algebras — first amongst which is the following:

Definition 5.1.12. Let A and B be $\Bbbk$-algebras. We say that a map $f : A \to B$ is a $\Bbbk$-algebra homomorphism if it preserves the algebra structure as follows:

$$f(\alpha x + \beta y) = \alpha f(x) + \beta f(y) \qquad f(xy) = f(x)f(y)$$

for all $\alpha, \beta \in \Bbbk$ and $x, y \in A$. We say that f is a a $\Bbbk$-algebra isomorphism if, in addition, f is bijective.

5.2 The Temperley–Lieb algebras

We now come to our first example of a bonafide diagram algebra. We define a Temperley–Lieb diagram of rank n to be a crossingless matching between n northern and n southern points in a framed rectangular region. Examples of such a diagrams are as follows

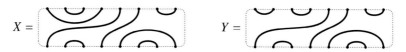

both of which are of rank 8. Now fix $\delta \in \Bbbk \setminus \{0\}$. A pair of Temperley–Lieb diagrams X, Y of rank n can be "multiplied" by way of vertical concatenation. That is, we place X above Y as follows

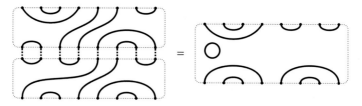

however, we notice that such a vertical concatenate of diagrams may contain a total of $t \in \mathbb{Z}_{\geq 0}$ "floating" closed loops, thus breaking the definition of a Temperley–Lieb diagram; we remove these loops to obtain a new Temperley–Lieb diagram, Z; we define the product $X * Y$ to be the $\delta^t Z$, hence "remembering" these closed loops. Thus in our example, the product $X * Y$ is as follows:

5.2 The Temperley–Lieb algebras

Definition 5.2.1. The Temperley–Lieb algebra $TL_n(\delta)$ is the $\mathbb{k}$-vector space spanned by all Temperley–Lieb diagrams of rank n together with the operation $*$.

Example 5.2.2. The algebra $TL_3(\delta)$ consists of precisely 5 diagrams. We record these diagrams, together with the multiplication rule, in Figure 5.1.

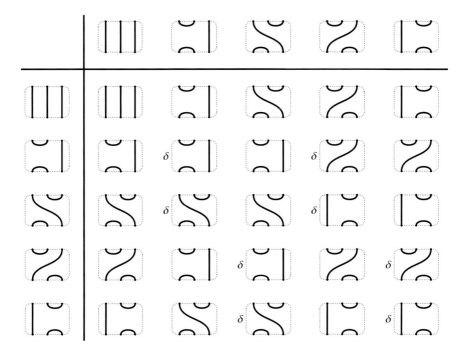

Fig. 5.1: The multiplication table for $TL_3(\delta)$.

We now fix some notation by setting e_i for $1 \leq i < n$ to be the diagram with $n - 2$ vertical strands, and a northern arc (respectively southern arc) connecting the ith and $(i+1)$th northern (respectively southern) vertices. For example the diagrams $e_1, e_2, e_3 \in TL_5(\delta)$ are as follows:

We refer to a strand from the northern edge to the southern edge of a Temperley–Lieb diagram as a **propagating strand**. We refer to a strand which starts and ends on the northern edge (or starts and ends on the southern edge) as a **northern arc** (or a **southern arc**). The number of Temperley–Lieb diagrams of rank n is equal to the nth Catalan number

$$C_n = \frac{1}{n+1}\binom{2n}{n}.$$

The Catalan numbers enumerate lots of other mathematical objects and are of significant interest in algebraic combinatorics [Sta15].

We are now ready to provide a presentation of the Temperley–Lieb algebra. The following proof is based on ideas of Kaufmann [Kau90]; this was later much improved upon by Ridout and Saint Aubin [RSA14]; the version presented here is due to Bowman, De Visscher, Farrell, Hazi, and Norton [BDF⁺].

Theorem 5.2.3. *Fix $\delta \in \Bbbk \setminus \{0\}$. The Temperley–Lieb algebra $\mathrm{TL}_n(\delta)$ is generated by $e_1, \ldots, e_{n-1}$ subject to the rules $e_i e_{i\pm 1} e_i = e_i$, $e_i e_j = e_j e_i$, and $e_i^2 = \delta e_i$ for all admissible $1 \leq i, j < n$ with $|i - j| > 1$.*

Proof. We need only show that any Temperley–Lieb diagram d can be written as a product of the diagrams $e_1, \ldots, e_{n-1}$. The "removing closed loops" rule follows from $e_i^2 = \delta e_i$. That we only care about the underlying matching of nodes (rather than any meaningless "wiggles" we draw within a given strand) follows from $e_i e_{i\pm 1} e_i = e_i$. We write d as a product of the generators via an algorithm, which we now detail. We first number the northern (respectively southern) vertices of d by the integers $1, \ldots, n$ (respectively $1', \ldots, n'$). Given a diagram d, we construct a sequence as follows: we set

$$\pi(k) = \begin{cases} \text{NE} & \text{if the vertex } k \text{ is connected to a vertex weakly to its right;} \\ \text{SE} & \text{otherwise} \end{cases}$$

$$\pi(k') = \begin{cases} \text{NW} & \text{if the vertex } k' \text{ is connected to a vertex strictly to its right;} \\ \text{SW} & \text{otherwise.} \end{cases}$$

We let $\Pi(d) = \{\pi(k)\}_{k=1,2\ldots,n,n',\ldots,2,1}$ be the path which moves one unit in each prescribed direction in the prescribed order of $\Pi(d)$. By definition, this path carves out a polygon in $\mathbb{R}^2$, which we can tile with (1×1)-squares; we fill each square with an e_i and hence obtain the required factorisation of d as a product of the generators. □

Example 5.2.4. For $a \in \mathrm{TL}_9(\delta)$ as in Figure 5.2, we have that $\Pi(a)$ is the sequence

{NE, NE, SE, SE, SE, SE, NE, NE, SE, SW, SW, SW, SW, NW, NW, NW, NW},

see Figure 5.3 for the resulting tiling.

Exercise 5.2.5. Construct an isomorphism between $G_3(\Bbbk)$ and $\mathrm{TL}_n(2)$.

5.3 Idempotents and things

Fig. 5.2: Examples of Temperley–Lieb diagrams $a, b, c \in \mathrm{TL}_9(\delta)$.

Fig. 5.3: The tilings of the diagrams from Figure 5.2. In each case the path begins at the western-most point, which is denoted with a circle; the path then follows the orientation depicted on the diagram. The colours have been chosen to distinguish between the e_i for $1 \leq i < 8$ and will be important later.

In the proof of Theorem 5.2.3 we rewrote a diagram as a product of the generators, or if you prefer: as a "word" in the "alphabet" $e_1, \ldots, e_{n-1}$. We notice that any two such "words" differ only by application of the commuting relation $e_i e_j = e_j e_i$ for $|i - j| > 1$. Indeed, these products were of *of minimal length* — that is, we cannot rewrite them as a shorter product of elements using the algebra relations. By way of contrast, the product e_i^2 (of length 2 in the generators) can be rewritten as a shorter product of generators using the relations as δe_i (of length 1 in the generators).

5.3 Idempotents and things

We will attempt to understand the "structure" or "shape" of a $\Bbbk$-algebra via its representation theory. The most important elements of a $\Bbbk$-algebra, from a representation theory point of view, are its idempotents, which we now define.

Definition 5.3.1. We say that an element $\varepsilon \in A$, is an **idempotent** if $\varepsilon^2 = \varepsilon$. We say that two idempotents $\varepsilon_1, \varepsilon_2 \in A$ are **orthogonal** if $\varepsilon_1 \varepsilon_2 = \varepsilon_2 \varepsilon_1 = 0$.

Example 5.3.2. We have that $\frac{1}{2}(\mathrm{id} + r)$ and $\frac{1}{2}(\mathrm{id} - r)$ in $\mathbb{C}C_2$ are both idempotents, for example

$$\tfrac{1}{2}(\mathrm{id} - r)^2 = \tfrac{1}{4}(\mathrm{id} - r)(\mathrm{id} - r) = \tfrac{1}{4}(\mathrm{id} - r - r + r^2) = \tfrac{1}{4}(\mathrm{id} - r - r + \mathrm{id}) = \tfrac{1}{2}(\mathrm{id} - r).$$

Exercise 5.3.3. Verify that the idempotents $\frac{1}{2}(\mathrm{id} + r)$ and $\frac{1}{2}(\mathrm{id} - r)$ are orthogonal.

Example 5.3.4. The Temperley–Lieb algebra has an abundance of easily constructible idempotents, for example

$$E_1 := \frac{1}{\delta} \quad \cdots \quad E_2 := \frac{1}{\delta^2} \quad \cdots$$

are both easily seen to be idempotents in $\mathrm{TL}_5(\delta)$ as follows:

$$E_1^2 = \frac{1}{\delta^2} \quad \cdots \quad = \frac{1}{\delta} \quad \cdots \quad = E_1$$

(check for yourself that E_2 is an idempotent).

More generally, any Temperley–Lieb diagram which is symmetric through its horizontal axis can be rescaled to be an idempotent. To see this, note that squaring any of these elements results in the same diagram, but with some number of closed loops in the middle, and that the power of δ can be chosen so as to cancel-out with the number of closed loops. For $0 \leqslant \ell \leqslant n/2$, we set

$$E_\ell := \frac{1}{\delta^\ell} \underbrace{\quad \cdots \quad}_{\ell \text{ arcs}} \underbrace{\quad \cdots \quad}_{n - 2\ell \text{ strands}} \tag{5.3.5}$$

to be the idempotent with ℓ adjacent northern and southern arcs at the left of the diagram and $(n - 2\ell)$ vertical strands at the right of the diagram.

Remark 6. What are the most obvious things we can say about general idempotents of an arbitrary (unital) $\Bbbk$-algebra, A?

- the identity element, $\mathrm{id} \in A$, is always an idempotent;
- for each idempotent $\varepsilon \in A$, we have that $\mathrm{id} - \varepsilon \in A$ is an orthogonal idempotent;
- for two orthogonal idempotents $\varepsilon_1, \varepsilon_2 \in A$, their sum $\varepsilon_1 + \varepsilon_2 \in A$ is also an idempotent.

We have already mentioned that idempotents are incredibly important in representation theory. One of the most powerful aspects of *categorical representation theory* is that it brings previously difficult-to-construct idempotents to the foreground, by

5.3 Idempotents and things

way of intuitive diagrammatic presentations (we will see this later in the book!). We finish this short section with a few more definitions which will be useful within this Chapter.

Definition 5.3.6. We say that $f \in A$ is unipotent if $f^k = 1$ for some $k \geqslant 0$. We say that an element $f \in A$ is nilpotent if $f^k = 0$ for some $k \geqslant 0$. We say that an element $f \in A$ is a quasi-idempotent if $f^2 = \alpha f$ for some $\alpha \in \Bbbk$.

Example 5.3.7. Recall that $\mathbb{F}_2$ is the field of 2 elements. We have that $\mathrm{id} + r \in \mathbb{F}_2 C_2$ is nilpotent, to see this note that

$$(\mathrm{id} + r)^2 = (\mathrm{id} + r)(\mathrm{id} + r) = \mathrm{id} + r + r + \mathrm{id} = 2(\mathrm{id} + r) = 0$$

since $1 + 1 = 0 \in \mathbb{F}_2$.

Example 5.3.8. The diagrams in Example 5.3.4 are quasi-idempotent if $\delta \in \mathbb{C}$ is specialised to be a root of unity. (For example $i \in \mathbb{C}$ is a 4th root of unity.)

Exercise 5.3.9. Verify that the matrix $\sum_{i-j=k} E_{i,j} \in \mathrm{Mat}_n(\Bbbk)$ for $k > 0$ is nilpotent. (Hint: start with the $k = n - 1$ case.)

Before embarking on the representation theory of $\Bbbk$-algebras we will require a notion of self-symmetry (generalising the idea of "inverses" from group theory).

Definition 5.3.10. We say that a map $\iota : A \to A$ is an anti-involution providing that

$$\iota(a * b) = \iota(b) * \iota(a)$$

for all $a, b \in A$.

Example 5.3.11. For the Temperley–Lieb algebras, the anti-involution is given by "flip through the horizontal axis". For example

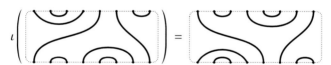

Exercise 5.3.12. Verify that the anti-involution property holds for the map ι defined as flipping a Temperley–Lieb diagram through the horizontal axis.

For diagrammatic algebras, the anti-involution is always easily pictured as "flip through the horizontal axis". For symmetric groups, we have already seen that flipping a strand diagram corresponds to the map $: g \mapsto g^{-1}$ and, indeed, this idea of inversion is what we are trying to replicate in the context of $\Bbbk$-algebras. Of course, it is easy to see that "flip and then concatenate" is the same as "concatenate and flip" and so this clearly defines an anti-involution for any sensible diagram algebra (whose multiplication is based on vertical stacking).

Exercise 5.3.13. Let G be any finite group and $\Bbbk$ be a field. Verify that the linear map determined by $\iota : \Bbbk G \to \Bbbk G$, $\iota(g) = g^{-1}$ is indeed an anti-involution.

5.4 Gradings and zig-zag algebras

We now define another family of algebras, the (extended) zig-zag algebras ZZ_n, via generators and relations. We will realise the zig-zag algebras as our prototypical examples of $\mathbb{Z}$-graded algebras and illustrate how this grading provides us with richer algebraic structures.

Definition 5.4.1. For $n \geq 2$ we define the (extended) zigzag algebra, ZZ_n, to be the $\Bbbk$-algebra generated by $\langle e_j, a^j_{j+1}, a^{j+1}_j \mid 0 \leq j < n \rangle$ subject to the idempotent relations
$$\text{id} = \sum_{0 \leq j < n} e_j \quad e_i e_j = \delta_{i,j} e_j \quad e_j a^j_{j\pm1} e_{j\pm1} = a^j_{j\pm1}$$

together with
$$a^j_{j\pm1} a^{j\pm1}_{j\pm2} = 0 \quad a^j_{j+1} a^{j+1}_j = -a^j_{j-1} a^{j-1}_j \quad a^0_1 a^1_0 = 0$$

for all admissible j. We formally set $a^j_j = e_j$ for $0 \leq j < n$.

Whilst defining a $\Bbbk$-algebra via generators and relations is very useful, it does not immediately give the reader a "feeling" for the algebra itself. For example, it is often difficult to see whether a $\Bbbk$-algebra defined in such a manner is finite or infinite-dimensional, or even if it is 0-dimensional! We begin by looking at small rank examples of zig-zag algebras, and then apply what we learn to understand the general case.

Example 5.4.2. Let $n = 2$. The algebra, ZZ_2, is generated by $\langle e_0, e_1, a^0_1, a^1_0 \rangle$ modulo the relations
$$1 = e_0 + e_1, \quad e_0^2 = e_0, \quad e_1^2 = e_1, \quad e_1 e_0 = 0 = e_0 e_1, \quad e_0 a^0_1 e_1 = a^0_1, \quad a^0_1 a^1_0 = 0.$$

We think of the idempotents as "staying still" (note that multiplying an idempotent e_i by itself over and over again produces nothing new — we do not "go anywhere") and we imagine that the a^0_1, a^1_0 elements are "stepping between the idempotents". This can be visualised as walking on the following graph:

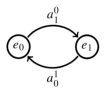

this allows us to think of the idempotents of ZZ_2 as vertices and the non-idempotent elements as paths in this graph. We hence think of the relation $a^0_1 a^1_0 = 0$ as saying "any path from vertex e_0 to itself is equal to 0". In this manner, we see that any path of length 3 must also be zero simply because it must be of the form

5.4 Gradings and zig-zag algebras

$$\ldots a_1^0 a_0^1 \ldots$$

We hence realise that ZZ_2 is 5-dimensional with basis

$$\{e_0, e_1, a_1^0, a_0^1, a_0^1 a_1^0\}.$$

Now that we have a presentation *together with* a basis for our algebra ZZ_2, we have a very concrete understanding of our algebra. In particular, we are able to provide a multiplication table for this algebra with respect to this basis is as follows:

	e_0	e_1	a_1^0	a_0^1	$a_0^1 a_1^0$
e_0	e_0	·	a_1^0	·	·
e_1	·	e_1	·	a_0^1	$a_0^1 a_1^0$
a_1^0	·	a_1^0	·	·	·
a_0^1	a_0^1	·	$a_0^1 a_1^0$	·	·
$a_0^1 a_1^0$	·	$a_0^1 a_1^0$	·	·	·

Armed with the multiplication table, we are now able to concretely realise ZZ_2 as a matrix algebra. For $p = 2$ we define an isomorphism from ZZ_2 to an $\mathbb{F}_2$-subalgebra of $\text{Mat}_3(\mathbb{F}_2)$ by first defining the map on idempotents as follows:

$$e_0 \mapsto \begin{bmatrix} 1 & 0 & 0 \\ 0 & 0 & 0 \\ 0 & 0 & 0 \end{bmatrix} \quad e_1 \mapsto \begin{bmatrix} 0 & 0 & 0 \\ 0 & 1 & 0 \\ 0 & 0 & 1 \end{bmatrix}$$

and for the remaining, non-idempotent elements, as follows:

$$a_1^0 \mapsto \begin{bmatrix} 0 & 1 & 1 \\ 0 & 0 & 0 \\ 0 & 0 & 0 \end{bmatrix} \quad a_0^1 \mapsto \begin{bmatrix} 0 & 0 & 0 \\ 1 & 0 & 0 \\ 1 & 0 & 0 \end{bmatrix} \quad a_0^1 a_1^0 \mapsto \begin{bmatrix} 0 & 0 & 0 \\ 0 & 1 & 1 \\ 0 & 1 & 1 \end{bmatrix}.$$

This subalgebra of $\text{Mat}_3(\mathbb{F}_2)$ is 5-dimensional and we visualise its basis as follows:

Exercise 5.4.3. We define a map from ZZ_n to itself by setting $\iota(e_k) = e_k$, $\iota(a_k^{k+1}) = a_{k+1}^k$, and $\iota(a_{k+1}^k) = a_k^{k+1}$ for all admissible k. Verify that ι is an anti-involution.

Definition 5.4.4. Let $\Bbbk$ be a field and A be a $\Bbbk$-algebra. A $\mathbb{Z}$-grading on A is a decomposition of A as a direct sum $A = \oplus_{i \in \mathbb{Z}} A_i$ such that if $a \in A_i$ and $b \in A_j$, then $ab \in A_{i+j}$. A graded algebra is an algebra with a specified $\mathbb{Z}$-grading. If $a \in A_i$, we say a is homogeneous of degree i and with $\deg(a) = i$. We define the **graded dimension** of A to be the formal power series

$$\dim_q(A) = \sum_{i \in \mathbb{Z}} \dim_\Bbbk(A_i) q^i.$$

Example 5.4.5. Our path-theoretic way of thinking about $\mathbb{ZZ}_2$ allows us to $\mathbb{Z}$-grade the algebra according to the length of the paths. In other words,

$$\deg(e_0) = 0 = \deg(e_1), \quad \deg(a_1^0) = 1 = \deg(a_0^1), \quad \deg(a_0^1 a_1^0) = 2$$

and so $\dim_q(\mathbb{ZZ}_2) = 2 + 2q + q^2$.

Exercise 5.4.6. Building on Example 5.1.8 and Exercise 5.1.11, verify that $\deg(E_{i,j}) = i - j$ defines a $\mathbb{Z}$-grading on $\operatorname{Mat}_n(\Bbbk)$ and calculate the resulting graded dimension of the algebra.

Definition 5.4.7. Let A and B be $\mathbb{Z}$-graded $\Bbbk$-algebras. We say that a $\Bbbk$-algebra homomorphism $f : A \to B$ is a $\mathbb{Z}$-graded $\Bbbk$-algebra homomorphism if $f(A_i) \subseteq B_i$ for all $i \in \mathbb{Z}$. We say that f is a a $\mathbb{Z}$-graded $\Bbbk$-algebra isomorphism if, in addition, f is bijective.

Example 5.4.8. Let $n = 3$. The algebra, $\mathbb{ZZ}_3$, is generated by $\langle e_0, e_1, e_2, a_1^0, a_0^1, a_2^1, a_1^2 \rangle$ modulo the relations

$$\mathrm{id} = \sum_{0 \leqslant k \leqslant 2} e_k, \quad e_i e_j = \delta_{i,j} e_j, \quad e_j a^j_{j \pm 1} e_{j \pm 1} = a^j_{j \pm 1}, \quad a_0^1 a_1^0 = -a_2^1 a_1^2,$$

$$a_1^0 a_2^1 = a_1^2 a_0^1 = a_1^0 a_0^1 = 0,$$

together with their flips under the anti-involution ι. We continue to think of the idempotents as "staying still" (again noting that multiplying an idempotent e_i by itself over and over again produces nothing new) and we imagine that the $a^j_{j \pm 1}$ elements are "stepping between the idempotents". This can be visualised as walking on the following graph:

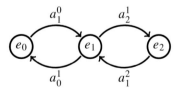

and we hence think of the relation $a_1^0 a_0^1 = 0$ as saying "any path from vertex e_0 to itself is equal to 0" and the relation $a_0^1 a_1^0 = -a_2^1 a_1^2$ as saying "the two paths from vertex e_1 to itself are equal, up to sign". Similarly we can think of $a_1^0 a_2^1 = 0$ as saying "any path of length 2 from vertex e_0 to vertex e_2 is equal to 0". We hence realise that $\mathbb{ZZ}_3$ has basis

$$\{e_0, e_1, e_2, a_1^0, a_0^1, a_2^1, a_1^2, a_0^1 a_1^0, a_1^2 a_2^1\}.$$

5.4 Gradings and zig-zag algebras

In particular, all paths of length 3 are zero; for example

$$a_1^2 a_2^1 a_1^2 = a_1^2(a_2^1 a_1^2) = -a_1^2(a_0^1 a_1^0) = -(a_1^2 a_0^1)a_1^0 = 0.$$

Grading the algebra again by the lengths of these paths as before, we have that

$$\dim_q(ZZ_3) = 3 + 4q + 2q^2.$$

The multiplication table for this algebra is as follows:

	e_0	e_1	e_2	a_1^0	a_0^1	a_2^1	a_1^2	$a_0^1 a_1^0$	$a_1^2 a_2^1$
e_0	e_0	·	·	a_1^0	·	·	·	·	·
e_1	·	e_1	·	·	a_0^1	a_2^1	·	$a_0^1 a_1^0$	·
e_2	·	·	e_2	·	·	·	a_1^2	·	$a_1^2 a_2^1$
a_1^0	·	a_1^0	·	·	·	·	·	·	·
a_0^1	a_0^1	·	·	$a_0^1 a_1^0$	·	·	·	·	·
a_2^1	·	·	a_2^1	·	·	·	$-a_0^1 a_1^0$	·	·
a_1^2	·	a_1^2	·	·	$a_1^2 a_2^1$	·	·	·	·
$a_0^1 a_1^0$	·	$a_0^1 a_1^0$	·	·	·	·	·	·	·
$a_1^2 a_2^1$	·	·	$a_1^2 a_2^1$	·	·	·	·	·	·

We can treat the algebra ZZ_n for $n > 3$ in a similar fashion. For each $0 < j \leq n-1$ there is a pair of degree 2 paths from vertex e_j to itself, which we identify up to sign (and any path from vertex e_0 to itself is set to be equal to zero), any path of length 2 from a vertex to a distinct vertex is zero. Putting all this together one deduces that any path of length 3 is always set to zero (as above). Therefore we arrive at the following:

Proposition 5.4.9. *For $n \geq 2$, the algebra ZZ_n is a $\mathbb{Z}$-graded $\Bbbk$-algebra with the grading defined on the generators as follows:*

$$\deg(e_j) = 0 \quad \deg(a_{j\pm 1}^j) = 1$$

for all admissible j. The algebra ZZ_n has graded basis

$$\{e_j \mid 0 \leq j < n\} \cup \{a_{j+1}^j, a_j^{j+1} \mid 0 \leq j < n-1\} \cup \{a_{j-1}^j a_j^{j-1} \mid 0 < j \leq n-1\}$$

and therefore ZZ_n has graded dimension

$$\dim_q(ZZ_n) = n + (2n-2)q + (n-1)q^2.$$

Exercise 5.4.10. Construct a $\mathbb{k}$-algebra isomorphism from $\mathbb{ZZ}_2$ to the 5-dimensional $\mathbb{F}_3$-subalgebra of $\text{Mat}_4(\mathbb{F}_3)$ with basis visualised as follows

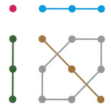

Generalise this construction to all primes $p \geqslant 3$.

Remark 7. In this section we spoke of the algebras $\mathbb{ZZ}_2$ and $\mathbb{ZZ}_3$ in terms of paths in an oriented graph or "quiver". This visualisation is very important in representation theory, we refer the reader to [EH18] for more details on representations of quivers.

5.5 Oriented Temperley–Lieb algebras

We now introduce our first example of an infinite dimensional $\mathbb{k}$-algebra. We let q denote a parameter. For $m + n \in \mathbb{N}$, we define an (m, n)-**weight** to be a diagram, λ, obtained by taking $m + n$ vertices and labelling m of these vertices with an $\wedge$ and n of these vertices with a $\vee$. We let $\Lambda_{m,n}$ denote the set of all (m, n)-weights. For example, for $m = 2$ and $n = 3$ we have the following elements

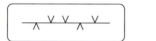 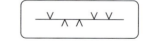

are elements of $\Lambda_{2,3}$. An (m, n)-orientation on a Temperley–Lieb diagram is a pair of (m, n)-weights $\lambda, \mu \in \Lambda_{m,n}$ placed on the top and bottom edges of the Temperley–Lieb diagram in such a manner that the arrows on propagating strands match, whereas the arrows on arcs are the opposites of each other. Given d a Temperley–Lieb diagram of rank $m + n$, we let d_μ^λ be the oriented diagram obtained by placing λ on the northern edge of d and placing μ on the southern edge of d providing that the resulting diagram is oriented, and we leave d_μ^λ undefined otherwise. Two oriented Temperley–Lieb diagrams for $m = 5$ and $n = 4$ are depicted in Figure 5.4.

Given pair of oriented Temperley–Lieb diagrams a_μ^λ and b_π^ρ for $\lambda, \mu, \rho, \pi \in \Lambda_{m,n}$, we let c_π^λ be the diagram obtained by vertically concatenating and removing all closed loops (if $\rho = \mu$) and be undefined otherwise. We define the product $a_\mu^\lambda * b_\pi^\rho$ as follows, by setting

$$a_\mu^\lambda * b_\pi^\rho = \begin{cases} q^{\#\{\text{clockwise closed loops}\} - \#\{\text{anti-clockwise closed loops}\}} c_\pi^\lambda & \text{if } \rho = \mu \\ 0 & \text{otherwise.} \end{cases}$$

5.5 Oriented Temperley–Lieb algebras

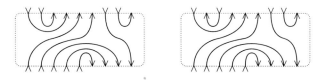

Fig. 5.4: A pair of oriented Temperley–Lieb diagrams with $m = 5$ and $n = 4$.

Definition 5.5.1. The oriented Temperley–Lieb algebra $\text{TL}_{m,n}^{\uparrow\downarrow}(q)$ is the $\Bbbk$-algebra generated by the (m, n)-oriented Temperley–Lieb diagrams under the $*$ operation.

The algebra $\text{TL}_{m,n}^{\uparrow\downarrow}(q)$ has basis given by

$$\{q^i a_\mu^\lambda \mid a_\mu^\lambda \text{ is an oriented Temperley–LIeb diagram}, i \in \mathbb{Z}\}$$

and in particular, this algebra is infinite-dimensional. A grading on the oriented Temperley–Lieb algebra can be defined as follows:

Definition 5.5.2. We set $\deg(q) = 1$ and $\deg(q^{-1}) = -1$. We define the degree of an oriented Temperley–Lieb diagram, d_μ^λ, as follows

$\sharp\{\text{clockwise oriented northern arcs}\} - \sharp\{\text{anti-clockwise oriented southern arcs}\}$.

Diagrammatically, we record this grading as follows

$$\deg(d_\mu^\lambda) = \sharp\left\{\;\smile\;\right\} - \sharp\left\{\;\frown\;\right\}. \tag{5.5.3}$$

Example 5.5.4. We have that

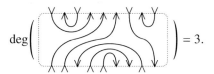

Proposition 5.5.5. *The algebra $\text{TL}_{m,n}^{\uparrow\downarrow}(q)$ is a $\mathbb{Z}$-graded $\Bbbk$-algebra.*

Proof. The multiplication removes clockwise closed loops (of degree $0+1$) and anti-clockwise loops (of degree $0-1$) and replaces them with q and q^{-1} respectively. This honours the grading, as required. All other relations preserve the arcs and strands of the diagram and so respect the grading. □

Exercise 5.5.6. Define an anti-involution on the oriented Temperley–Lieb algebra.

Exercise 5.5.7. Construct an example of an element of $\text{TL}_{4,5}^{\uparrow\downarrow}(q)$ and degree -8.

5.6 The binary Schur algebra

In this section we discuss one of the most beautiful objects in Lie theory, the binary Schur algebra. This algebra has an elementary presentation and has truly beautiful structure: this algebra represents the forefront of our knowledge for a whole host of important problems across representation theory, knot theory, physics...

We let $\mathbb{T}_r$ be the set of all sequences of length r consisting solely of the numbers 1 and 2. For example, $\mathbb{T}_3 = \{111, 112, 121, 122, 221, 212, 122, 222\}$. We define the weight of such a sequence, $\pi \in \mathbb{T}_r$, to the pair $\lambda = (\lambda_1, \lambda_2) \in \mathbb{N}^2$ where λ_1 (respectively λ_2) is the total number of 1s (respectively 2s) in the sequence π. We let Λ_r be the set of all possible weights of sequences from $\mathbb{T}_r$.

We wish to consider all maps between linear combinations of these sequences given by "raising" and "lowering" operators. The raising operator $f := f_{1 \to 2}$ takes a sequence π to the sum over all sequences which can be obtained from π by replacing a 1 with a 2. In fact, we further require the operator $f^{[k]} := f^{[k]}_{1 \to 2}$ takes a sequence π to the sum over all sequences which can be obtained from π by replacing k of the 1s with 2s. The lowering operators $e^{[k]} := e^{[k]}_{2 \to 1}$ take a sequence π to the sum over all sequences which can be obtained from π by replacing k of the 2s with 1s. We also require the idempotent maps 1_λ which send a sequence π of weight λ to itself and annihilate all sequences of weight $\mu \neq \lambda$.

Example 5.6.1. We have that

$(11122)e = 1111\underline{1}2 + 1112\underline{1}$ $(11122)f = \underline{2}1122 + 1\underline{2}122 + 11\underline{2}22$
$(11122)e^{[2]} = 111\underline{11}$ $(11122)f^{[2]} = \underline{22}122 + 1\underline{2}2\underline{2}2 + \underline{2}12\underline{2}2$
$(11122)1_{(3,2)} = 11122$ $(11122)1_{(4,1)} = 0$

here we have underlined the replaced entries in each term, for the reader's benefit.

Example 5.6.2. For $r = 2$, the set of binary sequences is given by 11, 12, 21, 22. The first sequence is of weight $(2, 0)$, the middle two sequences are of weight $(1, 1)$ and the final sequence is of weight $(0, 2)$. There are a total of 10 linearly independent maps which we can construct by composing the e, f, and $1_{(2,0)}$, $1_{(1,1)}$ and $1_{(0,2)}$ maps above; these are visualised as $(\mathbb{T}_2 \times \mathbb{T}_2)$-matrices as follows:

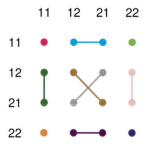

5.6 The binary Schur algebra

For example, the maps $1_{(2,0)}$, $1_{(1,1)}$, $1_{(0,2)}$ are the 3 maps along the diagonal given by $E_{11,11}$, $E_{12,12} + E_{21,21}$, $E_{22,22}$. The map $1_{(1,1)}e$ kills every sequence except for 12 and 21, both of which it sends to 11; in other words $1_{(1,1)}e = E_{12,11} + E_{21,11}$. The map $1_{(0,2)}e$ kills every sequence except 22 which is takes to 12 + 21 and therefore $1_{(0,2)}e = E_{22,12} + E_{22,21}$. We have that $1_{(0,2)}e^{[2]}$ kills every sequence except 22, which is takes to 11; in other words $e^{[2]}1_{(0,2)} = E_{22,11}$. (We note that $1_{(0,2)}ee = 2 \times 1_{(0,2)}e^{[2]}$.) Now, we have that the map

$$1_{(1,1)}ef = E_{12,12} + E_{21,21} + E_{21,12} + E_{12,21},$$

to see this, note that $1_{(1,1)}$ kills all sequences except 12 and 21; then e sends both these sequences to 11; and finally f sends 11 to the sum 12 + 21. We have not depicted $1_{(1,1)}ef$ in our matrix above (this involves a sum over four elementary matrices) and we have instead depicted the pair of maps

$$1_{(1,1)} = E_{12,12} + E_{21,21}$$

(the idempotent already considered above) together with

$$1_{(1,1)}(ef - 1) = E_{12,21} + E_{21,12}$$

(the conjoined off-diagonal dots in the middle of the matrix). One can continue in this manner until one finds all 10 basis elements. As this is our first example of a binary Schur algebra, we write the 10 basis elements of this algebra explicitly, as follows

$$\begin{bmatrix} 1 & 0 & 0 & 0 \\ 0 & 0 & 0 & 0 \\ 0 & 0 & 0 & 0 \\ 0 & 0 & 0 & 0 \end{bmatrix} \begin{bmatrix} 0 & 0 & 0 & 0 \\ 0 & 1 & 0 & 0 \\ 0 & 0 & 1 & 0 \\ 0 & 0 & 0 & 0 \end{bmatrix} \begin{bmatrix} 0 & 0 & 0 & 0 \\ 0 & 0 & 0 & 0 \\ 0 & 0 & 0 & 0 \\ 0 & 0 & 0 & 1 \end{bmatrix} \begin{bmatrix} 0 & 0 & 0 & 0 \\ 0 & 0 & 0 & 0 \\ 0 & 0 & 0 & 0 \\ 1 & 0 & 0 & 0 \end{bmatrix} \begin{bmatrix} 0 & 0 & 0 & 1 \\ 0 & 0 & 0 & 0 \\ 0 & 0 & 0 & 0 \\ 0 & 0 & 0 & 0 \end{bmatrix} \quad (5.6.3)$$

$$\begin{bmatrix} 0 & 1 & 1 & 0 \\ 0 & 0 & 0 & 0 \\ 0 & 0 & 0 & 0 \\ 0 & 0 & 0 & 0 \end{bmatrix} \begin{bmatrix} 0 & 0 & 0 & 0 \\ 1 & 0 & 0 & 0 \\ 1 & 0 & 0 & 0 \\ 0 & 0 & 0 & 0 \end{bmatrix} \begin{bmatrix} 0 & 0 & 0 & 0 \\ 0 & 0 & 1 & 0 \\ 0 & 1 & 0 & 0 \\ 0 & 0 & 0 & 0 \end{bmatrix} \begin{bmatrix} 0 & 0 & 0 & 0 \\ 0 & 0 & 0 & 1 \\ 0 & 0 & 0 & 1 \\ 0 & 0 & 0 & 0 \end{bmatrix} \begin{bmatrix} 0 & 0 & 0 & 0 \\ 0 & 0 & 0 & 0 \\ 0 & 0 & 0 & 0 \\ 0 & 1 & 1 & 0 \end{bmatrix} \quad (5.6.4)$$

and we invite the reader to compare these with the coloured entries above (this will help in understanding the next couple of examples).

Exercise 5.6.5. Construct the (10×10)-multiplication table for $S^{\Bbbk}(2)$. Hint: The idempotents $1_{(2,0)}$, $1_{(1,1)}$, $1_{(0,2)}$ break the matrices down into rows/columns whose pairwise distinct products are zero and this accounts for many of the 100 entries you need to compute.

Exercise 5.6.6. Construct an injective $\mathbb{F}_2$-algebra homomorphism : $\mathbb{ZZ}_2 \to S^{\Bbbk}(2)$.

Exercise 5.6.7. Rewrite every matrix in (5.6.3) and (5.6.4) as a product of the generators $e^{[i]}, f^{[j]}, 1_\lambda$ for $\lambda \in \mathbb{T}_2$ and $0 \leqslant i, j \leqslant 2$.

In order to illustrate the above construction on a slightly larger scale, we now consider the $r = 3$ and 4 cases. This will provide us with enough insight to define the general binary Schur algebras via generators and relations.

Example 5.6.8. We let $r = 3$, so that the set of binary sequences is given by 111, 112, 121, 122, 221, 212, 122, and 222. There are a total of 20 linearly independent maps which we can construct by composing the e, f, and $1_{(3,0)}$, $1_{(2,1)}$, $1_{(1,2)}$ and $1_{(0,3)}$ maps defined above. These are visualised as $(\mathbb{T}_3 \times \mathbb{T}_3)$-matrices as follows:

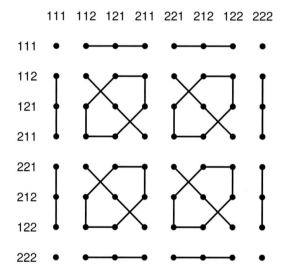

This matrix is not much harder to calculate than our previous example. For an example, the "top-left hexagon" in this matrix is given by

$$1_{(2,1)}(ef - 1) = E_{112,121} + E_{112,211} + E_{121,112} + E_{121,211} + E_{211,112} + E_{211,121}.$$

To see this, note that $1_{(2,1)}$ kills all sequences except 112 and 121 and 211; then e sends all three of these sequences to 111; and finally f sends 111 to the sum $112 + 121 + 211$.

Example 5.6.9. Let $\pi = 1112 \in \mathbb{T}_4$. We have that

$$\begin{aligned}(1112)1_{(3,1)}fe &= (2112 + 1212 + 1122)e \\ &= (1112 + 2111) + (1112 + 1211) + (1112 + 1121)) \\ &= (1112 + 1121 + 1211 + 2111) + 2(1112)1_{(3,1)} \\ &= (1112)ef + 2(1112)\end{aligned}$$

and so we see that $1_{(3,1)}fe = 1_{(3,1)}(ef + 2)$.

5.6 The binary Schur algebra

Example 5.6.10. For $r = 4$ there are a total of 34 linearly independent maps which we can construct by composing the e, f, and $1_{(4,0)}$, $1_{(3,1)}$, $1_{(2,2)}$, $1_{(1,3)}$ and $1_{(0,4)}$ maps. These are visualised as $(\mathbb{T}_4 \times \mathbb{T}_4)$-matrices as follows:

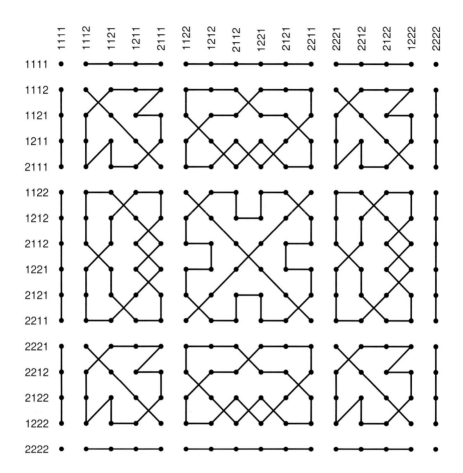

We now consider the binary Schur algebras of arbitrary rank. It is clear from the construction of the maps e, f, and 1_λ that the former two maps are nilpotent (there are only finitely many 2s to change into 1s and vice versa) and that the final map is idempotent. The idempotents 1_λ are free to pass through the e and f maps at the expense of their weight λ begin raised or lowered accordingly. The trivial map (which takes any sequence π to itself) is clearly given by the sum over all the idempotents 1_λ for $\lambda \in \Lambda_r$. Moreover, since the weight of a sequence is unique we have that $1_\lambda 1_\mu(\pi) = 0$ for any $\lambda \neq \mu$ and $\pi \in \mathbb{T}_r$.

Now we need to consider the effect of composing the e and f maps. We can break this calculation down according to weight spaces (since the idempotents labelled by these spaces sum to be the identity map). The maps ef and fe preserve weight spaces (i.e. send a sequence of weight λ to a linear combination of sequences of weight λ). Indeed, with a little work one can check that $1_\lambda fe = 1_\lambda(ef - \lambda_2 + \lambda_1)$.

Exercise 5.6.11. Calculate $(111122)1_{(4,2)}f^{[2]}e$ and rewrite $1_{(4,2)}f^{[2]}e$ in the form $\alpha ef^{[2]} + \beta f$ for $\alpha, \beta \in \mathbb{k}$.

For larger examples, we emphasise that our raising and lowering operators are acting on $\mathbb{T}_r$ by sums over all possible ways of turning 1s into 2s followed by 2s into 1s and vice versa, therefore we should expect arbitrary binomial coefficients to start appearing when we compose these maps. With a slightly larger example, we see that

$(11111122)f^{[2]}e^{[2]}$

$= (11112222 + 11121222 + 11211222 + 11221122 + \ldots)e^{[2]}$

$= (11112222)e^{[2]} + (11121222)e^{[2]} + (11211222)e^{[2]} + (11221122)e^{[2]} + \ldots$

$= (11111122 + 11111212 + 11112112 + 11112121 + 11112112 + 11112211) + \ldots$

is a sum over $\binom{6}{2} \times \binom{4}{2}$ terms, which we claim can be rewritten as follows:

$(11111122)f^{[2]}e^{[2]} = (11111122)e^{[2]}f^{[2]} + 4(11111122)ef + 6(11111122)1_{(6,2)}.$

We leave grouping together the terms and verifying the claim as an exercise for the reader. As a sanity check, we note that $(11111122)e^{[2]}f^{[2]}$ is a sum over all $\binom{8}{2}$ elements in $\mathbb{T}_8$; $(11111122)ef$ is a sum over $\binom{2}{1} \times \binom{7}{1}$ elements; and $(11111122)1_{(6,2)}$ is a single term; summing these altogether we obtain

$$1 \times \binom{8}{2} + 4 \times \binom{2}{1} \times \binom{7}{1} + 6 \times 1 = 90 = \binom{6}{2} \times \binom{4}{2}.$$

So we do, at least, obtain the correct number of terms. By working carefully with binomials coefficients, we can come up with a general rule for moving the raising and lowering operators past one another, in this way we arrive at the following definition...

Definition 5.6.12. Let $\mathbb{k}$ be a field and $r \geq 1$. We define the binary Schur algebra $S^{\mathbb{k}}(r)$ to be the algebra generated by the elements $f^{[k]}$, $e^{[k]}$, and 1_μ for $\mu \in \Lambda_r$ and $k \geq 0$ subject to the following relations. We have the idempotent relations

$$\sum_{\lambda \in \Lambda_r} 1_\lambda = 1$$

$$1_\lambda 1_\mu = \delta_{\lambda\mu} 1_\lambda$$

$$f^{[k]}1_{(\lambda_1, \lambda_2)} = 1_{(\lambda_1+k, \lambda_2-k)}f^{[k]}$$

$$e^{[k]}1_{(\lambda_1, \lambda_2)} = 1_{(\lambda_1-k, \lambda_2+k)}e^{[k]}$$

(with the convention that any 1_λ for $\lambda \notin \Lambda_r$ is zero), the index raising relations

$$e^{[a]}e^{[b]} = \frac{(a+b)!}{a!b!}e^{[a+b]} \qquad e^{[0]} = 1$$

$$f^{[a]}f^{[b]} = \frac{(a+b)!}{a!b!}f^{[a+b]} \qquad f^{[0]} = 1,$$

5.6 The binary Schur algebra

and the relations for passing divided powers through one another as follows:

$$f^{[a]}1_\lambda e^{[b]} = \sum_{k \geqslant 0} \binom{a+b+\lambda_1-\lambda_2}{k} e^{[b-k]}1_{(\lambda_1+a+b-k,\lambda_2-a-b+k)}f^{[a-k]} \quad (5.6.13)$$

$$e^{[b]}1_\lambda f^{[a]} = \sum_{k \geqslant 0} \binom{a+b-\lambda_1+\lambda_2}{k} f^{[a-k]}1_{(\lambda_1-a-b+k,\lambda_2+a+b-k)}e^{[b-k]}. \quad (5.6.14)$$

Exercise 5.6.15. Prove that the raising and lowering operators as defined on $\mathbb{T}_r$ do indeed satisfy relation (5.6.13). Hint: work over $\mathbb{Q}$ and afterwards restrict to $\mathbb{Z} \subseteq \mathbb{Q}$. For $\pi \in \mathbb{T}_r$, we have that $(\pi)f^{[a]} = \frac{1}{a!}ff\ldots f$ (applied a times) and so we can calculate $f^{[a]}1_\lambda e = \frac{1}{a}f(f^{[a-1]}1_\lambda e)$ by induction on $a \geqslant 1$. Similarly, $(\pi)e^{[b]} = \frac{1}{k!}ee\ldots e$ (applied b times) and so we can then use the previous step to calculate $f^{[a]}1_\lambda e^{[b]} = \frac{1}{b}(f^{[a]}1_\lambda e^{[b-1]})e$ by induction on $b \geqslant 1$.

Example 5.6.16. Over the rationals, the binary Schur algebras $S^\mathbb{Q}(r)$ have many idempotent elements. For example, all the elements 1_λ for $\lambda \in \Lambda_r$ are idempotents. Over the rationals, the following elements of $S^\mathbb{Q}(2)$ are all idempotents

$$\begin{bmatrix} 1 & 0 & 0 & 0 \\ 0 & 0 & 0 & 0 \\ 0 & 0 & 0 & 0 \\ 0 & 0 & 0 & 1 \end{bmatrix} \quad \begin{bmatrix} 0 & 0 & 0 & 0 \\ 0 & 1 & 0 & 0 \\ 0 & 0 & 1 & 0 \\ 0 & 0 & 0 & 0 \end{bmatrix} \quad \frac{1}{2}\begin{bmatrix} 0 & 0 & 0 & 0 \\ 0 & 1 & 1 & 0 \\ 0 & 1 & 1 & 0 \\ 0 & 0 & 0 & 0 \end{bmatrix}$$

where these elements are $1_{(2,0)} + 1_{(0,2)}$, $1_{(1,1)}$, and $\frac{1}{2}e1_{(2,0)}f$ respectively.

Exercise 5.6.17. Fix the field $\Bbbk$ to be of characteristic zero (for example, $\mathbb{Q}$ or $\mathbb{C}$). Prove that the relations (5.6.13) and (5.6.14) for $a, b \geqslant 1$ all follow from the $a = b = 1$ case. This vastly simplifies the presentation of the binary Schur algebra for these fields.

Proposition 5.6.18. *For $r > 2$ we have a $\Bbbk$-algebra homomorphism* $\det_r : S^\Bbbk(r) \to S^\Bbbk(r-2)$ *given by*

$$\det_r : e^{[k]} \mapsto e^{[k]} \qquad \det_r : f^{[k]} \mapsto f^{[k]}$$

$$\det_r : 1_{(\lambda_1,\lambda_2)} \mapsto \begin{cases} 1_{(\lambda_1-1,\lambda_2-1)} & \text{if } \lambda_1, \lambda_2 > 1 \\ 0 & \text{otherwise} \end{cases}$$

for $k \geqslant 0$. We refer to this as the **determinant homomorphism**.

Proof. The idempotent and the summing power relations are trivially preserved. To see that the divided power relations are preserved, we simply note that we have that $\lambda_1 - \lambda_2 = (\lambda_1 - 1) - (\lambda_2 - 1)$ and so the binomial coefficients in (5.6.13) and (5.6.14) are preserved, as required. □

Proposition 5.6.19. *For $r \geq 1$ the $\Bbbk$-algebra $S^{\Bbbk}(r)$ has basis*

$$\{e^{[j]}1_\lambda f^{[k]} \mid \lambda \in \Lambda_r, 0 \leq j, k \leq \lambda_1 - \lambda_2\}. \tag{5.6.20}$$

In particular

$$\dim(S^{\Bbbk}(r)) = \begin{cases} 1^2 + 3^2 + 5^2 + \cdots + (r+1)^2 & \text{if } r \text{ is even} \\ 2^2 + 4^2 + 6^2 + \cdots + (r+1)^2 & \text{if } r \text{ is odd}. \end{cases}$$

Proof. For $r = 1$ the result is trivial, for $r = 2$ see Example 5.6.2. We observe that $e^{[r+j]}1_{(r,0)} = 0 = 1_{(r,0)}f^{[r+k]}$ for $j, k > 0$, simply because the number of 1s or 2s in $\pi \in \mathbb{T}_r$ is less than or equal to r. Therefore

$$S^{\Bbbk}(r)1_{(r,0)}S^{\Bbbk}(r) = \Bbbk\{e^{[j]}1_{(r,0)}f^{[k]} \mid 0 \leq j, k \leq r\} \tag{5.6.21}$$

by relations (5.6.13) and the fact that $\sum_{\lambda \in \Lambda_r} 1_\lambda = 1$. Truncating this set on the left and right by $\mu, \nu \in \Lambda_r$ we have that

$$1_\mu S^{\Bbbk}(r)1_{(r,0)}S^{\Bbbk}(r)1_\nu = \Bbbk\{e^{[\mu_2]}1_{(r,0)}f^{[\nu_2]}\}$$

is 1-dimensional and therefore the set of elements in (5.6.21) is linearly independent, since the 1_λ for $\lambda \in \Lambda_r$ are a set of orthogonal idempotents. Finally, we observe that the map $\det_r$ for $r > 2$ is surjective with kernel

$$\ker(\det_r) = S^{\Bbbk}(r)(1_{(r,0)} + 1_{(0,r)})S^{\Bbbk}(r) = S^{\Bbbk}(r)1_{(r,0)}S^{\Bbbk}(r) \tag{5.6.22}$$

where the first equality follows by the definition of $\det_r$ and the second follows since $1_{0,r} = 1_{r,0}f^{[r]}$. The basis result follows by (5.6.21) (and the linear dependence of these elements) together with (5.6.22) and induction on $r \geq 1$. The dimension result follows then follows immediately. $\square$

Exercise 5.6.23. Check that the map $\iota : S^{\Bbbk}(r) \to S^{\Bbbk}(r)$ given by $\iota(e^{[m]}) = f^{[m]}, \iota(f^{[m]}) = e^{[m]}$ and $\iota(1_\lambda) = 1_\lambda$ for $\lambda \in \Lambda_r$ and $m \geq 1$ is an anti-involution.

Exercise 5.6.24. Calculate $(\pi)e^{[j]}1_{(r,0)}f^{[k]}$ for arbitrary $0 \leq j, k \leq r$ and $\pi \in \mathbb{T}_r$.

Exercise 5.6.25. Construct an injective $\mathbb{F}_3$-algebra homomorphism : $\mathbb{Z}\mathbb{Z}_3 \to S^{\Bbbk}(3)$.

Remark 8. It is worth noting that the determinant homomorphism is so-named because the binary Schur algebra was originally defined as the image of $\mathrm{GL}_2(\Bbbk)$ acting on $\mathbb{T}_r$. In this manner, the homomorphism $\det_r$ can be seen as an incarnation of the matrix determinant map

$$: \begin{bmatrix} a & b \\ c & d \end{bmatrix} \mapsto (ad - bc)^{-1}.$$

5.7 Hidden gradings on Schur algebras in positive characteristic

The binary Schur algebra is not a $\mathbb{Z}$-graded algebra in any obvious fashion (indeed, the discovery of such gradings was a huge surprise to the experts in the field). However, for small examples we are able to construct gradings rather easily by hand. This is an insightful exercise as it hints at the hidden connections with the Kazhdan–Lusztig theoretic diagram algebras we will encounter later in the book. The gradings on $S^{\Bbbk}(r)$ will depend heavily on the field $\Bbbk$ and they incorporate a great deal of structural information (as we shall see in the next chapter).

Example 5.7.1. We let $\Bbbk = \mathbb{F}_2$ be the field of two elements. We claim that the algebra $S^{\Bbbk}(2)$ is a $\mathbb{Z}$-graded $\Bbbk$-algebra and that we can picture the grading on basis elements as follows. First we have the identity element which is clearly of degree zero

$$\begin{bmatrix} 0 & 0 & 0 & 0 \\ 0 & 1 & 0 & 0 \\ 0 & 0 & 1 & 0 \\ 0 & 0 & 0 & 0 \end{bmatrix}$$

then we colour the remaining elements as follows

$$\begin{bmatrix} 1 & 0 & 0 & 0 \\ 0 & 0 & 0 & 0 \\ 0 & 0 & 0 & 0 \\ 0 & 0 & 0 & 0 \end{bmatrix} \quad \begin{bmatrix} 0 & 1 & 1 & 0 \\ 0 & 0 & 0 & 0 \\ 0 & 0 & 0 & 0 \\ 0 & 0 & 0 & 0 \end{bmatrix} \quad \begin{bmatrix} 0 & 0 & 0 & 1 \\ 0 & 0 & 0 & 0 \\ 0 & 0 & 0 & 0 \\ 0 & 0 & 0 & 0 \end{bmatrix}$$

$$\begin{bmatrix} 0 & 0 & 0 & 0 \\ 1 & 0 & 0 & 0 \\ 1 & 0 & 0 & 0 \\ 0 & 0 & 0 & 0 \end{bmatrix} \quad \begin{bmatrix} 0 & 0 & 0 & 0 \\ 0 & 1 & 1 & 0 \\ 0 & 1 & 1 & 0 \\ 0 & 0 & 0 & 0 \end{bmatrix} \quad \begin{bmatrix} 0 & 0 & 0 & 0 \\ 0 & 0 & 0 & 1 \\ 0 & 0 & 0 & 1 \\ 0 & 0 & 0 & 0 \end{bmatrix}$$

$$\begin{bmatrix} 0 & 0 & 0 & 0 \\ 0 & 0 & 0 & 0 \\ 0 & 0 & 0 & 0 \\ 1 & 0 & 0 & 0 \end{bmatrix} \quad \begin{bmatrix} 0 & 0 & 0 & 0 \\ 0 & 0 & 0 & 0 \\ 0 & 0 & 0 & 0 \\ 0 & 1 & 1 & 0 \end{bmatrix} \quad \begin{bmatrix} 0 & 0 & 0 & 0 \\ 0 & 0 & 0 & 0 \\ 0 & 0 & 0 & 0 \\ 0 & 0 & 0 & 1 \end{bmatrix}$$

where the degree of each corner pink element is zero, each blue element is one, and the degree of the unique green element in the centre is two.

To verify the claim, we simply multiply the matrices together: we notice that any non-zero product of any two pink elements is again pink, as required; any non-zero product of the two blue matrices is equal to the green matrix as required, for example we have the following non-zero product:

$$\begin{bmatrix} 0 & 1 & 1 & 0 \\ 0 & 0 & 0 & 0 \\ 0 & 0 & 0 & 0 \\ 0 & 0 & 0 & 0 \end{bmatrix} \begin{bmatrix} 0 & 0 & 0 & 0 \\ 1 & 0 & 0 & 0 \\ 1 & 0 & 0 & 0 \\ 0 & 0 & 0 & 0 \end{bmatrix} = \begin{bmatrix} 0 & 0 & 0 & 0 \\ 0 & 1 & 1 & 0 \\ 0 & 1 & 1 & 0 \\ 0 & 0 & 0 & 0 \end{bmatrix}.$$

Finally, the product of any element of degree 1 or 2 with the green matrix is always equal to zero (which is necessary, as there are no element of degree 3 or 4), for example

$$\begin{bmatrix} 0 & 0 & 0 & 0 \\ 0 & 1 & 1 & 0 \\ 0 & 1 & 1 & 0 \\ 0 & 0 & 0 & 0 \end{bmatrix} \begin{bmatrix} 0 & 0 & 0 & 0 \\ 0 & 1 & 1 & 0 \\ 0 & 1 & 1 & 0 \\ 0 & 0 & 0 & 0 \end{bmatrix} = 2 \begin{bmatrix} 0 & 0 & 0 & 0 \\ 0 & 1 & 1 & 0 \\ 0 & 1 & 1 & 0 \\ 0 & 0 & 0 & 0 \end{bmatrix} = 0$$

and so this does indeed define a graded structure on the Schur algebra, but *only over fields of characteristic 2!*

Exercise 5.7.2. Let $\Bbbk$ be a field of characteristic 3. Define a grading structure on $S^{\Bbbk}(3)$ analogous to the one we defined in Example 5.7.1. Hint: the resulting graded dimension of the algebra should be equal to $8 + 8q + 4q^2$.

Further Reading. The Temperley–Lieb algebras were first defined in the context of "Potts"-like problems in statistical mechanics, where they arise as the transfer matrices acting on electron-spin state space [TL71]. The Temperley–Lieb algebras were later rediscovered by Jones and famously put-to-use in his definition of the *Jones polynomial* of knot theory [Jon85, Jon87]. From our perspective, the most important appearance of Temperley–Lieb algebras is via their Schur–Weyl duality with the binary Schur algebra, which we will discuss in Part 5 of this book.

The first use of zig-zag algebras (of which the author is aware) was in Khovanov–Seidel's work on Floer cohomology [KS02] The zig-zag algebra and its many generalisations appear now appear in various places, such as symplectic geometry [EL17], the block-wise study of symmetric groups and their quiver Hecke algebras [EK18, KM19], and 2-representation theory [MT19].

The material on (binary) Schur algebras was inspired by Doty–Giaquinto's treatment of these algebras [DG02]. We have approached these algebras in this manner because these idempotent-rich presentations are similar in spirit to the definition of the diagrammatic algebras $\mathscr{H}_{(W,P)}$ via generators and relations. Our term "binary" Schur algebra (motivated by its definition in terms of sequences of 1s and 2s) is not used in the literature; it is usually defined as via the action of $GL_2(\Bbbk)$ on tensor space. For a more classical treatment of the Schur algebra we refer to [Gre07, Mar93].

Chapter 6
Representation theory

The previous chapters have introduced some of our favourite mathematical objects: the symmetric groups, Temperley–Lieb algebras, and the (binary) Schur algebras. We are now going to delve into the structure of these mathematical objects. We do this by "representing them" concretely as matrices.

Our first aim as representation theorists are to classify and construct the smallest such representations (that is the "simple representations"). This chapter introduces the framework of cellularity as an effective general theory for tackling this problem. We provide the reader with the representation theoretic tools that are necessary to understand and calculate Williamson's counter examples to Lusztig's conjecture. In this chapter we demonstrate the power of these tools by constructing the simple modules of binary Schur algebras in terms of the combinatorics of coloured Pascal triangles. As also develop concrete ways of visualising the submodule structures of more complicated representations.

6.1 Examples of simple modules and representations

We defined a $\Bbbk$-algebra to be an n-dimensional $\Bbbk$-vector space (with basis $\mathcal{B} = \{b_1, \ldots, b_n\}$, say) together with a multiplication (which our axioms ensured was "compatible" with the vector space structure). Thus, we can realise an abstractly defined $\Bbbk$-algebra A concretely as a matrix algebra as follows: suppose that

$$b_j x = \sum_{1 \leqslant k \leqslant n} \alpha_{j,k}^x b_k$$

for $x \in A$, $1 \leqslant j \leqslant n$, $\alpha_{j,k}^x \in \Bbbk$; we define a map

$$\varphi_{\mathcal{B}} : A \to \mathrm{Mat}_n(\Bbbk), \quad \varphi_{\mathcal{B}}(x) = \sum_{1 \leqslant j,k \leqslant n} \alpha_{j,k}^x E_{j,k}.$$

We refer to $\varphi_{\mathscr{B}}$ as the **natural representation** of A (as a matrix algebra) with respect to the basis $\mathscr{B}$.

Example 6.1.1. Let $G = C_3$ the cyclic group of order 3. We pick $\mathscr{B} = \{\mathrm{id}, r, r^2\}$ as a basis of the group algebra $\Bbbk C_3$ for $\Bbbk$ any field. The natural representation $\varphi_{\mathscr{B}}$ is given by

$$\varphi_{\mathscr{B}}(\mathrm{id}) = \begin{bmatrix} 1 & 0 & 0 \\ 0 & 1 & 0 \\ 0 & 0 & 1 \end{bmatrix} \quad \varphi_{\mathscr{B}}(r) = \begin{bmatrix} 0 & 1 & 0 \\ 0 & 0 & 1 \\ 1 & 0 & 0 \end{bmatrix} \quad \varphi_{\mathscr{B}}(r^2) = \begin{bmatrix} 0 & 0 & 1 \\ 1 & 0 & 0 \\ 0 & 1 & 0 \end{bmatrix}.$$

Example 6.1.2. We consider the Temperley–Lieb algebra $\mathrm{TL}_3(\delta)$. We pick

(6.1.3)

as a basis, $\mathscr{B}$, of $\mathrm{TL}_3(\delta)$. The natural representation is given by

$$\varphi_{\mathscr{B}}(e_1) = \begin{bmatrix} 0 & 1 & 0 & 0 & 0 \\ 0 & \delta & 0 & 0 & 0 \\ 0 & 0 & 0 & 1 & 0 \\ 0 & 0 & 0 & \delta & 0 \\ 0 & 1 & 0 & 0 & 0 \end{bmatrix} \quad \varphi_{\mathscr{B}}(e_2) = \begin{bmatrix} 0 & 0 & 1 & 0 & 0 \\ 0 & 0 & 0 & 0 & 1 \\ 0 & 0 & \delta & 0 & 0 \\ 0 & 0 & 1 & 0 & 0 \\ 0 & 0 & 0 & 0 & \delta \end{bmatrix}.$$

We note that e_1 and e_2 generate $\mathrm{TL}_3(\delta)$ as a unital $\Bbbk$-algebra and so this determines the representation on the remaining basis elements $\varphi_{\mathscr{B}}(e_1 e_2)$ and $\varphi_{\mathscr{B}}(e_2 e_1)$ (and $\varphi_{\mathscr{B}}(\mathrm{id}) = \mathrm{id}_5$).

When studying $\Bbbk$-algebras, the substructures of interest are the vector spaces preserved by our algebra multiplication. This leads us to the following definition:

Definition 6.1.4. Given $(A, *)$ a $\Bbbk$-algebra, we say that a $\Bbbk$-vector space $M \subseteq A$ is a left (respectively right) A-submodule if $a * m \in M$ (respectively $m * a \in M$) for all $a \in A$ and $m \in M$.

Definition 6.1.5. We say that M is an ideal if it is both a left and a right A-submodule.

We let A be a $\Bbbk$-algebra with basis $\mathscr{B} = \{b_1, \ldots, b_n\}$ and multiplication rule $b_j x = \sum_{1 \leq k \leq n} \alpha_{j,k}^x b_k$ for $x \in A$, $1 \leq j \leq n$, $\alpha_{j,k}^x \in \Bbbk$. We suppose that the A-submodule $M \subseteq A$ has basis $\mathscr{B}_M = \{b_1, \ldots, b_m\} \subseteq \{b_1, \ldots, b_n\} = \mathscr{B}$ for $m \leq n$. We let

$$\varphi_{\mathscr{B}_M} : A \to \mathrm{Mat}_m(\Bbbk), \quad \varphi_{\mathscr{B}_M}(x) = \sum_{1 \leq j,k \leq m} \alpha_{j,k}^x E_{j,k}.$$

We refer to $\varphi_{\mathscr{B}_M}$ as the **representation corresponding to the A-submodule M with basis $\mathscr{B}_M$**.

6.1 Examples of simple modules and representations

Example 6.1.6. The algebra $TL_3(\delta)$ has basis as in (6.1.3). For any $\delta \neq 1$, the element

$$\mathbb{1} = \left[\text{|||} \right] - \frac{\delta}{\delta^2 - 1} \left(\left[\smile\atop\frown\right] + \left[\smile\atop\frown\right] \right) + \frac{1}{\delta^2 - 1} \left(\left[\text{crossing}\right] + \left[\text{crossing}\right] \right)$$

spans a 1-dimensional right (and left) submodule of $TL_3(\delta)$. The identity diagram acts as on $\mathbb{1}$ as scalar multiplication by 1; all non-identity diagrams from 6.1.3 act as zero (on both the left and the right). In particular, the corresponding representation is determined by

$$\varphi_{\mathbb{1}}(\text{id}) = (1) \qquad \varphi_{\mathbb{1}}(e_1) = (0) \qquad \varphi_{\mathbb{1}}(e_2) = (0).$$

Example 6.1.7. Consider the right $TL_3(\delta)$-submodule $M = e_1 TL_3(\delta) \subseteq TL_3(\delta)$. We pick a basis of M as follows

$$\mathcal{B}_M = \{e_1, e_1 e_2\} = \left\{ \left[\smile\atop\frown\right], \left[\text{crossing}\right] \right\}. \tag{6.1.8}$$

With respect to this basis, the representation $\varphi_{\mathcal{B}_M}$ is determined on the generators as follows

$$\varphi_{\mathcal{B}_M}(\text{id}) = \begin{bmatrix} 1 & 0 \\ 0 & 1 \end{bmatrix}, \quad \varphi_{\mathcal{B}_M}(e_1) = \begin{bmatrix} \delta & 0 \\ 1 & 0 \end{bmatrix}, \quad \varphi_{\mathcal{B}_M}(e_2) = \begin{bmatrix} 0 & 1 \\ 0 & \delta \end{bmatrix}.$$

Exercise 6.1.9. Let A and B be $\Bbbk$-algebras and suppose that $f : A \to B$ is a $\Bbbk$-algebra homomorphism. Prove that $\ker(\varphi)$ is an ideal of A.

Definition 6.1.10. Let N, M be A-modules and suppose that $N \subseteq M$, we say that N is an A-submodule of M. We define the quotient A-module to be the vector space $M/N = \{m + N \mid m \in M\}$ together with multiplication $a(m + N) = am + N$.

Example 6.1.11. Consider the group algebra $\mathbb{F}_2 C_2$. Pick the following basis, $\mathcal{B}$, of $\mathbb{F}_2 C_2$

$$x_0 = \text{id} \qquad x_1 = \text{id} + r.$$

The action of the generators with respect to this basis is:

$$\varphi_{\mathcal{B}}(\text{id}) = \begin{bmatrix} 1 & 0 \\ 0 & 1 \end{bmatrix} \qquad \varphi_{\mathcal{B}}(r) = \begin{bmatrix} 1 & 1 \\ 0 & 1 \end{bmatrix}.$$

The sub module $\mathbb{F}_2\{x_1\}$ is 1-dimensional and the quotient module $\mathbb{F}_2\{x_0\}/\mathbb{F}_2\{x_1\}$ is also 1-dimensional.

In what follows, we will often refer to an A-(sub)quotient module simply as an A-module. We say that a submodule $N \subseteq M$ is a **proper submodule** if $N \neq \emptyset$ and $N \neq M$.

Definition 6.1.12. We say that an A-module, M, is simple (or irreducible) if it contains no A-submodules (apart from the submodules $\{0\} \subseteq M$ and $M \subseteq M$ itself). We say that the module is reducible otherwise.

Suppose that M is reducible, so that there is a proper A-submodule $N \subseteq M$ with $0 < \dim(N) < \dim(M)$. Take a basis $\mathcal{B}_N$ and choose $\mathcal{B}_M$ a basis of M such that $\mathcal{B}_M = \mathcal{B}_N \sqcup \mathcal{B}_{M/N}$ for some basis $\mathcal{B}_{M/N}$ of the quotient module M/N. For any $a \in A$, we have that

$$\varphi_{\mathcal{B}_M}(a) = \left[\begin{array}{c|ccc} & * & \cdots & * \\ \varphi_{\mathcal{B}_{M/N}}(a) & \vdots & & \vdots \\ & * & \cdots & * \\ \hline 0 & \multicolumn{3}{c}{\varphi_{\mathcal{B}_N}(a)} \end{array}\right] \tag{6.1.13}$$

where the south-westerly matrix is zero, but the north-easterly matrix can be arbitrary.

- If the bases can be chosen so that the north-easterly matrix in (6.1.13) is zero for all $a \in A$, then we say M splits or decomposes as a direct sum $N/M \oplus N$ (and we note that M/N is a submodule of M in this case, with N the corresponding quotient module).
- If there exists no basis $\mathcal{B}_M$ such that the north-easterly matrix in (6.1.13) is zero for all $a \in A$, then we say that M is a non-split extension of N and M/N (and we note that M/N is *not* a submodule in this case).

Example 6.1.14. Consider the group algebra $\mathbb{C}C_3$ and let ξ be a primitive 3rd root of unity. We define

$$x_0 = \text{id} + r + r^2 \qquad x_1 = \text{id} + \xi r + \xi^2 r^2 \qquad x_2 = \text{id} + \xi^2 r + \xi r^2$$

in $\mathbb{C}C_3$. We have that $(x_k)r = \xi^k x_k$ for $k \geq 0$ and so we have that $\mathbb{C}C_3$ splits/decomposes as a direct sum of three simple modules

$$\mathbb{C}C_3 = \mathbb{C}\{x_0\} \oplus \mathbb{C}\{x_1\} \oplus \mathbb{C}\{x_2\}. \tag{6.1.15}$$

To see this decomposition, we set $\mathcal{B} = \{x_0, x_1, x_2\}$ and we note that

$$\varphi_{\mathcal{B}}(\text{id}) = \begin{bmatrix} 1 & 0 & 0 \\ 0 & 1 & 0 \\ 0 & 0 & 1 \end{bmatrix} \qquad \varphi_{\mathcal{B}}(r) = \begin{bmatrix} 1 & 0 & 0 \\ 0 & \xi^2 & 0 \\ 0 & 0 & \xi \end{bmatrix} \qquad \varphi_{\mathcal{B}}(r^2) = \begin{bmatrix} 1 & 0 & 0 \\ 0 & \xi & 0 \\ 0 & 0 & \xi^2 \end{bmatrix}.$$

To see that the modules are simple, we simply note that each module $\mathbb{C}\{x_k\}$ for $k = 0, 1, 2$ is 1-dimensional (and so has no proper submodules).

6.1 Examples of simple modules and representations

Example 6.1.16. Consider the right 2-dimensional $TL_3(\delta)$-module M from Example 6.1.7. Any submodule of the 2-dimensional, M, must be 1-dimensional. Thus we consider a general element of the form $\alpha_1 e_1 + \alpha_2 e_1 e_2 \in M$ and we ask if/when this can form a submodule. The generators $e_1, e_2 \in TL_3(\delta)$ act as follows

$$\left(\alpha_1 \begin{bmatrix} \cup \\ \cap \end{bmatrix} + \alpha_2 \begin{bmatrix} \diagdown \\ \diagup \end{bmatrix}\right) * \begin{bmatrix} \cup \\ \cap \end{bmatrix} = (\delta\alpha_1 + \alpha_2) \begin{bmatrix} \cup \\ \cap \end{bmatrix}$$

and

$$\left(\alpha_1 \begin{bmatrix} \cup \\ \cap \end{bmatrix} + \alpha_2 \begin{bmatrix} \diagdown \\ \diagup \end{bmatrix}\right) * \begin{bmatrix} | \\ \cap \end{bmatrix} = (\alpha_1 + \delta\alpha_2) \begin{bmatrix} \diagdown \\ \diagup \end{bmatrix}$$

and so there exists a proper submodule (spanned by $\alpha_1 e_1 + \alpha_2 e_1 e_2$) if and only if

$$\delta\alpha_1 + \alpha_2 = 0 \quad \text{and} \quad \alpha_1 + \delta\alpha_2 = 0.$$

In other words, when we set $\delta = \pm 1$ we obtain a unique (irreducible) proper submodule

$$N_{\pm 1} = \mathbb{C}\left\{\begin{bmatrix} \cup \\ \cap \end{bmatrix} \mp \begin{bmatrix} \diagdown \\ \diagup \end{bmatrix}\right\} \subseteq \mathbb{C}\left\{\begin{bmatrix} \cup \\ \cap \end{bmatrix}, \begin{bmatrix} \diagdown \\ \diagup \end{bmatrix}\right\} = M$$

and for all all other values of δ the 2-dimensional module M has no proper submodules. We now fix $\delta = 1$ and consider the structure of M in more detail. We choose a new basis of M to be given by $\mathscr{B}'_M = \{e_1, e_1 - e_1 e_2\}$ pictured as follows

$$e_1 = \begin{bmatrix} \cup \\ \cap \end{bmatrix} \qquad e_1 - e_1 e_2 = \begin{bmatrix} \cup \\ \cap \end{bmatrix} - \begin{bmatrix} \diagdown \\ \diagup \end{bmatrix}$$

and, with respect to this basis, the action of the generators of $TL_3(1)$ is given by

$$\varphi_{\mathscr{B}'_M}(\mathrm{id}) = \begin{pmatrix} 1 & 0 \\ 0 & 1 \end{pmatrix} \quad \varphi_{\mathscr{B}'_M}(e_1) = \begin{pmatrix} 1 & 0 \\ 0 & 0 \end{pmatrix} \quad \varphi_{\mathscr{B}'_M}(e_2) = \begin{pmatrix} 1 & -1 \\ 0 & 0 \end{pmatrix}$$

and in particular, this module is a *non-split* extension of two 1-dimensional modules. (To see that the extension is non-split, we note that we have already shown that N_{+1} is the unique non-zero $TL_3(1)$-submodule of M, as above.)

Example 6.1.17. Revisiting Example 6.1.16, we have already shown that the $TL_3(\delta)$-module M is simple for $\delta \notin \{\pm 1\}$. Clearly, the 1-dimensional $TL_3(1)$-modules N_{+1} and M/N_{+1} are irreducible (because any 1-dimensional module must be simple).

Example 6.1.18. For G any finite group, we have that there is always a 1-dimensional simple module $\mathbb{C}\{\sum_{g \in G} g\} \subseteq \mathbb{C}G$. Of course, any 1-dimensional A-module is simple. For example, the modules in Example 6.1.6 and the three direct summands in (6.1.15) of Example 6.1.14 are all simple modules.

We now have a notion of the smallest possible A-modules (the simples ones). The principal aims of representation theory are (1) to classify and construct these simple A-modules and (2) to understand how they "fit together". This is what we mean when we say we want to "understand the substructures" of a $\Bbbk$-algebra. This is analogous to classifying and constructing atoms and then seeking to understand how these atoms fit together into molecules. The first step in classifying these simple constituents is a notion of "sameness".

Definition 6.1.19. Let M, N be right A-modules. We define an A-module homomorphism $f : M \to N$ to be a $\Bbbk$-linear map such that $f(ma) = f(m)a$ for all $m \in M$ and $a \in A$. We say that f is an A-module isomorphism if, in addition, this map is bijective.

In what follows, we will always regard isomorphic modules as "the same module".

Example 6.1.20. Revisiting Example 6.1.16, we see that the 1-dimensional modules N_{+1} and M/N_{+1} are non-isomorphic simply by noting that the elements e_1, e_2 both act on M/N_{+1} as the scalar 1, whereas they act on the submodule N_{+1} as the scalar 0.

We say that a module is **semisimple** if it is a direct sum of its simple constituents (that is, every submodule is a quotient module). We say that an algebra is **semisimple** if its natural module is semisimple. We have already seen that $\mathbb{C}C_3$ is a semisimple algebra (in fact, a theorem of Maschke states that every group algebra over the complex field is semisimple!) whereas $TL_3(\delta)$ is non-semisimple for $\delta = \pm 1$. Modular representation theorists regard semisimple algebras as boring and they will not be given much consideration in this book.

For a given $\Bbbk$-algebra, A, the goal of representation theory is to understand the A-module substructures of its natural module. For $\mathbb{C}C_3$ we have already completely determined everything about its submodule structure in Example 6.1.14. We now consider the non-semisimple example of $\mathbb{F}_3 C_3$.

Example 6.1.21. We consider the group algebra $\mathbb{F}_3 C_3$. We pick as our basis

$$\mathscr{B} = \{\mathrm{id}, r + 2\mathrm{id}, \mathrm{id} + r + r^2\}$$

and we note that, with respect to this basis, the action on the natural module is as follows

$$\varphi_{\mathscr{B}}(\mathrm{id}) = \begin{bmatrix} 1 & 0 & 0 \\ 0 & 1 & 0 \\ 0 & 0 & 1 \end{bmatrix} \quad \varphi_{\mathscr{B}}(r) = \begin{bmatrix} 1 & 1 & 0 \\ 0 & 1 & 1 \\ 0 & 0 & 1 \end{bmatrix} \quad \varphi_{\mathscr{B}}(r^2) = \begin{bmatrix} 1 & 2 & 1 \\ 0 & 1 & 2 \\ 0 & 0 & 1 \end{bmatrix}$$

(note that these $\varphi_{\mathscr{B}}(r)^3 = \mathrm{id}_3$ since $3 \equiv 0$ in our field $\mathbb{F}_3$). In particular, we have a chain of submodules coming from the upper uni-triangularity

$$\mathbb{F}_3\{\mathrm{id} + r + r^2\} \subseteq \mathbb{F}_3\{\mathrm{id} + r + r^2, r + 2\mathrm{id}\} \subseteq \mathbb{F}_3\{\mathrm{id} + r + r^2, r + 2\mathrm{id}, \mathrm{id}\}.$$

6.1 Examples of simple modules and representations

The quotient module at each stage is isomorphic to the trivial module (simply because all the diagonal entries of these matrices are equal to 1). In particular, $\mathbb{F}_3 C_3$ has precisely one simple module (the trivial module) which appears three times as a subquotient of $\mathbb{F}_3 C_3$.

We now consider a slightly more complicated type of group (dihedral, rather than cyclic) and completely determine its representation theory in the (boring!) semisimple case.

Example 6.1.22. Consider the group algebra $\mathbb{C} D_6 = \langle r, f \mid r^3 = \mathrm{id} = f^2, r^k f = fr^{3-k}\rangle$ and let ξ be a primitive 3rd root of unity. We define

$$\begin{aligned} x_0 &= \mathrm{id} + r + r^2 & y_0 &= f + fr + fr^2 \\ x_1 &= \mathrm{id} + \xi r + \xi^2 r^2 & y_1 &= f + \xi fr + \xi^2 fr^2 \\ x_2 &= \mathrm{id} + \xi^2 r + \xi r^2 & y_2 &= f + \xi^2 fr + \xi fr^2 \end{aligned}$$

in $\mathbb{C} D_6$. We have that

$$(x_k)r = \xi^k x_k \qquad (y_k)r = \xi^{-k} y_k \qquad (x_k)f = y_k \qquad (y_k)f = x_k \qquad (6.1.23)$$

for $0 \leqslant k \leqslant 2$ and so $\mathbb{C} D_6$ decomposes as a direct sum of right submodules as follows

$$\mathbb{C} D_{2n} = \mathbb{C}\{x_0, y_0\} \oplus \mathbb{C}\{x_1, y_1\} \oplus \mathbb{C}\{x_2, y_2\}.$$

We notice that the $\mathbb{C} D_6$-module $\mathbb{C}\{x_0, y_0\}$ has two 1-dimensional submodules

$$\mathbb{C}\{x_0, y_0\} \cong \mathbb{C}\{x_0 + y_0\} \oplus \mathbb{C}\{x_0 - y_0\}$$

where f acts as $+1$ on the former and -1 on the latter module; r acts by $+1$ on both submodules. In fact, these are the only 1-dimensional $\mathbb{C} D_6$-modules. To see this, we simply note that (i) $f^2 = \mathrm{id}$ and so f must act as ± 1 on any 1-dimensional module (ii) $r^3 = \mathrm{id}$ and so f must act as by multiplication by ξ^i for $0 \leqslant i \leqslant 2$ on any 1-dimensional module (iii) $r^k f = fr^{-k}$ and so r^k must act as a real number on any 1-dimensional module. Putting this altogether implies that r must act as 1 and f must act as ± 1. Thus any 1-dimensional module is isomorphic to one of the two 1-dimensional modules constructed above.

We claim that the modules $\mathbb{C}\{x_1, y_1\}$ and $\mathbb{C}\{x_2, y_2\}$ are simple. We have seen in (6.1.23) that r acts on $\mathbb{C}\{x_1, y_1\}$ with eigenvalues ξ and ξ^2. Therefore if $\mathbb{C}\{x_1, y_1\}$ is not a simple module, it must contain a 1-dimensional submodule and a 1-dimensional quotient module, each of which is isomorphic to one of these eigenvectors for r. However, we have seen that the *only* 1-dimensional $\mathbb{C} D_6$-representations are the eigenvectors for r with eigenvalue 1, we thus obtain a contradiction. Thus $\mathbb{C}\{x_1, y_1\}$ is simple.

Finally, we note that $\mathbb{C}\{x_1, y_1\}$ and $\mathbb{C}\{x_2, y_2\}$ are isomorphic via the map $f(x_1) = y_2$ and $f(y_1) = x_2$. Therefore $\mathbb{C}\{x_2, y_2\}$ is also simple.

Exercise 6.1.24. Decompose the algebra $\mathbb{C} D_{2n}$ for $n > 3$ as a direct sum of simple modules. Hint: the arguments and notation we have used in order to understand $\mathbb{C} D_6$

are readily generalised to all $\mathbb{C}D_{2n}$ for $n \geqslant 3$. Be careful, as the n is odd and even cases are slightly different (there are four 1-dimensional modules of $\mathbb{C}D_{2n}$ for n even).

Exercise 6.1.25. We have seen that $\mathbb{C}D_6$ is decomposable. Show that every $\mathbb{C}G$ is decomposable for G any non-trivial group. Hint: check that $\mathbb{C}\{(g - \mathrm{id}_G) \mid g \in G\}$ is an ideal of $\mathbb{C}G$ and construct a 1-dimensional module which does not belong to this ideal.

We now turn to the representation theory of the binary Schur algebra (of rank 4). These Schur algebras will be a focal point of the rest of the chapter, and so we first spend some time considering one illustrative module in great detail (in both the semisimple and non-semsimple cases).

Example 6.1.26. We consider the algebra $S^{\Bbbk}(4)$ concretely as the matrix algebra pictured in Example 5.6.10. We consider the right submodule $\Delta^{\Bbbk}(4) \subseteq S^{\Bbbk}(4)$ generated by the element $1_{(4,0)} \in S^{\Bbbk}(4)$ where

$$(\pi)1_{(4,0)} = (\pi)E_{1111,1111} = \begin{cases} \pi & \text{if } \pi = 1111 \\ 0 & \text{otherwise.} \end{cases}$$

We have that $(1111)e = 0$ (as there are no 2s in the sequence). Thus the right submodule $1_{(4,0)}S^{\Bbbk}(4) = E_{1111,1111}S^{\Bbbk}(4)$ has spanning set given by

$$\{1_{(4,0)} f^{[k]} \mid 0 \leqslant k \leqslant 4\}$$

(where we note that $1_{(4,0)} f^{[5]} = 0$ as no sequence $\pi \in \mathbb{T}_4$ has five 2s in it). We visualise these maps concretely as elements of $(\mathbb{T}_4 \times \mathbb{T}_4)$-matrices as follows:

$$(E_{1111,1111})1_{(4,0)} = E_{1111,1111}$$
$$(E_{1111,1111})f = E_{1111,1112} + E_{1111,1121} + E_{1111,1211} + E_{1111,2111}$$
$$(E_{1111,1111})f^{[2]} = E_{1111,1122} + E_{1111,1212} + E_{1111,2112}$$
$$+ E_{1111,1221} + E_{1111,2121} + E_{1111,2211}$$
$$(E_{1111,1111})f^{[3]} = E_{1111,2221} + E_{1111,2212} + E_{1111,2122} + E_{1111,1222}$$
$$(E_{1111,1111})f^{[4]} = E_{1111,2222}.$$

For example, we have that

$$(\pi)1_{(4,0)} f^{[2]} = \begin{cases} 1122 + 1212 + 2112 + 1221 + 1221 + 2211 & \text{if } \pi = 1111; \\ 0 & \text{otherwise} \end{cases}$$

and so $1_{(4,0)} f^{[2]}$ is indeed equal to the claimed $(\mathbb{T}_4 \times \mathbb{T}_4)$-matrix. The matrix notation is useful and very explicit, but rather cumbersome in two ways. Firstly, we note that the basis vectors above are all of the form $E_{1111,\pi} \in \mathrm{Mat}_{\mathbb{T}_4}(\Bbbk)$ for some $\pi \in \mathbb{T}_4$; therefore we might as well identify this elementary matrix simply with the sequence

6.1 Examples of simple modules and representations

$\pi \in \mathbb{T}_4$. Secondly, we note that each $1_{(4,0)} f^{[k]}$ is a sum of all possible permutations the sequence

$$\underbrace{1\ldots 1}_{4-k}\underbrace{2\ldots 2}_{k}$$

we encode this by denoting these orbit-sums for $k = 0, 1, 2, 3, 4$ by the representatives

$$\boxed{1\,|\,1\,|\,1\,|\,1}\quad \boxed{1\,|\,1\,|\,1\,|\,2}\quad \boxed{1\,|\,1\,|\,2\,|\,2}\quad \boxed{1\,|\,2\,|\,2\,|\,2}\quad \boxed{2\,|\,2\,|\,2\,|\,2}.$$

For example, putting the above points together we have the following shorthand:

$$\boxed{1\,|\,1\,|\,1\,|\,2} \leftrightarrow 1112 + 1121 + 1211 + 2111$$
$$\leftrightarrow E_{1111,1112} + E_{1111,1121} + E_{1111,1211} + E_{1111,2111} = 1_{(4,0)} f^{[1]}.$$

In summary, $S^k(4)$ has a 5-dimensional (right) submodule $\Delta^k(4)$ with basis

$$\left\{ \boxed{1\,|\,1\,|\,1\,|\,1},\ \boxed{1\,|\,1\,|\,1\,|\,2},\ \boxed{1\,|\,1\,|\,2\,|\,2},\ \boxed{1\,|\,2\,|\,2\,|\,2},\ \boxed{2\,|\,2\,|\,2\,|\,2} \right\}.$$

Example 6.1.27. We now consider the action of $S^Q(4)$ on the 5-dimensional module $\Delta^k(4)$. This can be done very explicitly using the generators and relations or, equivalently, the matrix representation. For example, we have that

$$\left(\boxed{1\,|\,1\,|\,1\,|\,2} \right) f = (1112)f + (1121)f + (1211)f + (2111)f$$
$$= (1122 + 1212 + 2112) + (1122 + 1221 + 2121)$$
$$+ (1212 + 1221 + 2211) + (2112 + 2121 + 2211)$$
$$= 2 \times (1122 + 1212 + 1221 + 2112 + 2121 + +2211)$$
$$= 2 \times \left(\boxed{1\,|\,1\,|\,2\,|\,2} \right)$$

In a similar manner, the action of e and f on $\Delta^k(4)$ can be encoded in the matrices

$$f = \begin{bmatrix} 0 & 1 & 0 & 0 & 0 \\ 0 & 0 & 2 & 0 & 0 \\ 0 & 0 & 0 & 3 & 0 \\ 0 & 0 & 0 & 0 & 4 \\ 0 & 0 & 0 & 0 & 0 \end{bmatrix} \quad e = \begin{bmatrix} 0 & 0 & 0 & 0 & 0 \\ 4 & 0 & 0 & 0 & 0 \\ 0 & 3 & 0 & 0 & 0 \\ 0 & 0 & 2 & 0 & 0 \\ 0 & 0 & 0 & 1 & 0 \end{bmatrix} \quad (6.1.28)$$

and of course the matrices for $e^{[k]} = \frac{1}{k}e^k$ and $f^{[k]} = \frac{1}{k}f^k$ can be constructed by taking powers of these and dividing by the appropriate scalars. The idempotent elements simply act as the scalar 1 (if the weights match) or as zero (otherwise). Thus, over the rationals (or more generally, any ring in which 2, 3, 4 are invertible) we can encode the representing matrices diagrammatically as follows:

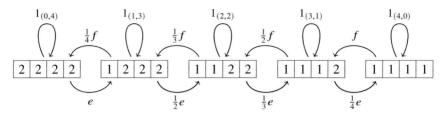

We can now easily verify that this module is simple providing that the scalars 2, 3 (and therefore 4) are invertible in our field $\Bbbk$. To see this, note that we can get from any vector to any other vector using the operators $f, \frac{1}{2}f, \frac{1}{3}f, \frac{1}{4}f$ and $e, \frac{1}{2}e, \frac{1}{3}e, \frac{1}{4}e$ and the idempotents in the manner illustrated in the diagram (and applying relation (5.6.13)).

Exercise 6.1.29. Let $\Delta^{\Bbbk}(6, 0)$ be the 7-dimensional right $S^{\Bbbk}(6)$-module $1_{(6,0)} S^{\Bbbk}(6)$. Write down the action of the e and f operators on this module by analogy with Example 6.1.27.

Example 6.1.30. Let $\Bbbk = \mathbb{F}_3$ be the field of 3 elements. We now consider the $S^{\Bbbk}(4)$-module $\Delta^{\Bbbk}(4)$. We claim that there is a submodule spanned by

$$\boxed{1\ 1\ 2\ 2} = 1122 + 1212 + 2112 + 1221 + 2121 + 2211.$$

This is essentially because there are $6 = 3 \times 2$ vectors in this sum, and so when we apply e or f to each of them in turn, we obtain

$$\left(\boxed{1\ 1\ 2\ 2}\right)e = (1122)e + (1212)e + (2112)e + (1221)e + (2121)e + (2211)e$$
$$= (1112 + 1121) + (1112 + 1211) + (1112 + 2111)$$
$$\quad + (1121 + 1211) + (1121 + 2111) + (1211 + 2111)$$
$$= 3 \times (2221 + 2212 + 2122 + 1222)$$
$$= 3\left(\boxed{1\ 1\ 1\ 2}\right) \equiv 0$$

in characteristic 3. One can similarly verify that

$$\left(\boxed{1\ 1\ 2\ 2}\right)f = 3\left(\boxed{1\ 2\ 2\ 2}\right) \equiv 0$$

in characteristic 3. Thus we have a 1-dimensional submodule

$$M = \Bbbk\left\{\boxed{1\ 1\ 2\ 2}\right\} \subseteq \Delta^{\Bbbk}(4)$$

and the representing matrices for the action on this submodule are

$$1_{(2,2)} \mapsto (1) \qquad 1_\mu \mapsto (0) \qquad e^{[k]} \mapsto (0) \qquad f^{[k]} \mapsto (0)$$

6.2 Temperley–Lieb algebras as the prototypical cellular algebras

for $\mu \neq (2, 2)$ and $k \geqslant 0$. Now, the resulting quotient of $\Delta^k(4)$ by this 1-dimensional submodule is a 4-dimensional module (obviously!) with the action of e and f given by

$$f = \begin{bmatrix} 0 & 1 & 0 & 0 \\ 0 & 0 & 0 & 0 \\ 0 & 0 & 0 & 1 \\ 0 & 0 & 0 & 0 \end{bmatrix} \quad e = \begin{bmatrix} 0 & 0 & 0 & 0 \\ 1 & 0 & 0 & 0 \\ 0 & 0 & 0 & 0 \\ 0 & 0 & 1 & 0 \end{bmatrix}.$$

We now check that this 4-dimensional quotient module is simple. By relation (5.6.13)), it suffices to check that each of

$$\boxed{1\,|\,1\,|\,1\,|\,1}, \boxed{1\,|\,1\,|\,1\,|\,2}, \boxed{1\,|\,2\,|\,2\,|\,2}, \boxed{2\,|\,2\,|\,2\,|\,2}$$

generates the whole of $\Delta^k(4)/M$ (and hence generates $\Delta^k(4)$). We can encode this action on the quotient $\Delta^k(4)/M$ as follows

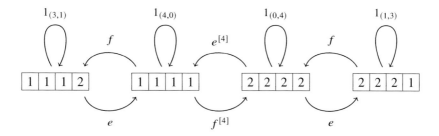

and so every vector does indeed generate the whole module, as required. To check that the scalar on the leftmost left-to-right arrow is 1 (as claimed) we note that

$$\left(\boxed{1\,|\,1\,|\,1\,|\,2} \right) e = 4 \left(\boxed{1\,|\,1\,|\,1\,|\,1} \right) \equiv \boxed{1\,|\,1\,|\,1\,|\,1}$$

by (6.1.28) followed by the fact that $4 \equiv 1$ in $\Bbbk = \mathbb{F}_3$.

6.2 Temperley–Lieb algebras as the prototypical cellular algebras

In this section we introduce the framework of "cellularity" as a way of understanding the representation theory of the algebras of interest in this book. We begin with the Temperley–Lieb algebra, because it conveys all the important axioms of cellularity in a relatively straight-forward manner.

Definition 6.2.1. We recall that a propagating strand of a Temperley–Lieb diagram is any strand which connects a northern vertex to a southern vertex. Given a diagram, $d \in \mathrm{TL}_n(\delta)$, we define the propagating number, $\sharp(d)$, to be the total number of propagating strands in d.

Concatenation of diagrams can only reduce the number of propagating strands, that is
$$\sharp(d \circ d') \leqslant \min\{\sharp(d), \sharp(d')\}.$$
Thus the idempotents $\{E_k \mid 0 \leqslant k \leqslant \lfloor n/2 \rfloor\}$ of (5.3.5) give rise to a chain of ideals
$$0 \subset \text{TL}_n(\delta) E_{\lfloor n/2 \rfloor} \text{TL}_n(\delta) \subset \cdots \subset \text{TL}_n(\delta) E_1 \text{TL}_n(\delta) \subset \text{TL}_n(\delta).$$
The kth layer of this stratification
$$\text{TL}_n(\delta) E_k \text{TL}_n(\delta) / \text{TL}_n(\delta) E_{k+1} \text{TL}_n(\delta)$$
has basis consisting of the diagrams, d, with exactly $n - 2k$ propagating strands (that is, $\sharp(d) = n - 2k$). This forms both a left and right module for the algebra. We choose to work with right modules and fix our choice for the top of the Temperley–Lieb diagram. For $0 \leqslant k \leqslant \lfloor n/2 \rfloor$ we define
$$\Delta^k(k) = E_k \text{TL}_n(\delta) / ((\text{TL}_n(\delta) E_{k+1} \text{TL}_n(\delta)) \cap E_k \text{TL}_n(\delta))$$
and this (apparently abstract) module is readily constructed via a natural and easy to describe basis. This basis consists of all diagrams with precisely k adjacent northern horizontal arcs at the top-left of the diagram.

Example 6.2.2. The $\text{TL}_5(\delta)$-modules $\Delta_5(1)$, and $\Delta_5(2)$ are 4- and 5-dimensional with bases depicted in Figures 6.1 and 6.2. The trivial $\text{TL}_5(\delta)$-module $\Delta_5(0)$ is the 1-dimensional module spanned by the identity element, with all non-trivial elements acting as zero. The corresponding layers of the stratification of $\text{TL}_5(\delta)$ are 1^2-, 4^2-, and 5^2-dimensional with bases given by "choosing both the top and bottom freely".

Fig. 6.1: The basis of the $\text{TL}_5(\delta)$-module $\Delta_5(1)$.

Example 6.2.3. The action of the generators of the $\text{TL}_5(\delta)$ on the module $\Delta_5(1)$ is given by

$$e_1 \mapsto \begin{bmatrix} \delta & 0 & 0 & 0 \\ 1 & 0 & 0 & 0 \\ 0 & 0 & 0 & 0 \\ 0 & 0 & 0 & 0 \end{bmatrix} \quad e_2 \mapsto \begin{bmatrix} 0 & 1 & 0 & 0 \\ 0 & \delta & 0 & 0 \\ 0 & 1 & 0 & 0 \\ 0 & 0 & 0 & 0 \end{bmatrix} \quad e_3 \mapsto \begin{bmatrix} 0 & 0 & 0 & 0 \\ 0 & 0 & 1 & 0 \\ 0 & 0 & \delta & 0 \\ 0 & 0 & 1 & 0 \end{bmatrix} \quad e_4 \mapsto \begin{bmatrix} 0 & 0 & 0 & 0 \\ 0 & 0 & 0 & 0 \\ 0 & 0 & 0 & 1 \\ 0 & 0 & 0 & \delta \end{bmatrix}.$$

For example, let's calculate the effect of applying each of the e_k for $1 \leqslant k \leqslant 4$ generators to the third diagram in Figure 6.1 (or equivalently, let's calculate the third row of each of the above matrices)...

6.2 Temperley–Lieb algebras as the prototypical cellular algebras

For $k = 1, 2$ the products e_k with the third basis element of Figure 6.1 are calculated as follows:

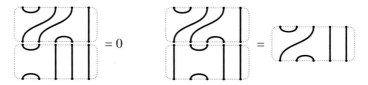

(where the zero is in the module $\Delta_5(1)$, because the concatenated diagram has fewer propagating strands, that is it belongs to the ideal $\mathrm{TL}_5(\delta)E_2\mathrm{TL}_5(\delta)$). For e_k with $k = 3, 4$ we have that

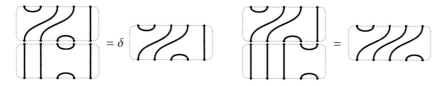

Exercise 6.2.4. Calculate the action of the generators of $\mathrm{TL}_5(\delta)$ on the basis of the module $\Delta_5(2)$ pictured in Figure 6.2.

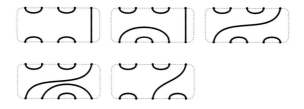

Fig. 6.2: The basis of the $\mathrm{TL}_5(\delta)$-module $\Delta_5(2)$.

So far, we have learnt that the Temperley–Lieb algebra has a chain of 2-sided ideals, whose layers provide us with a family of easily constructible modules. We are now ready to axiomatise this idea:

Definition 6.2.5. Let A be a $\Bbbk$-algebra with anti-involution $*$. Let $(\Lambda^+, <)$ be a totally ordered set, $\lambda_1 < \lambda_2 < \cdots < \lambda_t$ indexing a set of quasi-idempotents $x_{\lambda_k} = x^*_{\lambda_k} \in A$ for $1 \leq k \leq t$ and for which we have a chain of inclusions

$$0 \subset Ax_{\lambda_1}A \subset Ax_{\lambda_2}A \subset \cdots \subset Ax_{\lambda_t}A = A.$$

For $1 \leq k \leq t$, suppose there exist $d_S \in A$ indexed by $S \in \mathrm{Std}(\lambda_k)$ such that

$$\{x_{ST} := d^*_S x_{\lambda_k} d_T \mid S, T \in \mathrm{Std}(\lambda_k)\} \tag{6.2.6}$$

is a $\Bbbk$-basis of $Ax_{\lambda_k}A/Ax_{\lambda_{k-1}}A$. In which case, we say that A is a **cellular algebra** and that the basis in (6.2.6) is a **cellular basis**.

We often refer to the indexing elements $S \in \mathrm{Std}(\lambda_k)$ for $1 \leqslant k \leqslant t$ as **standard tableaux**, generalising the combinatorial notion of tableaux for symmetric groups (which we will soon discuss).

Example 6.2.7. Fix $\delta \neq 0$. The Temperley–Lieb algebra, $\mathrm{TL}_n(\delta)$, is cellular with respect to the poset
$$\Lambda^+ = \{\lfloor n/2 \rfloor \prec \cdots \prec 2 \prec 1 \prec 0\}.$$
We claim that the quasi-idempotents are given by $\{E_k \mid 0 \leqslant k \leqslant \lfloor n/2 \rfloor\}$ and the layer
$$\{x_{\mathsf{ST}} := d_{\mathsf{S}}^* x_{\lambda_k} d_{\mathsf{T}} \mid \mathsf{S}, \mathsf{T} \in \mathrm{Std}(\lambda_k)\}$$
consists of the n-strand Temperley–Lieb diagrams elements with precisely $n - 2k$ propagating strands. To see this, simply rewrite any $d \in \mathrm{TL}_n(\delta)$ with $n - 2k$ propagating strands in the required form $d = d_1 E_k d_2$ in the obvious fashion, see for example Figure 6.3.

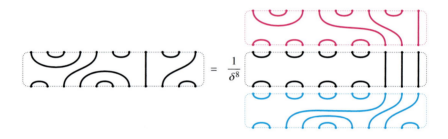

Fig. 6.3: Rewriting a Temperley–Lieb diagram in the form $d_{\mathsf{S}}^* E_4 d_{\mathsf{T}} \in \mathrm{TL}_{11}(\delta)$.

Definition 6.2.8. For $1 \leqslant k \leqslant t$ and $\lambda_k \in \Lambda^+$, we set $A^{\leqslant \lambda_k} = A x_{\lambda_k} A$ and similarly we set $A^{<\lambda_k} = A x_{\lambda_{k-1}} A$.

With this notation in place, we can drop the subscripts $1 \leqslant k \leqslant t$ and simply write $\lambda \in \Lambda^+$.

Definition 6.2.9. For any $\mathsf{S} \in \mathrm{Std}(\lambda)$ we set $x_{\mathsf{S}} := x_\lambda d_{\mathsf{S}} + A^{<\lambda}$.

Definition 6.2.10. We let A a cellular algebra as in Definition 6.2.5. For $\lambda \in \Lambda^+$, we define the right cell module to be
$$\Delta^{\Bbbk}(\lambda) = x_\lambda A / (A^{<\lambda} \cap x_\lambda A)$$
with basis given by $\{x_{\mathsf{S}} \mid \mathsf{S} \in \mathrm{Std}(\lambda)\}$. The left cell module can similarly be defined as the left A-module $A x_\lambda / (A^{<\lambda} \cap A x_\lambda)$.

In Example 6.1.16 we saw that the cell modules *can give us a handle* on the simple modules. In fact, cellularity provides a general machinery for constructing the simple modules, which we now recall.

6.2 Temperley–Lieb algebras as the prototypical cellular algebras

Definition 6.2.11. There exists a unique bilinear map $\langle -, - \rangle_\lambda^\Bbbk : \Delta^\Bbbk(\lambda) \times \Delta^\Bbbk(\lambda) \to \Bbbk$ such that $\langle x_S, x_T \rangle_\lambda^\Bbbk \in \Bbbk$ for $S, T \in \mathrm{Std}(\lambda)$ is given by

$$x_S x_T^* = x_\lambda d_S d_T^* x_\lambda + A^{<\lambda} = (\langle x_S, x_T \rangle_\lambda^\Bbbk) x_\lambda + A^{<\lambda}. \qquad (6.2.12)$$

We refer to $\langle -, - \rangle_\lambda^\Bbbk$ as the intersection form on the cell module $\Delta^\Bbbk(\lambda)$. We let $\mathrm{rad}\langle -, - \rangle_\lambda^\Bbbk = \{ x \in \Delta^\Bbbk(\lambda) \mid \langle x, y \rangle = 0 \text{ for all } y \in \Delta^\Bbbk(\lambda) \}$.

Remark 9. The first equality in (6.2.12) follows by Definition 6.2.9, we now justify the second equality. To see that $x_\lambda d_S d_T^* x_\lambda$ is indeed a scalar multiple of x_λ (the scalar $\langle x_S, x_T \rangle_\lambda^\Bbbk \in \Bbbk$) one need only note that $x_\lambda d_S d_T^* x_\lambda$ belongs to

$$(x_\lambda A \cap A x_\lambda) \setminus (x_\lambda A \cap A x_\lambda \cap A^{<\lambda}) = \Bbbk\{ x_\lambda d_S \mid S \in \mathrm{Std}(\lambda) \} \cap \Bbbk\{ d_T^* x_\lambda \mid T \in \mathrm{Std}(\lambda) \}$$

where the righthand-side is equal to $\Bbbk\{x_\lambda\}$ and so the equality follows.

Proposition 6.2.13. *For all $x, y \in \Delta^\Bbbk(\lambda)$ we have that*

(i) $\langle x, y \rangle_\lambda^\Bbbk = \langle y, x \rangle_\lambda^\Bbbk$.
(ii) $\langle ax, y \rangle_\lambda^\Bbbk = \langle x, ya^* \rangle_\lambda^\Bbbk$ for all $a \in A$.
(iii) $yx_{UV} = \langle y, x_U \rangle_\lambda^\Bbbk x_V$ for all $U, V \in \mathrm{Std}(\lambda)$.

In particular, $\mathrm{rad}\langle -, - \rangle_\lambda^\Bbbk$ *is an A-submodule of* $\Delta^\Bbbk(\lambda)$.

Proof. By bilinearity, it suffices to check (i), (ii), and (iii) for $x = x_S$ and $y = x_T$ for $S, T \in \mathrm{Std}(\lambda)$. By assumption $x_\lambda = x_\lambda^*$ and so we have that

$$\langle x_S, x_T \rangle_\lambda^\Bbbk x_\lambda = x_\lambda d_S d_T^* x_\lambda + A^{<\lambda} = (x_\lambda d_T d_S^* x_\lambda)^* + A^{<\lambda} = \langle x_T, x_S \rangle_\lambda^\Bbbk x_\lambda^* = \langle x_T, x_S \rangle_\lambda^\Bbbk x_\lambda$$

by definition chasing, and thus (i) follows. The proof of (ii) is similar. Substituting $y = x_T$ into (iii), we have that

$$x_T x_{UV} = (x_\lambda d_T)(d_U^* x_\lambda d_V) + A^{<\lambda}$$
$$= (x_\lambda d_T d_U^* x_\lambda) d_V + A^{<\lambda}$$
$$= \langle x_T, x_U \rangle_\lambda^\Bbbk x_\lambda d_V + A^{<\lambda}$$
$$= \langle x_T, x_U \rangle_\lambda^\Bbbk x_V$$

again by definition chasing, as required. □

We are now ready to construct the simple modules of a given cellular algebra, using only knowledge of the intersection forms.

Definition 6.2.14. We set $\Lambda_\Bbbk^0 = \{ \lambda \in \Lambda^+ \mid \mathrm{rad}\langle -, - \rangle_\lambda^\Bbbk \neq \Delta^\Bbbk(\lambda) \}$. For $\lambda \in \Lambda_\Bbbk^0$, we define

$$L^\Bbbk(\lambda) = \Delta^\Bbbk(\lambda) / \mathrm{rad}\langle -, - \rangle_\lambda^\Bbbk.$$

We now justify this suggestive notation by showing that these modules are, indeed, simple A-modules over the field $\Bbbk$.

Proposition 6.2.15. *Let $\Bbbk$ be a field. For $\lambda \in \Lambda_\Bbbk^0$ we have that $\mathrm{rad}\langle -, -\rangle_\lambda^\Bbbk$ is the unique maximal proper submodule of $\Delta^\Bbbk(\lambda)$. The quotient module $L^\Bbbk(\lambda) = \Delta^\Bbbk(\lambda)/\mathrm{rad}\langle -, -\rangle_\lambda^\Bbbk$ is a simple A-module.*

Proof. Let $z \in \Delta^\Bbbk(\lambda)/\mathrm{rad}\langle -, -\rangle_\lambda^\Bbbk$, so that $\langle z, y\rangle_\lambda^\Bbbk \neq 0 \in \Bbbk$ for some $y \in \Delta^\Bbbk(\lambda)$. Since $\Bbbk$ is a field, we may assume that $\langle z, y\rangle_\lambda^\Bbbk = 1$. We suppose $y = \sum_{S \in \mathrm{Std}(\lambda)} \alpha_S x_S \in \Delta^\Bbbk(\lambda)$ and we set $y_T = \sum_{S \in \mathrm{Std}(\lambda)} \alpha_S x_{ST} \in A$. By Proposition 6.2.13(*ii*), we have that

$$z y_T = \sum_{S \in \mathrm{Std}(\lambda)} \alpha_S z d_S^* x_\lambda d_T + A^{<\lambda} = \sum_{S \in \mathrm{Std}(\lambda)} \alpha_S \langle z, d_S\rangle x_\lambda d_T = \langle z, y\rangle x_T = x_T$$

and so by acting on the right with the various y_T we see that z generates the whole cell-module. Therefore every element $z \in \Delta^\Bbbk(\lambda)/\mathrm{rad}\langle -, -\rangle_\lambda^\Bbbk$ generates the quotient module $\Delta^\Bbbk(\lambda)/\mathrm{rad}\langle -, -\rangle_\lambda^\Bbbk$ and therefore it is a simple A-module, as required. □

Example 6.2.16. The Gram matrix of the $\mathrm{TL}_3(\delta)$-module $\Delta^\Bbbk(1)$ is given by

$$\begin{bmatrix} \delta & 1 \\ 1 & \delta \end{bmatrix}$$

and the radical is zero unless the determinant is zero, that is $\delta^2 - 1 = 0$. If $\delta = \pm 1$, then the Gram matrix has rank 1 and $\mathrm{rad}\langle -, -\rangle_1^\Bbbk$ is the 1-dimensional submodule spanned by

and the quotient module $L^\Bbbk(1)$ is 1-dimensional and therefore simple. The reader should compare this with Example 6.1.16.

Exercise 6.2.17. Continuing with Example 6.2.3, calculate the radical of the $\mathrm{TL}_5(\delta)$-module $\Delta^\mathbb{C}(1)$ for an arbitrary specialisation $\delta \in \mathbb{C} \setminus \{0\}$. Hence determine whether or not $\Delta^\mathbb{C}(1)$ is simple for specialisations $\delta \in \mathbb{C} \setminus \{0\}$.

Proposition 6.2.18. *Let $\Bbbk$ be a field and $\lambda \in \Lambda^+$, $\mu \in \Lambda_\Bbbk^0$. Let M be a proper submodule of $\Delta^\Bbbk(\mu)$ and suppose that $\vartheta : \Delta^\Bbbk(\lambda) \to \Delta^\Bbbk(\mu)/M$ is an A-module homomorphism.*

- *If ϑ is non-zero, then $\mu \leqslant \lambda$.*
- *If $\mu = \lambda$, then $\vartheta(x_\lambda) = \alpha_\vartheta x_\lambda + M$ for $\alpha_\vartheta \in \Bbbk$. Thus $\mathrm{Hom}_A(\Delta^\Bbbk(\lambda), \Delta^\Bbbk(\lambda)/M) \cong \Bbbk$.*

In particular, the modules $L^\Bbbk(\mu)$ for $\mu \in \Lambda_\Bbbk^0$ are pairwise non-isomorphic.

Proof. By assumption $\Delta^\Bbbk(\lambda)/\mathrm{rad}\langle -, -\rangle_\lambda^\Bbbk \neq 0$ and x_λ generates $\Delta^\Bbbk(\lambda)$; therefore $x_\lambda \notin \mathrm{rad}\langle -, -\rangle_\lambda^\Bbbk$ and there exists $y_\lambda \in \Delta^\Bbbk(\lambda)$ such that $\langle x_\lambda, y_\lambda\rangle = 1$, in other words $x_\lambda y_\lambda^* = x_\lambda$. We set $\vartheta(x_\lambda) = M + \sum_{S \in \mathrm{Std}(\mu)} \alpha_S x_S$ for some $\alpha_S \in \Bbbk$. Thus

6.2 Temperley–Lieb algebras as the prototypical cellular algebras

$$\vartheta(x_\lambda) = \vartheta(x_\lambda y_\lambda^*) = M + \left(\sum_{S\in \mathrm{Std}(\mu)} \alpha_S x_S\right) y_\lambda^*$$

where $y_\lambda \in A^{\leqslant \lambda}$ and hence the map is zero unless $\lambda \geqslant \mu$ (by the chain of ideals in the definition of cellularity) and thus (i) follows. Now suppose that $\lambda = \mu$ and we set $y_\lambda = \sum_{T\in \mathrm{Std}(\mu)} \beta_T x_\lambda d_T$. We have that

$$\vartheta(x_\lambda) = \vartheta(x_\lambda y_\lambda^*) = M + (\textstyle\sum_{S\in \mathrm{Std}(\mu)} \alpha_S x_S) y_\lambda^*$$
$$= M + (\textstyle\sum_{S\in \mathrm{Std}(\mu)} \alpha_S x_S)(\textstyle\sum_{T\in \mathrm{Std}(\mu)} \beta_T d_T^* x_\lambda)$$
$$= M + \textstyle\sum_{S,T\in \mathrm{Std}(\lambda)} \alpha_S \beta_T \langle x_S, x_T\rangle x_\lambda$$

where the first equality is by our definition of y_λ, the second equality is by the definition of ϑ, the third equality follows from the expansion of y_λ in the cell basis, the final equality follows from Proposition 6.2.13(iii). Therefore ϑ is simply the map given by scalar multiplication by $a_\vartheta = \sum_{S,T\in \mathrm{Std}(\lambda)} \alpha_S \beta_T$ (composed with the natural projection : $\Delta^{\Bbbk}(\lambda) \to \Delta^{\Bbbk}(\lambda)/M$ for M a submodule). $\square$

Lemma 6.2.19. *If λ is maximal in $\leqslant$, then $\Delta^{\Bbbk}(\lambda) = L^{\Bbbk}(\lambda)$.*

Proof. We set λ to be the maximal element in the total ordering $\leqslant$. Our aim is to show that $\mathrm{rad}\langle -, -\rangle_\lambda^{\Bbbk} = 0$. We have that the $1_A \notin A^{<\lambda}$ (as otherwise this would generate the whole algebra) but $1_A \in A^{\leqslant \lambda} = A$. Therefore we can assume that $x_\lambda = 1_A$. Since $1_A \cdot x \neq 0$ for any $0 \neq x \in A$ it follows that the radical is zero, as claimed. $\square$

Theorem 6.2.20. *Let $\Bbbk$ be a field and Λ^+ be finite. Then $\{L^{\Bbbk}(\mu) \mid \mu \in \Lambda_{\Bbbk}^0\}$ is a complete set of non-isomorphic A-modules over the field $\Bbbk$.*

Proof. In Proposition 6.2.18, we saw that the modules $\{L^{\Bbbk}(\mu) \mid \mu \in \Lambda_{\Bbbk}^0\}$ are pairwise non-isomorphic, thus we need only show that this list is exhaustive. By definition, the cellular structure provide a chain of ideals

$$0 \subset Ax_{\lambda_1} A \subset Ax_{\lambda_2} A \subset \cdots \subset Ax_{\lambda_t} A = A$$

filtering the algebra A. As a right A-module, each layer $Ax_{\lambda_k} A/Ax_{\lambda_{k-1}} A$ decomposes as $|\mathrm{Std}(\lambda_k)|$ copies of $\Delta^{\Bbbk}(\lambda_k)$ as follows:

$$Ax_{\lambda_k} A/Ax_{\lambda_{k-1}} A = \bigoplus_{S\in \mathrm{Std}(\lambda_k)} d_S^* x_\lambda A + A^{<\lambda_k}$$

where the isomorphism is given by $d_S^* x_\lambda d_T + A^{<\lambda_k} \mapsto x_\lambda d_T + A^{<\lambda_k}$. Hence (as a right A-module) the algebra A is filtered by cell modules. Thus having filtered the natural A-module by Δ-modules, we need only show that every simple composition factor of a $\Delta^{\Bbbk}(\lambda)$ for $\lambda \in \Lambda^+$ is of the claimed form.

We proceed by induction on the length of the cell chain. If λ_t is maximal, then $\lambda \in \Lambda_0$ and $\Delta^{\Bbbk}(\lambda_t) = L^{\Bbbk}(\lambda_t)$ (by Lemma 6.2.19) and so every composition factor is of the claimed form. Now suppose that $\lambda_k \in \Lambda^+$ for $1 \leqslant k < t$. We let L be any composition factor of $\Delta^{\Bbbk}(\lambda_k)$, by necessity either: $L \cong L^{\Bbbk}(\lambda_k)$ or L is a composition factor of $\mathrm{rad}\langle -, -\rangle^{\Bbbk}_{\lambda_k}$.

For any $a \in A^{\leqslant \lambda_k}$ we have that $\mathrm{rad}\langle -, -\rangle^{\Bbbk}_{\lambda_k} a = 0$ by the definition of the radical together with Proposition 6.2.13(*iii*). Thus every composition factor of $\mathrm{rad}\langle -, -\rangle^{\Bbbk}_{\lambda_k}$ is a composition factor of $A/Ax_{\lambda_k}A$. However, the algebra $A/Ax_{\lambda_k}A$ is itself a cellular algebra with respect to the basis

$$\{x_{\mathsf{ST}} := d^*_{\mathsf{S}} x_{\lambda_j} d_{\mathsf{T}} \mid \mathsf{S}, \mathsf{T} \in \mathrm{Std}(\lambda_j), k < j \leqslant t\}$$

which is of strictly smaller cell-chain-length, thus every simple $A/Ax_{\lambda_k}A$-module is a composition factor of some cell module of the algebra $A/Ax_{\lambda_k}A$. We note that the cell modules of A labelled by λ_j for $k < j \leqslant t$ are trivially obtained from inflating those of $A/Ax_{\lambda_k}A$ and so the result follows by induction. □

To summarise what we have learnt so far: the cell modules are easy to construct, and the simple modules can be defined as quotients of these cell modules using linear algebra. Therefore our aim is to understand the passage between the (easy to construct!) cell modules and the (more mysterious!) simple modules of our algebra A over a field $\Bbbk$. We can define a **composition series** of a cell module $\Delta^{\Bbbk}(\lambda)$ to be a filtration

$$0 \subset M_1 \subset M_2 \subset \cdots \subset M_n \subset \Delta^{\Bbbk}(\lambda)$$

such that each layer $M_k/M_{k-1} = L(\mu_k)$ for some $\mu_k \in \Lambda^0_{\Bbbk}$. Whilst composition series are not unique, one can prove that the multi-set of composition factors

$$\{L(\mu_k) \mid M_k/M_{k-1} = L(\mu_k) \text{ for } 1 \leqslant k \leqslant n\}$$

is unique, by an analogue of the Jordan–Hölder argument from the proof of Theorem 3.4.11. We write $[\Delta^{\Bbbk}(\lambda) : L^{\Bbbk}(\mu)]_A$ for the (well-defined) multiplicity of the simple A-module $L^{\Bbbk}(\mu)$ in a composition series of the cell-module $\Delta^{\Bbbk}(\lambda)$.

Definition 6.2.21. Let A be a cellular algebra over a field $\Bbbk$. We define the **decomposition matrix** of A over $\Bbbk$ to be the matrix recording the **decomposition numbers**

$$\mathbf{D} = (d_{\lambda\mu})_{\lambda \in \Lambda^+, \mu \in \Lambda^0_{\Bbbk}} \qquad d_{\lambda,\mu} = [\Delta^{\Bbbk}(\lambda) : L^{\Bbbk}(\mu)]_A.$$

By Proposition 6.2.18, we immediately deduce the following important fact about decomposition matrices:

Proposition 6.2.22. *The decomposition matrix is lower uni-triangular with respect to the ordering $\leqslant$. In more detail, we arrange the labels $\lambda \in \Lambda^+$ in descending order from top-to-bottom (along rows) and the labels $\mu \in \Lambda^0_{\Bbbk}$ in descending order from left-to-right along columns. The resulting matrix (with entries $(d_{\lambda\mu})_{\lambda \in \Lambda^+, \mu \in \Lambda^0_{\Bbbk}}$) has values 1 along the diagonal entries and with values 0 above the diagonal.*

6.3 The smallest non-semisimple Schur algebra

We will now establish the cellularity of the binary Schur algebra $S^k(2)$ and hence give a detailed account of its representation theory. This is not difficult, but serves as a warm-up to the ideas we will see later on. We set $\Lambda_2^+ = \{((2,0),(1,1)\}$ and we visualise and order the elements of Λ_2^+ as follows

$$(2,0) = \square\square \;<\; \genfrac{}{}{0pt}{}{\square}{\square} = (1^2).$$

Given $\lambda \in \Lambda_2^+$ we define $\mathrm{Std}(\lambda)$ to be given by the ways of filling the boxes of λ with 1s and 2s in such a manner that they weakly increase along rows and strictly increase along columns. That is $\mathrm{Std}(\square\square)$ consists of the three elements

$$\mathsf{T} = \boxed{1\;1} \quad \mathsf{U} = \boxed{1\;2} \quad \mathsf{V} = \boxed{2\;2}$$

and $\mathrm{Std}(\genfrac{}{}{0pt}{}{\square}{\square})$ consists of the single element

$$\mathsf{S} = \genfrac{}{}{0pt}{}{\boxed{1}}{\boxed{2}}.$$

Structurally, the 1s and 2s in these tableaux will correspond to weights in Λ_2^+ and hence to idempotents in $S^k(2)$.

We define the cellular basis of $S^k(2)$ simply by (re)labelling the basis elements of (5.6.3), (5.6.4) as follows. For the $\genfrac{}{}{0pt}{}{\square}{\square}$ layer, we set

$$x_{\mathsf{SS}} = \begin{bmatrix} 0 & 0 & 0 & 0 \\ 0 & 1 & 0 & 0 \\ 0 & 0 & 1 & 0 \\ 0 & 0 & 0 & 0 \end{bmatrix}$$

which is equal to the idempotent $1_{(1,1)}$. For the $\square\square$ layer, we set

$$x_{\mathsf{TT}} = \begin{bmatrix} 1 & 0 & 0 & 0 \\ 0 & 0 & 0 & 0 \\ 0 & 0 & 0 & 0 \\ 0 & 0 & 0 & 0 \end{bmatrix} \quad x_{\mathsf{TU}} = \begin{bmatrix} 0 & 1 & 1 & 0 \\ 0 & 0 & 0 & 0 \\ 0 & 0 & 0 & 0 \\ 0 & 0 & 0 & 0 \end{bmatrix} \quad x_{\mathsf{TV}} = \begin{bmatrix} 0 & 0 & 0 & 1 \\ 0 & 0 & 0 & 0 \\ 0 & 0 & 0 & 0 \\ 0 & 0 & 0 & 0 \end{bmatrix}$$

$$x_{\mathsf{UT}} = \begin{bmatrix} 0 & 0 & 0 & 0 \\ 1 & 0 & 0 & 0 \\ 1 & 0 & 0 & 0 \\ 0 & 0 & 0 & 0 \end{bmatrix} \quad x_{\mathsf{UU}} = \begin{bmatrix} 0 & 0 & 0 & 0 \\ 0 & 1 & 1 & 0 \\ 0 & 1 & 1 & 0 \\ 0 & 0 & 0 & 0 \end{bmatrix} \quad x_{\mathsf{UV}} = \begin{bmatrix} 0 & 0 & 0 & 0 \\ 0 & 0 & 0 & 1 \\ 0 & 0 & 0 & 1 \\ 0 & 0 & 0 & 0 \end{bmatrix}$$

$$x_{\mathsf{VT}} = \begin{bmatrix} 0 & 0 & 0 & 0 \\ 0 & 0 & 0 & 0 \\ 0 & 0 & 0 & 0 \\ 1 & 0 & 0 & 0 \end{bmatrix} \quad x_{\mathsf{VU}} = \begin{bmatrix} 0 & 0 & 0 & 0 \\ 0 & 0 & 0 & 0 \\ 0 & 0 & 0 & 0 \\ 0 & 1 & 1 & 0 \end{bmatrix} \quad x_{\mathsf{VV}} = \begin{bmatrix} 0 & 0 & 0 & 0 \\ 0 & 0 & 0 & 0 \\ 0 & 0 & 0 & 0 \\ 0 & 0 & 0 & 1 \end{bmatrix}$$

where we note that the top-left matrix is the idempotent $1_{(2,0)}$ and the bottom right matrix is the idempotent $1_{(0,2)}$.

Proposition 6.3.1. *The algebra $S^{\Bbbk}(2)$ is a cellular algebra with respect to the poset $(\Lambda_2^+, \leqslant)$, the basis*

$$\{x_{\mathsf{PQ}} \mid \mathsf{P}, \mathsf{Q} \in \mathrm{Std}(\lambda), \lambda \in \Lambda_2^+\}$$

and the anti-involution given by flipping a matrix through its principal diagonal.

Proof. That this is a basis and is preserved by the anti-involution $*$ is clear. One can check that the 9-dimensional layer labelled by $\square\square$ is closed under multiplication simply by taking all possible products with matrices from this layer. $\square$

The two cell modules of $S^{\Bbbk}(2)$ are 1- and 3-dimensional and can be explicitly constructed via their bases as follows:

$$\Delta^{\Bbbk}(\boxminus) = \Bbbk\{x_{\mathsf{S}}\} \qquad \Delta^{\Bbbk}(\square\square) = \Bbbk\{x_{\mathsf{T}}, x_{\mathsf{U}}, x_{\mathsf{V}}\}.$$

We now use our cellular structure in order to determine the simple modules of $S^{\Bbbk}(2)$ over an arbitrary field $\Bbbk$. In order to do this, we must first calculate the Gram matrices of their intersection forms.

Lemma 6.3.2. *The Gram matrices of the intersection forms of the cell modules $\Delta^{\Bbbk}(\boxminus)$ and $\Delta^{\Bbbk}(\square\square)$ are given by*

	S
S	1

	T	U	V
T	1	0	0
U	0	2	0
V	0	0	1

(6.3.3)

respectively.

Proof. For the $\boxminus$ cell module, we have that

$$\langle x_{\mathsf{S}}, x_{\mathsf{S}} \rangle x_{\mathsf{SS}} = x_{\mathsf{SS}} x_{\mathsf{SS}} = 1_{(1,1)} 1_{(1,1)} + S_{\Bbbk}^{<(1,1)}(2) = 1_{(1,1)} + S_{\Bbbk}^{<(1,1)}(2) = 1 \times x_{\mathsf{SS}}$$

and so the (S, S)-entry of the Gram matrix is equal to 1 as claimed. We now consider the $\square\square$ cell module. We first calculate the (T, T)-entry as follows,

$$\langle x_{\mathsf{T}}, x_{\mathsf{T}} \rangle x_{\mathsf{TT}} = x_{\mathsf{TT}} x_{\mathsf{TT}} = 1_{(1,1)} 1_{(1,1)} = 1_{(1,1)} = 1 \times x_{\mathsf{TT}}$$

and so the (T, T)-entry of the Gram matrix is equal to 1 as claimed. The (U, U)-entry can be calculated similarly.

6.3 The smallest non-semisimple Schur algebra

We now consider the off-diagonal (T, V)-entry (which is equal to the (V, T)-entry, by Proposition 6.2.13(*i*)). We have that

$$x_{TT} x_{VT} = \begin{bmatrix} 1 & 0 & 0 & 0 \\ 0 & 0 & 0 & 0 \\ 0 & 0 & 0 & 0 \\ 0 & 0 & 0 & 0 \end{bmatrix} \begin{bmatrix} 0 & 0 & 0 & 0 \\ 0 & 0 & 0 & 0 \\ 0 & 0 & 0 & 0 \\ 1 & 0 & 0 & 0 \end{bmatrix} = \begin{bmatrix} 0 & 0 & 0 & 0 \\ 0 & 0 & 0 & 0 \\ 0 & 0 & 0 & 0 \\ 0 & 0 & 0 & 0 \end{bmatrix} = 0 \cdot x_{TT}$$

and so the (T, V)-entry is 0, as claimed. The remaining off-diagonal entries in (6.3.3) are also all zero because the non-zero rows/columns in the products of matrices do not match-up (of, if you prefer, because $1_\mu 1_\nu = 0$ for $\mu \neq \nu$).

We now consider the (U, U)-entry. We have that

$$x_{TU} x_{UT} = \begin{bmatrix} 0 & 1 & 1 & 0 \\ 0 & 0 & 0 & 0 \\ 0 & 0 & 0 & 0 \\ 0 & 0 & 0 & 0 \end{bmatrix} \begin{bmatrix} 0 & 0 & 0 & 0 \\ 1 & 0 & 0 & 0 \\ 1 & 0 & 0 & 0 \\ 0 & 0 & 0 & 0 \end{bmatrix} = \begin{bmatrix} 2 & 0 & 0 & 0 \\ 0 & 0 & 0 & 0 \\ 0 & 0 & 0 & 0 \\ 0 & 0 & 0 & 0 \end{bmatrix} = 2 \cdot x_{TT}$$

as required (or, if you prefer, $1_{(2,0)} f e 1_{(2,0)} = 2 1_{(2,0)}$). Therefore the (U, U)-entry is equal to 2 as claimed. $\square$

With the Gram matrices of the intersection forms calculated, we are now able to construct the simple modules for $S^{\Bbbk}(2)$.

Lemma 6.3.4. *Over* $\Bbbk = \mathbb{F}_p$ *for* $p > 2$ *the algebra* $S^{\Bbbk}(2)$ *is semisimple with* $\Delta^{\Bbbk}(\text{╞})$ *and* $\Delta^{\Bbbk}(\square)$ *providing a complete set of non-isomorphic simple modules.*

Proof. The Gram matrices in (6.3.3) are diagonal matrices with non-zero entries along the diagonal, and so are of (full) rank 1 and 3, respectively. Therefore the radicals of these cell modules are 0-dimensional and the claim follows by Proposition 6.2.15. $\square$

Lemma 6.3.5. *Over* $\Bbbk = \mathbb{F}_2$ *the algebra* $S^{\Bbbk}(2)$ *is non-semisimple with* $L^{\Bbbk}(\text{╞})$ *and* $L^{\Bbbk}(\square)$ *providing a complete set of non-isomorphic simple modules. The decomposition matrix is as follows*

	$L^{\Bbbk}(\text{╞})$	$L^{\Bbbk}(\square)$
$\Delta^{\Bbbk}(\text{╞})$	1	0
$\Delta^{\Bbbk}(\square)$	1	1

Proof. We trivially have that $\Delta^{\Bbbk}(\text{╞}) = L^{\Bbbk}(\text{╞})$ as its Gram matrix is unchanged by reduction modulo 2. Thus the first row of the decomposition matrix is of the claimed form.

Over $\Bbbk = \mathbb{F}_2$ we see that the latter matrix in (6.3.3) has diagonal entries 1, 0, 1 and so this matrix is of rank 2; therefore

$$\dim(L^\Bbbk(\Box\Box)) = 2 \qquad \dim(\text{rad}(\langle -, -\rangle_\Box^\Bbbk)) = 3 - 2 = 1$$

and hence $\Delta^\Bbbk(\Box\Box)$ has a proper 1-dimensional submodule. By Proposition 6.2.22, there is only one possible candidate for such a submodule, namely $L^\Bbbk(\Box)$. Thus the second row of the decomposition matrix is of the claimed form. □

6.4 The symmetric group on 3 letters

In this section, we give a detailed account of the representation theory of $\Bbbk\mathfrak{S}_3$ over an arbitrary field, using the technology from cellular algebras. We hope this might help group theorists to ground themselves in the material. We let $\Lambda^+ = \{(3), (2, 1), (1^3)\}$ and we visualise and order these elements as follows

$$\Box\text{\tiny{}} < \Box\Box\text{\tiny{}} < \Box\Box\Box$$

and we set the x_λ to be the quasi-idempotents

$$\text{id} - (1, 2) - (2, 3) - (1, 3) + (1, 2, 3) + (1, 3, 2), \qquad \text{id} - (1, 2), \qquad \text{id},$$

respectively. We define $\text{Std}(\lambda)$ for $\lambda \in \Lambda^+$ to be given by the ways of filling these boxes with the numbers 1, 2, and 3 so that they strictly increase along rows and columns. That is, the set $\text{Std}(\lambda)$ for $\lambda \in \Lambda$ consists of the elements

$$\mathsf{U} = \begin{array}{|c|}\hline 1 \\\hline 2 \\\hline 3 \\\hline\end{array} \quad \mathsf{S} = \begin{array}{|c|c|}\hline 1 & 3 \\\hline 2 \\\cline{1-1}\end{array} \quad \mathsf{T} = \begin{array}{|c|c|}\hline 1 & 2 \\\hline 3 \\\cline{1-1}\end{array} \quad \mathsf{V} = \begin{array}{|c|c|c|}\hline 1 & 2 & 3 \\\hline\end{array}.$$

and we set $d_\mathsf{T} = (2, 3)$ and $d_\mathsf{S} = d_\mathsf{U} = d_\mathsf{V} = 1$. (Notice that S and T differ by swapping the entries 2 and 3.) With the above definitions in place, the cellular basis is as follows:

$\Box\Box\Box$	$x_\mathsf{VV} = x_{(3)}$
$\Box\Box\text{\tiny{}}$	$x_\mathsf{TT} = x_{(2,1)}$ $\quad x_\mathsf{TS} = x_{(2,1)}(2, 3)$ $x_\mathsf{ST} = (2, 3)x_{(2,1)}$ $\quad x_\mathsf{SS} = (2, 3)x_{(2,1)}(2, 3)$
$\Box\text{\tiny{}}$	$x_\mathsf{UU} = x_{(1^3)}$

6.4 The symmetric group on 3 letters

Note that we have highlighted the stratification of the basis into the the three layers and emphasised the left versus right construction of the 4-dimensional layer (using columns versus rows). We now use our cellular structure in order to determine the simple modules of $\Bbbk \mathfrak{S}_3$ over a field $\Bbbk = \mathbb{F}_p$.

Lemma 6.4.1. *The Gram matrices of the intersection forms of the cell modules are given by*

$$\begin{array}{c|c} & \mathsf{V} \\ \hline \mathsf{V} & 1 \end{array} \qquad \begin{array}{c|cc} & \mathsf{S} & \mathsf{T} \\ \hline \mathsf{S} & 2 & -1 \\ \mathsf{T} & -1 & 2 \end{array} \qquad \begin{array}{c|c} & \mathsf{U} \\ \hline \mathsf{U} & 6 \end{array} \qquad (6.4.2)$$

Proof. We proceed from left to right. Thus we first consider the cell module labelled by $(3) \in \Lambda^+$. We have that

$$x_{\mathsf{VV}} x_{\mathsf{VV}} = 1 \times 1 = 1 = 1 \cdot x_{\mathsf{VV}}$$

and so the Gram matrix is of the claimed form.

Next, we consider the cell module labelled by $(2, 1) \in \Lambda^+$. We first calculate the diagonal entries

$$x_{\mathsf{SS}} x_{\mathsf{SS}} = (1 - (1,2)) \times (1 - (1,2)) = 2(1 - (1,2)) = 2 x_{\mathsf{SS}}$$
$$x_{\mathsf{ST}} x_{\mathsf{TS}} = (1 - (1,2))(2,3) \times (2,3)(1 - (1,2)) = 2(1 - (1,2)) = 2 x_{\mathsf{SS}}$$

and so the diagonal entries are both equal to 2. We now calculate the off-diagonal entries as follows

$$\begin{aligned} x_{\mathsf{ST}} x_{\mathsf{SS}} &= (1 - (1,2))(2,3) \times (1 - (1,2)) \\ &= (2,3) - (1,2,3) - (1,3,2) + (1,3) \\ &= (-1 + (1,2) + (2,3) + (1,3) - (1,2,3) - (1,3,2)) - (1 - (1,2)) \\ &= x_{\mathsf{UU}} - x_{\mathsf{SS}} \\ &\in -x_{\mathsf{SS}} + \Bbbk \mathfrak{S}_3^{<(2,1)} \end{aligned}$$

and so the off-diagonal entry in row T and column S is equal to -1. Therefore the second Gram matrix is of the claimed form.

Finally, we consider the cell module labelled by $(1^3) \in \Lambda^+$. We have that

$$\begin{aligned} x_{\mathsf{UU}} x_{\mathsf{UU}} &= (1 - (1,2) - (2,3) - (1,3) + (1,2,3) + (1,3,2))^2 \\ &= 6(1 - (1,2) - (2,3) - (1,3) + (1,2,3) + (1,3,2)) \\ &= 6 \cdot x_{\mathsf{UU}}. \end{aligned}$$

Thus the final Gram matrix is of the claimed form $\square$

Exercise 6.4.3. Let $\Bbbk = \mathbb{F}_p$ for $p > 3$. Prove that the algebra $\Bbbk \mathfrak{S}_3$ is semisimple and that the decomposition matrix is trivial.

Lemma 6.4.4. *Over* $\Bbbk = \mathbb{F}_2$ *the algebra* $\Bbbk \mathfrak{S}_3$ *has decomposition matrix*

	$L^\Bbbk(\square\square\square)$	$L^\Bbbk(\square\!\square)$
$\Delta^\Bbbk(\square\square\square)$	1	0
$\Delta^\Bbbk(\square\!\square)$	0	1
$\Delta^\Bbbk(\square\!\square)$	1	0

Proof. Over $\mathbb{F}_2$ the first two matrices in (6.4.2) are of full rank and hence the first two rows of the decomposition matrix are of the stated form. Over $\mathbb{F}_2$ the final matrix in (6.4.2) is the zero matrix (and therefore has rank 0). Thus (1^3) does not label a column in the decomposition matrix. Now, since the cell module $\Delta^\Bbbk(1^3)$ is 1-dimensional, it must be isomorphic to a 1-dimensional simple module; there is only one such simple module, namely $L^\Bbbk(3)$. Therefore $\Delta^\Bbbk(1^3) \cong L^\Bbbk(3)$ and the third row of the decomposition matrix is of the claimed form. $\square$

Lemma 6.4.5. *Over* $\Bbbk = \mathbb{F}_3$ *the algebra* $\Bbbk \mathfrak{S}_3$ *has decomposition matrix*

	$L^\Bbbk(\square\square\square)$	$L^\Bbbk(\square\!\square)$
$\Delta^\Bbbk(\square\square\square)$	1	0
$\Delta^\Bbbk(\square\!\square)$	1	1
$\Delta^\Bbbk(\square\!\square)$	0	1

Proof. Over $\mathbb{F}_3$ the first matrix in (6.4.2) is of full rank. Therefore the first row of the decomposition matrix is of the stated form. Over $\mathbb{F}_3$ we have that $2 = -1$ and so the second matrix in (6.4.2) is of rank 1. Therefore the 2-dimensional module $\Delta^\Bbbk(2, 1)$ has a 1-dimensional submodule (which by Proposition 6.2.22 must be isomorphic to $L^\Bbbk(3)$) and the resulting quotient is a 1-dimensional module $L^\Bbbk(2, 1)$. Thus the second row of the decomposition matrix is of the stated form.

Over $\mathbb{F}_3$ the third matrix in (6.4.2) is the zero matrix; thus (1^3) does not label a column in the decomposition matrix. Therefore $\Delta^\Bbbk(1^3)$ must be isomorphic to one of the two distinct 1-dimensional simple modules $L^\Bbbk(3)$ or $L^\Bbbk(2, 1)$. We will show that $\Delta^\Bbbk(1^3) \not\cong L^\Bbbk(3)$ and hence deduce the result. We have that

$$x_{\mathsf{UU}}(1, 2) = (1 - (1, 2) - (2, 3) - (1, 3) + (1, 2, 3) + (1, 3, 2))(1, 2)$$
$$= (1, 2) - 1 - (1, 3, 2) - (1, 2, 3) + (2, 3) + (1, 3) = -x_{\mathsf{UU}}$$

and so the generator $(1, 2)$ acts on $\Delta^\Bbbk(1^3)$ as the scalar -1. On the other hand,

$$x_{\mathsf{VV}}(1, 2) = 1 \cdot (1, 2) = (1, 2) = -1 + (1 - (1, 2)) = x_{\mathsf{VV}} + \Bbbk\mathfrak{S}_3^{<(3)}$$

and so the generator $(1, 2)$ acts on $\Delta^\Bbbk(1^3)$ as the scalar 1. The result follows. $\square$

6.5 Semisimple filtrations of non-semisimple modules

As a first approximation to understanding a non-semisimple algebra's structure, we construct filtrations of this algebra in which the resulting subquotient layers *are* semisimple. To this end, we define the Jacobson radical of a finite-dimensional A-module M, denoted $\mathrm{rad}(M)$, to be the smallest submodule of M such that the corresponding quotient is semisimple. When the context is clear, we refer to this simple as the radical of M. We then let $\mathrm{rad}^2 M = \mathrm{rad}(\mathrm{rad}M)$ and inductively define the radical series, $\mathrm{rad}^i M$, of M by $\mathrm{rad}^{i+1} M = \mathrm{rad}(\mathrm{rad}^i M)$. We have a finite chain

$$M \supset \mathrm{rad}(M) \supset \mathrm{rad}^2(M) \supset \cdots \supset \mathrm{rad}^i(M) \supset \mathrm{rad}^{i+1}(M) \supset \cdots \supset \mathrm{rad}^s(M) = 0$$

and we let $\mathrm{rad}_i(M) = \mathrm{rad}^i(M)/\mathrm{rad}^{i+1}(M)$ denote the semisimple layers. We have already used the term "radical" in the context of bi-linear forms, fortunately we note that the Jacobson radical of the cell module $\Delta^{\Bbbk}(\lambda)$ is equal to the radical of the bilinear form $\mathrm{rad}(\langle -, - \rangle_\lambda^{\Bbbk})$ by Proposition 6.2.15 and so the terminology does not conflict.

Example 6.5.1. We have already seen an example of a 2-dimensional $\mathrm{TL}_3(1)$-module in 6.1.8 which has a unique (semi)simple 1-dimensional submodule and therefore a unique (semi)simple 1-dimensional quotient. Thus the module of (6.1.8) has two radical layers.

We now provide our first example of a cell module with three radical layers.

Example 6.5.2. Let $\Bbbk = \mathbb{F}_2$ and consider the $S^{\Bbbk}(4)$-module $\Delta^{\Bbbk}(4, 0) = 1_{(4,0)} S^{\Bbbk}(4)$. We claim that the radical series is of length three, with (semisimple) radical layers

$$\mathrm{rad}_0(\Delta^{\Bbbk}(4)) = \Bbbk \left\{ \boxed{1\,|\,1\,|\,1\,|\,1}\,,\, \boxed{2\,|\,2\,|\,2\,|\,2} \right\}$$

$$\mathrm{rad}_1(\Delta^{\Bbbk}(4)) = \Bbbk \left\{ \boxed{1\,|\,1\,|\,2\,|\,2} \right\}$$

$$\mathrm{rad}_2(\Delta^{\Bbbk}(4)) = \Bbbk \left\{ \boxed{1\,|\,1\,|\,1\,|\,2}\,,\, \boxed{1\,|\,2\,|\,2\,|\,2} \right\}.$$

We will now prove this claim. We first establish that

$$N = \Bbbk \left\{ \boxed{1\,|\,1\,|\,1\,|\,2}\,,\, \boxed{1\,|\,2\,|\,2\,|\,2} \right\}$$

is a submodule of $\Delta^{\Bbbk}(4)$ by noting that

$$\left(\boxed{1\,|\,1\,|\,1\,|\,2} \right) f = 2 \times \boxed{1\,|\,1\,|\,2\,|\,2} = 0 \qquad (6.5.3)$$

$$\left(\boxed{1\,|\,1\,|\,1\,|\,2} \right) e = 4 \times \boxed{1\,|\,1\,|\,1\,|\,1} = 0 \qquad (6.5.4)$$

$$\left(\boxed{1\,|\,2\,|\,2\,|\,2} \right) e = 2 \times \boxed{1\,|\,1\,|\,2\,|\,2} = 0 \qquad (6.5.5)$$

$$\left(\boxed{1\,|\,2\,|\,2\,|\,2} \right) f = 4 \times \boxed{2\,|\,2\,|\,2\,|\,2} = 0 \qquad (6.5.6)$$

and furthermore that N is a *simple* submodule by noting that

$$\left(\boxed{1\,|\,2\,|\,2\,|\,2}\right) e^2 = 3 \times \boxed{1\,|\,1\,|\,1\,|\,2} = \boxed{1\,|\,1\,|\,1\,|\,2} \tag{6.5.7}$$

$$\left(\boxed{1\,|\,1\,|\,1\,|\,2}\right) f^2 = 3 \times \boxed{1\,|\,2\,|\,2\,|\,2} = \boxed{1\,|\,2\,|\,2\,|\,2} \tag{6.5.8}$$

We now establish that

$$M = \Bbbk\left\{\boxed{1\,|\,1\,|\,1\,|\,2}, \boxed{1\,|\,1\,|\,2\,|\,2}, \boxed{1\,|\,2\,|\,2\,|\,2}\right\}. \tag{6.5.9}$$

is a submodule of $\Delta^{\Bbbk}(4)$ by noting that

$$\left(\boxed{1\,|\,1\,|\,2\,|\,2}\right) f^{[2]} = 4\left(\boxed{2\,|\,2\,|\,2\,|\,2}\right) \equiv 0 \tag{6.5.10}$$

$$\left(\boxed{1\,|\,1\,|\,2\,|\,2}\right) e^{[2]} = 4\left(\boxed{1\,|\,1\,|\,1\,|\,1}\right) \equiv 0 \tag{6.5.11}$$

(together with our previous observations (6.5.3) to (6.5.6)). Of course, we have already checked that N is simple; the quotient M/N is 1-dimensional and therefore also simple. Furthermore we deduce that the 3-dimensional module M is a non-split extension of N and M/N by putting together

$$\left(\boxed{1\,|\,1\,|\,2\,|\,2}\right) e = 3 \times \boxed{1\,|\,2\,|\,2\,|\,2} \equiv \boxed{1\,|\,2\,|\,2\,|\,2}$$

$$\left(\boxed{1\,|\,1\,|\,2\,|\,2}\right) f = 3 \times \boxed{1\,|\,1\,|\,1\,|\,2} \equiv \boxed{1\,|\,1\,|\,1\,|\,2}$$

with (6.5.3), (6.5.5). Finally, we will now establish that the quotient module $\Delta^{\Bbbk}(4)/M$ is simple by noting that

$$\left(\boxed{1\,|\,1\,|\,1\,|\,1}\right) f^{[4]} = \boxed{2\,|\,2\,|\,2\,|\,2} \qquad \left(\boxed{2\,|\,2\,|\,2\,|\,2}\right) e^{[4]} = \boxed{1\,|\,1\,|\,1\,|\,1}.$$

To summarise, we have a chain of submodules

$$0 \subseteq N \subseteq M \subseteq \Delta^{\Bbbk}(4)$$

which are non-split extensions and the resulting subquotient modules are simple; hence this is the radical filtration of $\Delta^{\Bbbk}(4)$. The subquotients have the claimed bases and so the claim follows.

Example 6.5.12. Let $\Bbbk = \mathbb{F}_3$ be the field of 3 elements. Prove that the $S^{\Bbbk}(4)$-module $\Delta^{\Bbbk}(4)$ has 2 radical layers. Hint: we have already seen that it has two simple composition factors, therefore it remains to show that the module is non-split.

Exercise 6.5.13. Check that every non-simple cell module for $S^{\Bbbk}(2)$ and $\Bbbk\mathfrak{S}_3$ has precisely two radical layers.

Exercise 6.5.14. Let $\Bbbk = \mathbb{F}_2$ be the field of 2 elements. Construct the radical filtration of the $S^{\Bbbk}(6)$-module $1_{(6,0)} S^{\Bbbk}(6)$.

6.6 Weighted cellular algebras and gradings

Two of the most powerful ideas in Lie theory are that of *highest weight theory* and of *gradings*, which we now incorporate into our cellular framework. Both of these ideas were key tools in the construction of Williamson's counterexamples to Lusztig's conjecture. Essentially, incorporating gradings and highest weight theory into our cellular basis allows us to decompose the Gram matrices of intersection forms with veritable ease (as we shall see in the next section).

Definition 6.6.1. Let A be a $\Bbbk$-algebra with anti-involution $*$ and a set of of pairwise orthogonal idempotents: 1_λ for $\lambda \in \Lambda$ which are fixed by $*$. Let $\Lambda^+ \subseteq \Lambda$ be a subset on which we have a total ordering $\lambda_1 < \lambda_2 < \cdots < \lambda_t$ and for which we have a chain of inclusions
$$0 \subset A1_{\leq \lambda_1} A \subset A1_{\leq \lambda_2} A \subset \cdots \subset A1_{\leq \lambda_t} A = A$$
where $1_{\leq \lambda_k} = \sum_{1 \leq j \leq k} 1_{\lambda_j}$. For $1 \leq k \leq t$, suppose there exist $d_\mathsf{T} \in 1_{\lambda_k} A1_\nu$ indexed by $\mathsf{T} \in \mathrm{SStd}(\lambda_k, \nu)$ such that
$$\{x_\mathsf{ST} := d_\mathsf{S}^* 1_{\lambda_k} d_\mathsf{T} \mid \mathsf{S} \in \mathrm{SStd}(\lambda_k, \mu), \mathsf{T} \in \mathrm{SStd}(\lambda_k, \nu), \mu, \nu \in \Lambda\}$$
is a $\Bbbk$-basis of $A1_{\leq \lambda_k} A / A1_{\leq \lambda_{k-1}} A$. In which case, we say that A is a **weighted cellular algebra**. For $\lambda \in \Lambda^+$, we often write $\mathrm{Std}(\lambda)$, $\mathrm{SStd}(\lambda)$, or $\mathrm{SStd}(\lambda, -)$ for the set $\cup_{\mu \in \Lambda} \mathrm{SStd}(\lambda, \mu)$.

We refer to the $\mathsf{S} \in \mathrm{SStd}(\lambda)$ as semistandard tableaux, generalising the classical combinatorial notion of semistandard tableaux for binary Schur algebras (which we will soon consider).

Proposition 6.6.2. *Let $\Bbbk$ be a field and let A be a weighted cellular algebra. Then $\Lambda_\Bbbk^0 = \Lambda$ and the decomposition matrix is square uni-triangular.*

Proof. Since 1_λ for $\lambda \in \Lambda^+$ are idempotents by assumption, we have that
$$\langle 1_\lambda, 1_\lambda \rangle = 1_\lambda 1_\lambda + A^{<\lambda} = 1_\lambda + A^{<\lambda}$$
and so $1_\lambda \notin \mathrm{rad}(\Delta^\Bbbk(\lambda))$ and therefore 1_λ belongs to the (non-zero) quotient module $L^\Bbbk(\lambda)$, as required. $\square$

The above definition allows us to import ideas from "highest weight theory". The first of these ideas is that of a "formal character":

Definition 6.6.3. Let A be a weighted cellular algebra with respect to $\Lambda^+ \subseteq \Lambda$ and let M be an A-module. We define the λ-weight-space of $[M]$ to be the submodule $M1_\lambda$. We define the **formal character**, $[M]$, as follows
$$[M] = \sum_{\lambda \in \Lambda} \dim(M1_\lambda)[\lambda]$$
where $[\lambda]$ for $\lambda \in \Lambda$ are merely formal symbols.

Whilst the notion of a formal character might seem quite innocuous, it is actually a very powerful idea which we will make heavy use of later in this chapter. In particular, it allows us to encode much representation-theoretic information in terms of the combinatorics of semistandard tableaux, via the following observation:

$$[\Delta(\lambda)] = \sum_{\lambda \in \Lambda} |\text{SStd}(\lambda, \mu)|[\lambda]. \tag{6.6.4}$$

Example 6.6.5. For $\Bbbk = \mathbb{F}_2$ we have the following formal characters of $S^{\Bbbk}(2)$-modules

$$[L^{\Bbbk}(\square\square)] = [(2,0)] + [(0,2)]$$
$$[\Delta^{\Bbbk}(\square\square)] = [(2,0)] + [(1,1)] + [(0,2)]$$
$$[1_{(1,1)} S^{\Bbbk}(2)] = [(2,0)] + 2[(1,1)] + [(0,2)].$$

Exercise 6.6.6. Let $\Bbbk = \mathbb{F}_2$. Calculate the formal characters of $S^{\Bbbk}(4)$-modules $[\Delta^{\Bbbk}(4)]$, $[L^{\Bbbk}(4)]$, and $[1_{(2,2)} S^{\Bbbk}(4)]$. Which of these are independent of the field?

We now wish to introduce a grading into the picture above, by defining a graded weighted cellular algebra as follows:

Definition 6.6.7. Let A be a (weighted) cellular algebra with respect to basis

$$\{x_{ST} \mid S, T \in \text{Std}(\lambda), \lambda \in \Lambda^+\}.$$

We say that A is a $\mathbb{Z}$-graded (weighted) cellular algebra if there exists a map $\deg : \text{Std}(\lambda) \to \mathbb{Z}$ and so that $\deg(x_{ST}) = \deg(S) + \deg(T)$ defines a $\mathbb{Z}$-grading on the algebra A.

We now require a notion of "sameness" for $\mathbb{Z}$-graded representation theory.

Definition 6.6.8. Let A be a $\mathbb{Z}$-graded $\Bbbk$-algebra. We say that an A-module M is a $\mathbb{Z}$-graded A-module if we have a direct sum decomposition $M = \oplus_{i \in \mathbb{Z}} M_i$ such that $m_i a_j \in M_{i+j}$ for all $m_i \in M_i$ and $a_j \in A_j$. Let M, N be two $\mathbb{Z}$-graded A-modules, and let $f : M \to N$ be an A-module homomorphism. We say that f is a $\mathbb{Z}$-graded $\Bbbk$-algebra homomorphism if $f(A_i) \subseteq B_i$ for all $i \in \mathbb{Z}$ (and a $\mathbb{Z}$-graded $\Bbbk$-algebra isomorphism if this map is bijective).

One can incorporate the grading into the arguments of Section 6.2 in a routine fashion. In particular, we have the following:

Exercise 6.6.9. Prove that $\text{rad}\langle -, - \rangle_\lambda^{\Bbbk}$ is a graded submodule of $\Delta^{\Bbbk}(\lambda)$ for $\lambda \in \Lambda$.

Theorem 6.6.10. *The $\mathbb{Z}$-graded modules*

$$L^{\Bbbk}(\lambda) = \Delta^{\Bbbk}(\lambda) / \text{rad}\langle -, - \rangle_\lambda^{\Bbbk}$$

for $\lambda \in \Lambda_{\Bbbk}^0$ and their grading shifts $L^{\Bbbk}(\lambda)\langle k \rangle$ for $k \in \mathbb{Z}$ provide a complete set of non-isomorphic simple $\mathbb{Z}$-graded A-modules.

6.6 Weighted cellular algebras and gradings

We hence define the **graded decomposition numbers** to be

$$d_{\lambda\mu}(q) = \sum_{k\in\mathbb{Z}}[\Delta^{\Bbbk}(\lambda) : L^{\Bbbk}(\mu)\langle k\rangle]q^k$$

for $\lambda \in \Lambda$ and $\mu \in \Lambda^0_{\Bbbk}$.

Example 6.6.11. We now revisit the binary Schur algebra $S^{\Bbbk}(2)$ for $\Bbbk = \mathbb{F}_2$ (which we focussed on in detail in Section 6.3). We will now establish the graded weighted cellular structure on $S^{\Bbbk}(2)$. We set $\Lambda_2^+ = \{((2,0),(1,1)\} \subseteq \Lambda_2 = \{(2,0),(1,1),(0,2)\}$. Given $\lambda \in \Lambda_2^+$ and $\mu \in \Lambda_2$, we define $\mathrm{SStd}(\lambda,\mu)$ to be given by the ways of filling the boxes of λ with a total of μ_1 1s and a total of μ_2 2s in such a manner that they weakly increase along rows and strictly increase along columns. That is, the semistandard tableaux are as follows

$$S = \boxed{\begin{array}{c}1\\2\end{array}} \quad T = \boxed{1\;1} \quad U = \boxed{1\;2} \quad V = \boxed{2\;2}$$

of weights $\mu = (1,1), (2,0), (1,1)$ and $(0,2)$ respectively (the first has shape $(1,1)$ and the latter three have shape $(2,0)$). We define a grading on $\mathrm{SStd}(\lambda,-)$ by

$$\deg(S) = \deg(T) = \deg(V) = 0 \quad \deg(U) = 1$$

and we are able to lift this to a grading on the algebra *but only for fields of characteristic 2!* Setting $\deg(x_{PQ}) = \deg(P) + \deg(Q)$, we picture the grading on basis elements as follows:

$$\begin{array}{cc} & x_{TT}\;x_{TU}\;x_{TV} \\ x_{SS} & x_{UT}\;x_{UU}\;x_{UV} \\ & x_{VT}\;x_{VU}\;x_{VV} \end{array}$$

where the degree each pink element is zero, each blue element is one, and the degree of the unique green element is two. To verify the claim, simply compare with Example 5.7.1 of the previous chapter against the tableaux-theoretic definitions of basis elements in Section 6.3.

Definition 6.6.12. Let A be a graded weighted cellular algebra with respect to $\Lambda^+ \subseteq \Lambda$ and let M be an A-module. We define the **graded formal character**, $[M]$, as follows

$$[M]_q = \sum_{\lambda\in\Lambda}\sum_{k\in\mathbb{Z}}\dim_q(M1_\lambda)[\lambda]q^k$$

where $[\lambda]$ for $\lambda \in \Lambda$ are merely formal symbols.

Example 6.6.13. For $\Bbbk = \mathbb{F}_2$, we note that the graded formal character of some $S^{\Bbbk}(2)$-modules are given as follows:

$$[\Delta^{\Bbbk}(\square)]_q = [(2,0)] + q[(1,1)] + [(0,2)]$$
$$[1_{(1,1)}S^{\Bbbk}(2)]_q = q[(2,0)] + (1+q^2)[(1,1)] + q[(0,2)].$$
$$[S^{\Bbbk}(2))]_q = (2+q)[(2,0)] + (1+2q+q^2)[(1,1)] + (2+q)[(0,2)].$$

6.7 The grading and idempotent tricks for intersection forms

Let A, be a graded weighted cellular algebra. The intersection forms on cell modules of A can be refined in two distinct ways: via the grading and idempotent decompositions. While these ideas are quite simple, they will be hugely important in the disproof of Lusztig's conjecture and so we take the time now to justify them. We wish to consider when the intersection form vanishes, that is when

$$\langle x_S, x_T \rangle_\lambda^{\Bbbk} = 0$$

for $S, T \in \mathrm{SStd}(\lambda, -)$.

The idempotent trick. The first (rather trivial) observation is that since $1_\mu, 1_\nu \in A$ are orthogonal idempotents, fixed by the anti-involution, and such that

$$x_S 1_\mu = x_S \qquad x_T 1_\nu = x_T$$

for $S \in \mathrm{SStd}(\lambda, \mu)$ and $T \in \mathrm{SStd}(\lambda, \nu)$, this implies that

$$\langle x_S, x_T \rangle_\lambda^{\Bbbk} = 0$$

unless $\mu = \nu$. To see this, assume $\mu \neq \nu$ and observe that

$$\langle x_S, x_T \rangle_\lambda^{\Bbbk} = \langle x_S 1_\mu, x_T 1_\nu \rangle_\lambda^{\Bbbk} = \langle x_S, x_T 1_\nu 1_\mu \rangle_\lambda^{\Bbbk} = 0$$

by Proposition 6.2.13(ii), our assumption that $1_\mu^* = 1_\mu$, and that $1_\mu 1_\nu = 0$. This observation is called the idempotent trick, and despite its simplicity it is hugely useful (as we will see in the disproof of Lusztig's conjecture). The reason this is so useful is that diagrammatic algebras have idempotents baked-in to their presentations and this makes the idempotent trick incredibly easy to apply.

The grading trick. So far in this book, one could be forgiven for thinking that gradings only offer us extra structure, rather than any *actual new tools* for understanding our algebras (i.e, more work, but no reward!). We are now ready to correct this misconception by introducing a very powerful new tool for calculating intersection forms. (We will apply this tool to zigzag algebras in the next section and to disprove Lusztig's conjecture later in the book.) For $S, T \in \mathrm{SStd}(\lambda, -)$ we have that

$$\deg(1_\lambda d_S d_T^* 1_\lambda + A^{<\lambda}) = \deg(\langle x_S, x_T \rangle_\lambda^{\Bbbk} 1_\lambda) = 0$$

where the first equality holds by definition of the bi-linear form and the latter equality follows simply because any scalar multiple of an idempotent $1_\lambda \in A$ is homogeneous of degree 0. In summary, we have that

$$\langle x_S, x_T \rangle_\lambda^{\Bbbk} = 0 \qquad \text{if } \deg(S) \neq -\deg(T)$$

and we refer to this as the grading trick.

Remark 10. We notice that if $\Delta^k(\lambda)$ is non-negatively graded, then the grading trick immediately implies that the simple head of $\Delta^k(\lambda)$ is concentrated solely in degree zero (if $\deg(\mathsf{S}) = -\deg(\mathsf{T})$ and all tableaux are of non-negative degree, then $\deg(\mathsf{S}) = 0 = -\deg(\mathsf{T})$). This is a very powerful application of the grading trick, which we will utilise in Chapter 10.

Example 6.7.1. We observe that all the zero values in the table in (6.3.3) can be deduced using the idempotent trick.

Example 6.7.2. We have already seen an example of the grading trick in action for $S^{\Bbbk}(2)$ for $\Bbbk$ the field of 2 elements (although our story was somewhat backward in this regard: we calculated the radical of the cell modules before we defined the grading). We have that $x_\mathsf{U} \in \Delta^k(\square\square)$ is of positive degree and so we could have deduced $\langle x_\mathsf{U}, x_\mathsf{U} \rangle_{\square\square}^{\Bbbk} = 0$ using the grading trick (instead of calculating this directly).

Exercise 6.7.3. Building on your answer to Exercise 5.7.2, establish that the algebra $S^{\Bbbk}(3)$ for $\Bbbk = \mathbb{F}_3$ is a graded weighted cellular algebra. Use the grading and idempotent tricks to calculate the graded decomposition numbers of this algebra.

6.8 Simple modules of zig-zag algebras

We now bring to bear the tools of cellularity and the idempotent and grading tricks in order to completely determine the simple modules and graded decomposition matrices of zig-zag algebras.

For $0 \leqslant j, k < n$ we set $\mathrm{SStd}(k, j) = \emptyset$ unless $j \in \{k, k+1\}$ in which case $\mathsf{T}_j^k \in \mathrm{Path}(k, j)$ is unique and we define $d_{\mathsf{T}_j^k} := a_j^k \in \mathsf{ZZ}_n$.

Theorem 6.8.1. *The zig-zag algebra ZZ_n is a graded weighted cellular algebra with respect to the natural ordering*

$$0 < 1 < 2 < \cdots < n-1$$

and the pairwise orthogonal idempotents e_k for $0 \leqslant k < n$, the anti-involution of Exercise 5.4.3, and the basis

$$\{d_\mathsf{S}^* e_k d_\mathsf{T} \mid \text{for } 0 \leqslant i, j, k < n \text{ and } \mathsf{S} \in \mathrm{SStd}(k, i), \mathsf{T} \in \mathrm{SStd}(k, j)\} \quad (6.8.2)$$

Proof. We proved that the set in (6.8.2) was a basis in the previous chapter. The anti-involution is compatible with the basis, by construction. The action of the idempotents on the basis is given by

$$e_p(a_k^i e_k a_j^k) e_q = \delta_{i,p} \delta_{j,q} (a_k^i e_k a_j^k)$$

simply by the idempotent relations of the zig-zag algebra. It remains to check that $\mathsf{ZZ}_n^{\leqslant k}$ is an ideal for $0 \leqslant k < n$. For $i = k+1$ or $j = k+1$, we have that

$$a_i^{i+1}(a_k^i e_k a_j^k) = 0 \qquad (a_k^i e_k a_j^k)a_{j+1}^j = 0$$

respectively, by the relations $a_i^{i+1}a_{i-1}^i = 0$ and $a_j^{j-1}a_{j+1}^j = 0$. All other products with $a_k^i e_k a_j^k$ clearly belong to $ZZ_n^{\leqslant k}$. □

In particular, every right cell module $\Delta^k(k)$ for $0 \leqslant k < n$ has basis

$$\Delta^k(k) = \{d_T \mid T \in \text{SStd}(k,i) \text{ for } 0 \leqslant i \leqslant n-1\}$$

and has graded dimension

$$\dim_q(\Delta^k(k)) = \begin{cases} 1+q & \text{for } 0 \leqslant k < n-1 \\ 1 & \text{for } k = n-1. \end{cases}$$

This algebra is non-negatively graded with dimension

$$(n-1)(1+q)^2 + 1 = n + (2n-2)q + (n-1)q^2.$$

Therefore by the grading and idempotent tricks, the graded simple ZZ_n-modules are

$$L^k(k) = \{d_T + \text{rad}\langle -,-\rangle_k^k \mid T \in \text{SStd}(k,k)\}$$

for $0 \leqslant k < n$, up to grading shift. Or, if you prefer, we can explicitly check that $d_S \in \text{rad}\langle -,-\rangle_k^k$ for $S \in \text{SStd}(k,j)$ for $j \neq k$ as follows. For $S \in \text{SStd}(k,i)$, $T \in \text{SStd}(k,j)$ with $\{i,j\} \neq \{k,k\}$ we have that

$$\langle e_k a_i^k, e_k a_j^k \rangle = e_k a_i^k a_j^k e_k + ZZ_n^{<k} = 0 + ZZ_n^{<k}$$

using the relations $a_{k+1}^k a_k^{k+1} = a_{k-1}^k a_k^{k-1} \in ZZ_n^{\leqslant k-1}$ for $i = j = k+1$ and the idempotent relations for the other two cases. Either way, the simple modules are all 1-dimensional and the graded decomposition matrix of ZZ_n is as depicted in Table 6.1.

	$L^k(n-1)$	$L^k(n-2)$	$L^k(n-3)$	$L^k(n-4)$	
$\Delta^k(n-1)$	1	0	0	0	...
$\Delta^k(n-2)$	q	1	0	0	...
$\Delta^k(n-3)$	0	q	1	0	...
$\Delta^k(n-4)$	0	0	q	1	...
⋮	⋮	⋮	⋮	⋮	⋱

Table 6.1: The graded decomposition matrix of ZZ_n for $n \in \mathbb{Z}_{>0}$.

6.9 Simple modules of the binary Schur algebra

We now apply the techniques of this chapter in order to explicitly construct the simple modules of binary Schur algebras over $\mathbb{F}_p$. Here we encounter, for the first time, the beautiful p-fractal-like behaviour that is characteristic of modular Lie theory. We begin by establishing the cellularity of the binary Schur algebras. Our indexing set will be given by

$$\Lambda_r^+ = \{(r-k, k) \mid 0 \leq k \leq \lfloor r/2 \rfloor\}$$

and we order these elements according to the natural ordering on $0 \leq k \leq \lfloor r/2 \rfloor$. We picture these elements diagrammatically as in Section 6.3. For example

for $r = 4$. We define $\mathrm{SStd}(\lambda, \mu)$ for $\lambda \in \Lambda_r^+$ and $\mu \in \Lambda$ to be all ways of filling the boxes of λ with a total of μ_1 1s and a total of μ_2 2s in such a manner that they weakly increase along rows and strictly increase along columns.

Example 6.9.1. For $r = 4$ we have that

are a complete list of all $\mathrm{SStd}(\lambda, -)$ for $\lambda \in \Lambda_4^+$.

Definition 6.9.2. Given $\lambda \in \Lambda_r$ we define $w(\lambda) = \lambda_1 - \lambda_2$. For $\mathsf{S} \in \mathrm{SStd}(\lambda, \mu)$ we let $d(\mathsf{S}) = \frac{1}{2}(w(\lambda) - w(\mu))$ and we let

$$x_{\mathsf{ST}} = e^{[d(\mathsf{S})]} 1_\lambda f^{[d(\mathsf{T})]}$$

for $\mathsf{S} \in \mathrm{SStd}(\lambda, \mu), \mathsf{T} \in \mathrm{SStd}(\lambda, \nu)$.

Theorem 6.9.3. *The $\Bbbk$-algebra $S^\Bbbk(r)$ is a weighted cellular algebra with idempotents 1_μ for $\mu \in \Lambda_r$, and basis*

$$\{x_{\mathsf{ST}} \mid \mathsf{S} \in \mathrm{SStd}(\lambda, \mu), \mathsf{T} \in \mathrm{SStd}(\lambda, \nu), \lambda \in \Lambda_r^+, \mu, \nu \in \Lambda_r\} \tag{6.9.4}$$

and the anti-involution $*$ given by $(e^{[m]})^* = f^{[m]}, (f^{[m]})^* = e^{[m]}$ and $(1_\lambda)^* = 1_\lambda$.

Proof. In Proposition 5.6.19 we showed that the set in (6.9.4) is a basis of the algebra. That the idempotents act in the required manner (in order to satisfy the weighted cellular axioms) is immediate from the idempotent relations for the binary Schur algebra. The anti-involution is clearly compatible with the cellular basis, by construction. It only remains to check that we have the desired chain of ideals. To that end, we define

$$S_{\underline{k}}^{\leqslant \lambda}(r) = \{e^{[j]}1_\mu f^{[k]} \mid 0 \leqslant j, k \leqslant w(\mu) \text{ for } \mu \leqslant \lambda\}$$

and for $\lambda = (r - k, k)$ we have that

$$S_{\underline{k}}^{\leqslant (r-k,k)}(r) = \ker(\det_r \circ \det_{r-2} \circ \cdots \circ \det_{r-k})$$

for the composition of determinant homomorphisms from Proposition 5.6.18. □

We now wish to construct the simple modules of the binary Schur algebra. Key to this construction is our understanding of prime divisors of binomial coefficients. Recall that Pascal's triangle is the triangular array with entries $\binom{r}{k}$ as in Figure 6.4. We have drawn the array in Figure 6.4 in such a manner that the entry $\binom{r}{k}$ has y-coordinate $-r$ and x-coordinate $r - 2k$. (Equivalently, so that $\binom{0}{0}$ has coordinate $(0, 0)$ and so that the triangle is preserved through reflection through the line $x = 0$.)

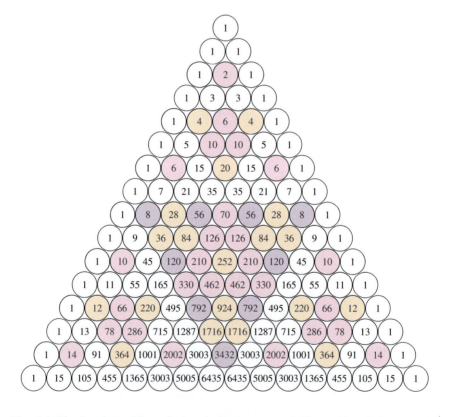

Fig. 6.4: The 2-coloured Pascal triangle for $0 \leqslant n < 16$. Here white circles are odd, pink circles are divisible by 2 (but not 4), orange circles are divisible by 4 (but not 8), and purple circles are divisible by 8 (but not 16).

6.9 Simple modules of the binary Schur algebra

We identify the $S \in \text{SStd}(\lambda, \mu)$ with the node in position $(w(\mu), -w(\lambda)) \in \mathbb{R}^2$ in Pascal's triangle. Thus S corresponds to the binomial coefficient $\binom{w(\lambda)}{d(S)}$. Given a fixed prime $p > 0$ we "p-colour" the entries of Pascal's triangle according to the highest power of p which divides the entry, as in Figure 6.4.

Example 6.9.5. We identify the elements of $S \in \text{SStd}((5, 1), -)$ (noting $4 = w(5, 1)$) with corresponding entries in a single row of Pascal's triangle as follows:

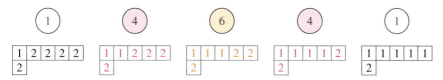

with $d(S)$ equal to $4, 3, 2, 1, 0$ respectively.

Exercise 6.9.6. Construct the 7 elements x_T for $T \in \text{SStd}((8, 2), -)$ and determine the colourings of the tableaux as in Example 6.9.5.

We are now ready to construct the simple modules of binary Schur algebras.

Theorem 6.9.7. *Let $\Bbbk$ be a field of characteristic $p > 0$. The algebra $S^{\Bbbk}(r)$ has simple modules labelled by $\lambda \in \Lambda_r^+$. The basis of the simple module $L^{\Bbbk}(\lambda)$ is given by*
$$\left\{ x_S + \text{rad}\langle -, -\rangle_\lambda^{\Bbbk} \mid p \text{ does not divide } \binom{w(\lambda)}{d(S)} \text{ for } S \in \text{SStd}(\lambda, -) \right\}.$$

Before embarking on the proof, we first consider an obvious corollary and an illustrative example.

Corollary 6.9.8. *Let $\Bbbk$ be a field of characteristic $p > r$. The cell modules $\Delta^{\Bbbk}(\lambda)$ for $\lambda \in \Lambda_r$ provide a complete set of non-isomorphic simple modules for the algebra $S^{\Bbbk}(r)$.*

Example 6.9.9. Let $\Bbbk = \mathbb{F}_2$. We fix $r = 4$ and construct the simple characters $L^{\Bbbk}(\lambda)$ by induction on $0 \leq w(\lambda) \leq 4$. We have that $\Delta^{\Bbbk}(4, 0)$ has basis indexed by

| 1 | 1 | 1 | 1 | | 1 | 1 | 1 | 2 | | 1 | 1 | 2 | 2 | | 1 | 2 | 2 | 2 | | 2 | 2 | 2 | 2 |

and we denote these tableaux by S_1, T_1, U, T_2, S_2 respectively. We have that
$$L^{\Bbbk}(4, 0) = \Bbbk\{x_{S_i} + \text{rad}\langle -, -\rangle_{(4)}^{\Bbbk} \mid i = 1, 2\} \qquad \text{rad}_{\Bbbk}(\Delta^{\Bbbk}(4, 0)) = \Bbbk\{x_{T_1}, x_U, x_{T_2}\}.$$

We have that $\Delta^{\Bbbk}(3, 1)$ has basis indexed by

| 1 | 1 | 1 | | 1 | 1 | 2 | | 1 | 2 | 2 |
| 2 | | | | 2 | | | | 2 | | |

and we denote these tableaux by P_1, Q, P_2 respectively. We have that
$$L^{\Bbbk}(3, 1) = \Bbbk\{x_{P_i} + \text{rad}\langle -, -\rangle_{(3,1)}^{\Bbbk} \mid i = 1, 2\} \qquad \text{rad}_{\Bbbk}(\Delta^{\Bbbk}(3, 1)) = \Bbbk\{x_Q\}.$$

We have that $\Delta^{\Bbbk}(2^2)$ is 1-dimensional with basis indexed by

$$\begin{array}{|c|c|} \hline 1 & 1 \\ \hline 2 & 2 \\ \hline \end{array}$$

and $L^{\Bbbk}(2^2) = \Delta^{\Bbbk}(2^2)$.

Exercise 6.9.10. Calculate the characters of $L^{\Bbbk}(p^3 + p, 0)$ for $\Bbbk = \mathbb{F}_p$ for $p = 2, 3,$ and 5. What do you notice?

We are now ready to prove Theorem 6.9.7. We break the proof up into the following two observations:

Proposition 6.9.11. *The Gram matrix for $\Delta^{\Bbbk}(\lambda)$ is simply the diagonal matrix whose (S, S)th entry for $S \in \text{Path}(\lambda, \mu)$ is given by*

$$\langle x_S, x_S \rangle = \binom{w(\lambda)}{d(S)}.$$

Proof. We let T_λ be the unique element of $SStd(\lambda, \lambda)$. We have that $x_{T_\lambda} e = 0$ by construction. Now, we have that the basis of $\Delta^{\Bbbk}(\lambda)$ is given by

$$x_S = x_{T_\lambda} f^{[d(S)]}$$

for $S \in SStd(\lambda, \mu)$. It remains to show that

$$\langle x_S, x_S \rangle = 1_\lambda f_S e_S 1_\lambda + S^{<\lambda}(r) = 1_\lambda f^{[d(S)]} e^{[d(S)]} 1_\lambda + S^{<\lambda}(r) = \binom{w(\lambda)}{d(S)} x_{T_\lambda}$$

where only the final equality is non-trivial. By (5.6.13), we have that

$$1_\lambda f^{[d(S)]} e^{[d(S)]} 1_\lambda = 1_\lambda f^{[d(S)]} 1_{(\lambda_1+d(S), \lambda_2-d(S))} e^{[d(S)]} 1_\lambda = 1_\lambda \binom{w(\lambda)}{d(S)} 1_\lambda + \ldots$$

where the ... terms are scalar multiples of $1_\lambda e^{[i]} f^{[i]} 1_\lambda \in S^{<\lambda}$ for $1 \leq i \leq d(S)$. The explicit coefficients can be obtained using relation (5.6.13). □

Corollary 6.9.12. *The radical of the $S^{\Bbbk}(r)$-module $\Delta^{\Bbbk}(\lambda)$ has basis*

$$\text{rad}(\Delta^{\Bbbk}(\lambda)) = \Bbbk \left\{ x_S \mid p \text{ divides } \binom{w(\lambda)}{d(S)} \right\}.$$

Proof. This follows immediately from the form of the entries in the diagonal Gram matrices in Corollary 6.9.12. □

We note that Corollary 6.9.12 immediately implies Theorem 6.9.7, simply by applying Theorem 6.6.10. The prime divisors of binomial coefficients actually satisfy a beautiful "p-fractal-like" property given by the following theorem (and further illustrated in Figure 6.5) which affords us a very explicit understanding of the bases of simple modules in Theorem 6.9.7 in terms of the combinatorics of Sierpinski's triangle.

6.9 Simple modules of the binary Schur algebra

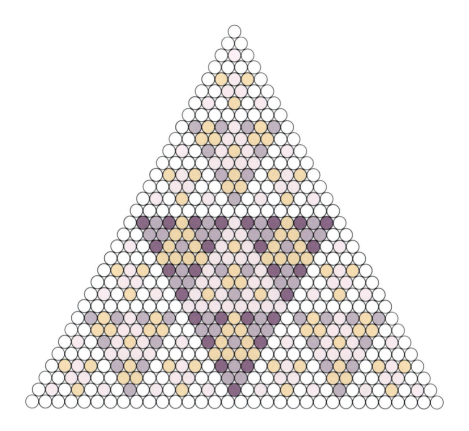

Fig. 6.5: The restriction of the 2-coloured Pascal triangle to the rows $0 \leqslant r < 32$. Here white circles are odd, pink circles are divisible by 2 (but not 4), orange circles are divisible by 4 (but not 8), light purple circles are divisible by 8 (but not 16), and dark purple circles are divisible by 16 (but not 32). Notice the rotational and fractional symmetries of the colouring; considering only the odd (white) circles as $r \to \infty$ one obtains the famous Sierpiński triangle.

Theorem 6.9.13 (Lucas' Theorem). *Given a pair $a, b \in \mathbb{N}$ with p-adic expansions*

$$a = a_k p^k + a_{k-1} p^{k-1} + \cdots + a_1 p + a_0 \qquad b = b_k p^k + b_{k-1} p^{p-1} + \cdots + b_1 p + b_0$$

we have that

$$\binom{a}{b} \equiv \prod_{j=0}^{k} \binom{a_j}{b_j} \mod p.$$

Remark 11. For $\lambda = (\lambda_1, \lambda_2) \in \Lambda_r^+$ and $\mu = (\mu_1, \mu_2) \in \Lambda_r$, Lucas' theorem states that the weight spaces $L^{\Bbbk}(\lambda) 1_\mu$ of $L^{\Bbbk}(\lambda)$ can be calculated in terms of the p-adic expansions of the integers $\lambda_1 - \lambda_2, \mu_1 - \mu_2$.

Example 6.9.14. Setting $w(\lambda) = p^k - 1$ for $k \geq 0$ we find that there are infinitely many simple modules which are also cell modules.

Remark 12. Let $\Bbbk = \mathbb{F}_p$. The formal character of the simple $S^{\Bbbk}(r)$-module $L^{\Bbbk}(\lambda)$ for $\lambda \in \Lambda_r^+$ is given by

$$[L^{\Bbbk}(\lambda)] = \sum_{\mu} \#\{S \in \mathrm{SStd}(\lambda, \mu) \mid p \text{ does not divide } \binom{w(\lambda)}{d(S)}\}[\mu].$$

Example 6.9.15. Let $\Bbbk = \mathbb{F}_2$. The dimension of all simple $S^{\Bbbk}(r)$-modules for $0 \leq r \leq 32$ can be read-off from Figure 6.5. For example, if $w(\lambda) = 6$ then we consider the 7th row of the diagram (as we label the rows of Pascal's triangle by the weights $0, 1, 2, \ldots, 6$) and we see that the white weights (for which 2 does not divide $\binom{6}{w(S)}$) are $w(S) = 6, 2, -2$, and -6 and therefore $L^{\Bbbk}(\lambda)$ is 4-dimensional for any λ such that $w(\lambda) = 6$. For example, $L^{\Bbbk}(6,0)$ and $L^{\Bbbk}(10,4)$ both have weight 6 and therefore are both 4-dimensional.

6.10 Decomposition numbers via highest weight theory

For weighted cellular algebras, the problem of determining the dimensions of simple modules is equivalent to determining the decomposition matrix. In fact, there is a simple algorithm for passing back-and-forth between these two equivalent problems, via "highest weight theory". This algorithm is best illustrated via an example. Having calculated the formal characters of the simple $S^{\Bbbk}(r)$-modules in the previous section, we now illustrate how to read off the composition factor multiplicities

$$d_{\lambda,\mu} = [\Delta^{\Bbbk}(\lambda) : L^{\Bbbk}(\mu)]$$

for $\lambda, \mu \in \Lambda_r^+$ by "highest weight theory". Namely, we notice that the "highest weight" of $L^{\Bbbk}(\mu)$ in the order $\leq$ is the weight μ itself (and this holds for all $\mu \in \Lambda_r^+$). Therefore we can "peal off" the weights of $\Delta^{\Bbbk}(\lambda)$ by induction *down* the ordering $\leq$ (starting with the cell module labelled by the maximal element of Λ_r^+) in a manner we now describe, by example.

Example 6.10.1. We fix $r = 4$. The formal characters of the cell-modules $\Delta^{\Bbbk}(\lambda)$ are trivially deduced from the weighted cellular basis and are as follows:

$$[\Delta^{\Bbbk}(4,0)] = [(4,0)] + [(3,1)] + [(2,2)] + [(1,3)] + [(0,4)],$$
$$[\Delta^{\Bbbk}(3,1)] = [(3,1)] + [(2,2)] + [(1,3)],$$
$$[\Delta^{\Bbbk}(2,2)] = [(2,2)].$$

In Remark 12, we calculated the formal characters of the simple modules $L^{\Bbbk}(\lambda)$ in terms of the 2-divisors of binomial coefficients; these formal characters are as follows,

6.10 Decomposition numbers via highest weight theory

$$[L^k(4,0)] = [(4,0)] + [(0,4)],$$
$$[L^k(3,1)] = [(3,1)] + [(1,3)],$$
$$[L^k(2,2)] = [(2,2)].$$

Our first observation is a powerful uni-triangularity property: the highest weight idempotent (under $\leqslant$) which does not kill the simple $L^k(\mu)$ is 1_μ. This means that we can "detect" composition factors of cell modules by working up the ordering $\leqslant$ and pealing off the formal characters of simple modules as we go, as follows. We first consider the 1-dimensional cell module,

$$[\Delta^k(2,2)] = [(2,2)] = [L^k(2,2)]$$

for which the composition series is trivial. Now, we know that $L^k(3,1)$ is a composition factor of $\Delta^k(3,1)$ and that we can peal it off as follows,

$$[\text{rad}\langle -,-\rangle^k_{(3,1)}] = [\Delta^k(3,1)] - [L^k(3,1)] = [(2,2)].$$

Now, the formal character $[\text{rad}(\Delta^k(3,1))]$ is equal to that $[L^k(2,2)]$ and so we have that

$$[\Delta^k(3,1)] = [L^k(3,1)] + [L^k(2,2)].$$

Finally, we consider $\Delta^k(4,0)$. We know that $L^k(4,0)$ is a composition factor of $\Delta^k(4,0)$ and so we can peal off this 2-dimensional module to leave

$$[\text{rad}(\Delta^k(2,2))] = [(3,1)] + [(2,2)] + [(1,3)].$$

Observe that the highest weight (under $\leqslant$ on Λ^+) is $(3,1)$. This implies (by uni-triangularity) that $L^k(3,1)$ must appear as a composition factor of $\Delta^k(4,0)$. Pealing off the character of $L^k(3,1)$ leaves us with the formal character $[(2,2)]$ and so we see that this implies that $L^k(2,2)$ must appear as the final composition factor. Therefore we we have that

$$[\Delta^k(4,0)] = [L^k(4,0)] + [L^k(3,1)] + [L^k(2,2)].$$

Thus the decomposition matrix of this algebra is as follows

	$L^k(2,2)$	$L^k(3,1)$	$L^k(4,0)$
$\Delta^k(2,2)$	1	·	·
$\Delta^k(3,1)$	1	1	·
$\Delta^k(4,0)$	1	1	1

Exercise 6.10.2. Calculate the decomposition matrix of $S^k(6)$, $S^k(8)$, and $S^k(10)$ for $k = \mathbb{F}_2$. Can you see any repeated patterns?

Example 6.10.3. We fix $r = 12$. One can calculate the composition factors of cell modules using the same highest weight theory arguments as in the previous example and hence obtain the following decomposition matrix:

	$L^{\Bbbk}(6,6)$	$L^{\Bbbk}(7,5)$	$L^{\Bbbk}(8,4)$	$L^{\Bbbk}(9,3)$	$L^{\Bbbk}(10,2)$	$L^{\Bbbk}(11,1)$	$L^{\Bbbk}(12,0)$
$\Delta^{\Bbbk}(6,6)$	1	·	·	·	·	·	·
$\Delta^{\Bbbk}(7,5)$	1	1	·	·	·	·	·
$\Delta^{\Bbbk}(8,4)$	1	1	1	·	·	·	·
$\Delta^{\Bbbk}(9,3)$	1	·	1	1	·	·	·
$\Delta^{\Bbbk}(10,2)$	1	·	1	1	1	·	·
$\Delta^{\Bbbk}(11,1)$	1	1	1	·	1	1	·
$\Delta^{\Bbbk}(12,0)$	1	1	·	·	1	1	1

Exercise 6.10.4. Calculate the final row of the decomposition matrices for $S^{\Bbbk}(p^3 + p, 0)$ for $\Bbbk = \mathbb{F}_p$ for $p = 2, 3,$ and 5. What do you notice? (Hint: you do not need to calculate the entire decomposition matrix!)

Exercise 6.10.5. Let A be an arbitrary weighted cellular algebra and suppose we know the formal characters of all simple A-modules for a fixed field $\Bbbk$. Write down an algorithm for calculating the decomposition matrix of A.

6.11 Alperin diagrams and submodule structures

It is worth emphasising that the highest weight arguments of the previous sections are not enough to determine the radical structures of these modules. In order to determine these, we have to "dig a little deeper" and examine the effect of applying the algebra generators to the composition factors of a given module. We provide a formal definition first, but the reader is invited to skip ahead to look at the examples (which are intuitive) and then come back to the definition.

Definition 6.11.1. We say that an A-module is multiplicity-free if no two distinct composition factors of M are isomorphic as A-modules.

Definition 6.11.2. Let M be a multiplicity-free A-module. The Alperin diagram of M is the graph whose nodes on the kth level are given by the direct summands of $\mathrm{rad}_k(M)$. We suppose that D is a summand of $\mathrm{rad}_j(M)$ and that D' is a summand of $\mathrm{rad}_k(M)$ for $j < k$. In other words, there exists $d \in \mathrm{rad}^j(M)$ such that $D \cong \langle d + \mathrm{rad}^{j+1}(M) \rangle$ and $d' \in \mathrm{rad}^k(M)$ such that $D' \cong \langle d' + \mathrm{rad}^{k+1}(M) \rangle$. We draw an edge, with label $a \in A$, from the node D to the node D' if, and only if, $ad = d'$.

Whenever there is a sequence of edges

$$D \xrightarrow{a} D' \xrightarrow{b} D''$$

we omit the edge $D \xrightarrow{ab} D''$ as this edge can be implicitly regarded as the composite of two smaller edges.

6.11 Alperin diagrams and submodule structures

Example 6.11.3. We continue with $r = 12$ from Example 6.10.3. We now construct the Alperin diagram for $\Delta^k(12, 0)$ whose composition factors are:

$$L^k(12,0) \quad L^k(11,1) \quad L^k(10,2) \quad L^k(7,5) \quad L^k(6,6)$$

which are generated as subquotients of $\Delta^k(12, 0)$ by the elements

$$1_{(12,0)} \quad 1_{(12,0)}f \quad 1_{(12,0)}f^{[2]} \quad 1_{(12,0)}f^{[5]} \quad 1_{(12,0)}f^{[6]},$$

respectively. We claim that the Alperin module for this diagram is the graph depicted in Figure 6.6.

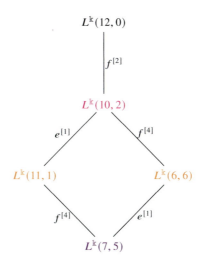

Fig. 6.6: The Alperin diagram of $\Delta^k(12, 0)$ in characteristic 2.

To understand the edges of the Alperin diagram, we consider the expansions in terms of elementary matrices

$$1_{(12,0)} f^{[k]} = \sum_\pi E_{111111111111,\pi} \tag{6.11.4}$$

with the sum over all π of weight $12 - 2k$ for $k = 0, 1, \ldots, 6$. There are precisely $\binom{12}{k}$ such terms in this sum, that is

$$1 \quad 12 \quad 66 \quad 792 \quad 924$$

for $k = 0, 1, 2, 5, 6$ respectively. Let's prove that there is an edge $L^k(10, 2) \to L^k(6, 6)$. By the presentation of $S^k(12)$ and the fact that $\Delta^k(12, 0)$ is a submodule of $S^k(12)$ we deduce that

$$(1_{(12,0)} f^{[2]}) f^{[4]} = \alpha 1_{(12,0)} f^{[6]}$$

for some scalar $\alpha \in \mathbb{k}$; thus, in order to conclude that the edge does exist we need only show that $\alpha \neq 0$.

Our first port of call is to explicitly record the expansions in terms of elementary matrices of the generators of these composition factors

$$1_{(12,0)} f^{[2]} = E_{11\ldots 1, 111111111122} + E_{11\ldots 1, 111111111212} + E_{11\ldots 1, 111111112112} + \ldots$$

where the sum is over 66 terms and

$$1_{(12,0)} f^{[6]} = E_{11\ldots 1, 111111222222} + E_{11\ldots 1, 111112122222} + E_{11\ldots 1, 111121122222} + \ldots$$

which is the sum over 924 terms. We now apply $f^{[4]}$ to $1_{(12,0)} f^{[2]} \in \Delta^{\mathbb{k}}(12, 0)$. We have that

$$(1_{(12,0)} f^{[2]}) f^{[4]}$$
$$= (E_{11\ldots 1, 111111111122} + E_{11\ldots 1, 111111111212} + E_{11\ldots 1, 111111112112} + \ldots) f^{[4]}$$
$$= E_{11\ldots 1, 111111111122} f^{[4]} + E_{11\ldots 1, 111111111212} f^{[4]} + E_{11\ldots 1, 111111112112} f^{[4]} + \ldots$$

and applying $f^{[4]}$ to each of the 66 terms on the right of the first equality gives a sum over $\binom{10}{4} = 210$ terms on the right of the second equality. Therefore,

$$(1_{(12,0)} f^{[2]}) f^{[4]} = \frac{210 \times 66}{924} 1_{(12,0)} f^{[6]} = 15 \times 1_{(12,0)} f^{[6]} \equiv 1_{(12,0)} f^{[6]}$$

over a field of characteristic 2. Thus there is indeed an edge $L^{\mathbb{k}}(10, 2) \to L^{\mathbb{k}}(6, 6)$ in the diagram in Figure 6.6, as claimed. The other edges can be calculated in a similar fashion.

Exercise 6.11.5. Show that $L^{\mathbb{k}}(11, 1)$ and $L^{\mathbb{k}}(6, 6)$ have no edge between them in the Alperin diagram of $L^{\mathbb{k}}(12, 0)$.

Exercise 6.11.6. Calculate the Alperin diagrams of $\Delta^{\mathbb{k}}(p^3 + p, 0)$ for $\mathbb{k}$ a field of characteristic $p = 2, 3$, and 5. What do you notice?

Exercise 6.11.7. Calculate the Alperin diagram of $\Delta^{\mathbb{k}}(11, 1)$. Hint: unlike in the previous examples, you will have to work modulo the non-zero ideal $S_{12}^{<(11,1)}$.

Theorem 6.11.8. *The Alperin diagram of the cell module $\Delta^{\mathbb{k}}(r, 0)$ has vertices as determined by the p-colouring on Pascal's triangle (see previous section) and edges determined as follows:*

$$L^{\mathbb{k}}(r - k, k) \xrightarrow{e^{[k-j]}} L^{\mathbb{k}}(r - j, j) \text{ if and only if } p \text{ does not divide } \binom{r-j}{r-k}$$

$$L^{\mathbb{k}}(r - j, j) \xrightarrow{f^{[k-j]}} L^{\mathbb{k}}(r - k, k) \text{ if and only if } p \text{ does not divide } \binom{k}{j}$$

for $j < k$.

6.11 Alperin diagrams and submodule structures

Proof. We know that

$$(1_{(r,0)} f^{[k]}) e^{[k-j]} = \alpha_{j,k} 1_{(r,0)} f^{[j]} \qquad (6.11.9)$$

$$(1_{(r,0)} f^{[j]}) f^{[k-j]} = \beta_{j,k} 1_{(r,0)} f^{[k]} \qquad (6.11.10)$$

for coefficients $\alpha_{j,k}$ and $\beta_{j,k} \in \Bbbk$. The edges

$$L^k(r-k,k) \xrightarrow{e^{[k-j]}} L^k(r-j,j)$$

correspond precisely to the non-zero coefficients $\alpha_{j,k}$ and the edges

$$L^k(r-j,j) \xrightarrow{f^{[k-j]}} L^k(r-k,k)$$

correspond to the non-zero scalars $\beta_{j,k}$. For $i \in \{j,k\}$, we consider the expansion in terms of elementary matrices

$$1_{(r,0)} f^{[i]} = \sum_\pi E_{111111111111,\pi} \qquad (6.11.11)$$

with the sum over all sequences $\pi \in \mathbb{T}_r$ of weight $r - 2i$ (there are precisely $\binom{r}{i}$ such terms in this sum). Putting together (6.11.9), (6.11.10) and (6.11.11) we have that

$$\alpha_{j,k} = \frac{\binom{r}{k}\binom{k}{k-j}}{\binom{r}{j}} = \left(\frac{r!}{k!(r-k)!}\right)\left(\frac{k!}{j!(k-j)!}\right)\left(\frac{j!(r-j)!}{r!}\right) = \binom{r-j}{r-k}.$$

The first equality follows from (6.11.9) and (6.11.10): on one hand $f^{[k]}$ turns k of the (r in total) 1s into 2s; then $e^{[k-j]}$ turns j of the (k in total) 2s back into 1s; whereas on the other hand $f^{[j]}$ turns j of the (r in total) 1s into 2s. (The second equality follows by cancelling terms in the numerator and denominator.) Similarly, we have that

$$\beta_{j,k} = \frac{\binom{r}{j}\binom{r-j}{k-j}}{\binom{r}{k}} = \left(\frac{r!}{j!(r-j)!}\right)\left(\frac{(r-j)!}{(r-k)!(k-j)!}\right)\left(\frac{k!(r-k)!}{r!}\right) = \binom{k}{j}.$$

This completes the proof. □

In fact, the structure of $\Delta^k(\lambda)$ for $\lambda \in \Lambda_r^+$ can be calculated in an identical fashion to Theorem 6.11.8 simply by replacing $(r,0)$ in the labels with $w(\lambda)$ and replacing the $(r-j,j), (r-k,k)$ with corresponding weights. We have only included the $\lambda = (r,0)$ case explicitly here because the statement and proof simplify nicely.

Exercise 6.11.12. Generalise the statement and proof of Theorem 6.11.8 to the case of $\Delta^k(\lambda)$ for general $\lambda \in \Lambda_r^+$. Hint: you can do this by making use of the relation 5.6.13 to rewrite products in the required form (or use the determinant homomorphism).

We say that a graded module is graded multiplicity-free if no two distinct composition factors of M are isomorphic *as $\mathbb{Z}$-graded modules*. We define a graded Alperin diagram of M exactly as in Definition 6.11.2, but under the assumption that the module M is *graded multiplicity free*.

Example 6.11.13. We now briefly revisit the graded $\Bbbk$-algebra $S^{\Bbbk}(2)$ for $\Bbbk = \mathbb{F}_2$. We consider the direct summands $1_\lambda S^{\Bbbk}(2)$ of $S^{\Bbbk}(2)$ for $\lambda \in \{(2,0), (1,1), (0,2)\}$. We have that each of these modules is *graded multiplicity-free*, indecomposable, and their graded Alperin diagrams are recorded in Figure 6.7.

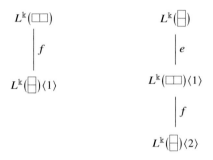

Fig. 6.7: The graded Alperin diagrams of the modules $1_{(2,0)} S^{\Bbbk}(2) \cong 1_{(0,2)} S^{\Bbbk}(2)$ and $1_{(1,1)} S^{\Bbbk}(2)$ respectively.

Exercise 6.11.14. We say that a module M is uniserial if it has a unique composition series, or equivalently, if every radical layer is simple (rather than *semi*simple). Using Theorem 6.11.8, classify the uniserial cell-modules of the binary Schur algebra. (This exercise is not easy!)

In general, the modules of the binary Schur algebra are not (graded) multiplicity-free and so we cannot define/construct Alperin diagrams for arbitrary $S^{\Bbbk}(r)$-modules (although, the fact that we could do this for the cell modules is already surprising — understanding cell modules of general $\Bbbk$-algebras is an impossibly difficult problem, as we will see later in the book). However, the zig-zag algebras have much simpler representation theoretic structure than the binary Schur algebras; we now construct the graded Alperin diagrams of the direct summands of the zigzag algebras.

Example 6.11.15. Consider the algebra ZZ_4 over an arbitrary field $\Bbbk$. We have already seen that this algebra has 4 distinct 1-dimensional simple modules $L^{\Bbbk}(k)$ for $0 \leq k < 4$, each generated by the idempotent e_k for $0 \leq k < 4$ with the representing matrices given by

$$\varphi_k(d) = \begin{cases} (1) & \text{if } d = e_k \\ (0) & \text{otherwise} \end{cases}$$

and decomposition matrix given by

6.11 Alperin diagrams and submodule structures

	$L^k(3)$	$L^k(2)$	$L^k(1)$	$L^k(0)$
$\Delta^k(3)$	1	0	0	0
$\Delta^k(2)$	q	1	0	0
$\Delta^k(1)$	0	q	1	0
$\Delta^k(0)$	0	0	q	1

We continue to bear in mind the visualisation of this algebra in terms of paths in the following graph

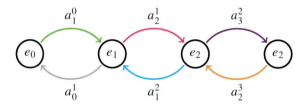

and recall that the non-idempotent relations are

$$a^j_{j\pm 1} a^{j\pm 1}_{j\pm 2} = 0 \qquad a^j_{j+1} a^{j+1}_j = -a^j_{j-1} a^{j-1}_j \qquad a^0_1 a^1_0 = 0 \qquad (6.11.16)$$

for all admissible j. We will use the above colouring when constructing graded Alperin diagrams, in order to emphasise the path visualisation. The Alperin diagrams of the cell modules are easy to construct:

$$
\begin{array}{cccc}
L^k(0) & L^k(1) & L^k(2) & L^k(3) \\
\downarrow a^0_1 & \downarrow a^1_2 & \downarrow a^2_3 & \\
L^k(1) & L^k(2) & L^k(3) &
\end{array}
$$

for $\Delta^k(k)$ with $k = 0, 1, 2, 3$. We claim that there are four direct summands of $\mathbb{Z}Z_4$ and that their graded Alperin diagrams are as follows:

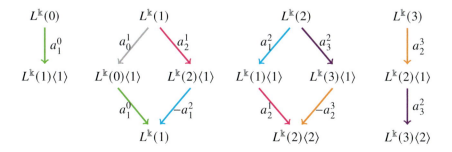

We now set about proving this claim. We first note that the algebra is non-negatively graded and that the elements of degree 0 are pairwise orthogonal idempotents. This implies that rad(ZZ_4) consists precisely of the elements of strictly positive degree. Thus, to deduce the graded Alperin diagrams, it suffices to consider the structure of $e_k ZZ_4$ for each $0 \leq k < 4$.

We now construct the graded Alperin diagram of $e_1 ZZ_4$ by exhaustion, the other cases are similar. By the idempotent relations, we know that

$$e_1 a_0^1, \quad e_1 a_2^1$$

are distinct elements of graded degree 1 in $e_1 ZZ_4$ and moreover, are the only such elements. Now, we have that

$$e_1 a_2^1 a_3^2 = 0 \quad e_1 a_0^1 a_1^0 = -e_1 a_2^1 a_1^2$$

are the only possible elements of degree 2 in $e_1 ZZ_4$ (by simply applying the relations in (6.11.16) above). There are no elements of degree 3 in ZZ_4 and so our exhaustive search is finished. Therefore $e_1 ZZ_4$ has the claimed graded Alperin diagram.

Exercise 6.11.17. Construct the graded Alperin diagrams of the direct summands of ZZ_n for $n \geq 2$.

Exercise 6.11.18. Let A be a graded weighted cellular algebra and suppose that A is generated (as a $\Bbbk$-algebra) by the set of all degree 0 and 1 elements in the grading. Further suppose that every element of degree 0 in the grading is an idempotent. Prove that

$$\mathrm{rad}_k(\Delta^k(\lambda)) = \{d_S \mid S \in \mathrm{SStd}(\lambda, -), \deg(S) = k\}.$$

6.12 Truncations, quotients, and saturation.

We begin by considering the effect of "truncating" an algebra by an idempotent. This allows us to introduce a notion of "equivalence" which is flabbier than the definition of isomorphism but which is still strong enough to preserve all the representation theoretic structures in which we are interested. This material is quite technical and we warmly invite the non-expert reader to skip to the next chapter (on first reading).

Definition 6.12.1. Let $\Bbbk$ be a field and A be a (graded) weighted cellular algebra with identity $1 = \sum_{\lambda \in \Lambda} 1_\lambda$. We suppose that $e_\Pi \in A$ is an idempotent of the form $e_\Pi = \sum_{\pi \in \Pi} 1_\pi$ for $\Pi \subseteq \Lambda$. The $\Bbbk$-algebra $A_\Pi := e_\Pi A e_\Pi$ is a cellular algebra with respect to the basis

$$\{x_{ST} := d_S^* 1_\lambda d_T \mid S, T \in \mathrm{SStd}(\lambda, -), \lambda \in \Lambda^+, \mu, \nu \in \Pi\}$$

and we say that the algebra A_Π is the truncation of A with respect to Π. We denote the cell modules of A_Π by $\Delta_\Pi^k(\lambda) := \Delta^k(\lambda) e_\Pi \neq 0$.

6.12 Truncations, quotients, and saturation. 151

The truncated algebra is cellular and so we can employ all the usual cellular machinery to the new basis. In particular, the cell modules $\Delta_\Pi^\Bbbk(\lambda)$ come equipped with bilinear forms, which can be easier to calculate *because they are usually smaller in rank* (as we have truncated, or "killed", some of the basis vectors of $\Delta^\Bbbk(\lambda)$).

Example 6.12.2. We set $\Pi = \{(2,0), (1,1)\} \subseteq \Lambda_4$. We truncate using the idempotent $e_\Pi = 1_{(2,0)} + 1_{(1,1)}$ in order to obtain the 5-dimensional algebra with cellular basis

$$\begin{bmatrix} 0 & 0 & 0 & 0 \\ 0 & 1 & 0 & 0 \\ 0 & 0 & 1 & 0 \\ 0 & 0 & 0 & 0 \end{bmatrix} \begin{bmatrix} 1 & 0 & 0 & 0 \\ 0 & 0 & 0 & 0 \\ 0 & 0 & 0 & 0 \\ 0 & 0 & 0 & 0 \end{bmatrix} \begin{bmatrix} 0 & 0 & 0 & 0 \\ 1 & 0 & 0 & 0 \\ 1 & 0 & 0 & 0 \\ 0 & 0 & 0 & 0 \end{bmatrix} \begin{bmatrix} 0 & 1 & 1 & 0 \\ 0 & 0 & 0 & 0 \\ 0 & 0 & 0 & 0 \\ 0 & 0 & 0 & 0 \end{bmatrix} \begin{bmatrix} 0 & 0 & 0 & 0 \\ 0 & 0 & 1 & 0 \\ 0 & 1 & 0 & 0 \\ 0 & 0 & 0 & 0 \end{bmatrix}$$

which we view as (4×4)-matrices in $S^\Bbbk(2)$ with non-zero entries only in the top leftmost (3×3) corner. For easy comparison with Example 5.6.2, we note that $e_\Pi S^\Bbbk(2) e_\Pi$ is the subset of matrices pictured as follows:

We note that $e_\Pi S^\Bbbk(2) e_\Pi$ and ZZ_2 are isomorphic as $\mathbb{F}_2$-algebras (see Example 5.4.2).

Example 6.12.3. We set $\Pi = \{(4,0), (3,1)\} \subseteq \Lambda_4$. The truncated algebra $e_\Pi S^\Bbbk(4) e_\Pi$ is 5-dimensional with (cellular) basis given by

$$\{x_{\mathsf{ST}} \mid \mathsf{S} \in \mathrm{SStd}(\lambda, \mu), \mathsf{S} \in \mathrm{SStd}(\lambda, \nu) \text{ for } \lambda, \mu, \nu \in \{(4,0), (3,1)\}\}.$$

These can be thought of as the elements from the "upper left corner" of the basis given in Example 5.6.10. Explicitly, the basis is given by the elements

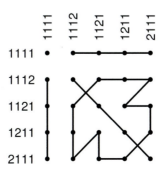

viewed as (16×16)-matrices with non-zero entries only in the (5×5) top left corner.

Example 6.12.4. We set $\Pi = \{(3,1),(2,2)\} \subseteq \Lambda_4 = \{(4,0),(3,1),(2,2),(1,3),(0,4)\}$. The truncated algebra $e_\Pi S^{\Bbbk}(4)e_\Pi$ is 9-dimensional with basis pictured as follows

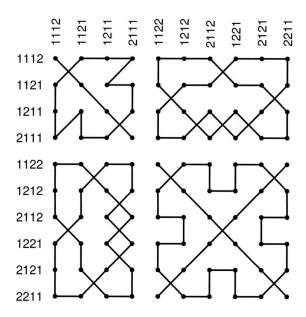

The cell modules of this algebra are the 1-dimensional module $\Delta^{\Bbbk}(2,2)e_\Pi$, the 2-dimensional module $\Delta^{\Bbbk}(3,1)e_\Pi$, and the 2-dimensional module $\Delta^{\Bbbk}(4,0)e_\Pi$.

Exercise 6.12.5. Let $\Bbbk = \mathbb{F}_2$ and let $\Pi = \{(4,0),(3,1)\} \subseteq \Lambda_4$. Calculate the Gram matrices of the cell modules and hence the decomposition matrix of $e_\Pi S^{\Bbbk}(4)e_\Pi$. Prove that $e_\Pi S^{\Bbbk}(2)e_\Pi$ and $\mathbb{Z}\mathbb{Z}_2$ are isomorphic as $\Bbbk$-algebras.

Exercise 6.12.6. Let $\Bbbk = \mathbb{F}_2$ and consider the subset

$$\Pi = \{(3,1),(2,2)\} \subseteq \Lambda_4 = \{(4,0),(3,1),(2,2),(1,3),(0,4)\}.$$

By calculating the intersection forms in the truncated cellular basis (and a little extra work) prove that the decomposition matrix of $e_\Pi S^{\Bbbk}(4)e_\Pi$ is given by

	$L^{\Bbbk}(2,2)e_\Pi$	$L^{\Bbbk}(3,1)e_\Pi$
$\Delta^{\Bbbk}(2,2)e_\Pi$	1	0
$\Delta^{\Bbbk}(3,1)e_\Pi$	1	1
$\Delta^{\Bbbk}(4,0)e_\Pi$	1	1

Show that (even though they have the same composition factors) the $e_\Pi S^{\Bbbk}(4)e_\Pi$-modules $\Delta^{\Bbbk}(3,1)e_\Pi$ and $\Delta^{\Bbbk}(4,0)e_\Pi$ are not isomorphic as A-modules.

6.12 Truncations, quotients, and saturation.

Our first port-of-call is to identify the simple modules of our truncated algebras (our truncation might kill some simples, as we have already seen). Of course, since our algebra is cellular, this simply means calculating the radicals of the *truncated* bilinear forms and applying the general theory of cellular algebras. For the remainder of this chapter, we focus on particularly nice truncations for which no such linear algebra calculations are required.

Definition 6.12.7. We say that a subset $\Pi \subseteq \Lambda$ is **saturated** if for all $\lambda \in \Lambda^+ \setminus (\Lambda^+ \cap \Pi)$ and $\mu \in \Pi$ we have that $\mathrm{SStd}(\lambda, \mu) = \emptyset$.

Theorem 6.12.8. *Let A be a weighted cellular algebra for some field $\Bbbk$ and totally ordered set $\Lambda^+ \subseteq \Lambda$. Given $\Pi \subseteq \Lambda$ a saturated subset, the modules*

$$\{L_\Pi^\Bbbk(\lambda)\langle j \rangle := L^\Bbbk(\lambda)\langle j \rangle e_\Pi \mid \lambda \in \Lambda^+ \cap \Pi, j \in \mathbb{Z}\}$$

form a complete set of non-isomorphic graded simple A_Π-modules.

Proof. By the definition of a saturated set, we have that $\Delta^\Bbbk(\lambda)e_\Pi = 0$ for $\lambda \in \Lambda^+ \setminus (\Lambda^+ \cap \Pi)$ and in particular $L^\Bbbk(\lambda)e_\Pi = 0$ for $\lambda \notin \Pi$. Now, since $1_\lambda e_\Pi = 1_\lambda \in \Delta_\Pi^\Bbbk(\lambda)$ for $\lambda \in \Lambda^+ \cap \Pi$, we have that $\mathrm{rad}(\Delta_\Pi^\Bbbk(\lambda)) \neq \Delta_\Pi^\Bbbk(\lambda)$ and the result follows. □

Corollary 6.12.9. *Let A be a (graded) weighted cellular algebra for some field $\Bbbk$ and totally ordered set $\Lambda^+ \subseteq \Lambda$. Let $\Pi \subseteq \Lambda$ be a saturated set and let $e_\Pi = \sum_{\pi \in \Pi} 1_\pi$. Let M be an A-module and*

$$0 \subset M_1 \subset M_2 \subset \cdots \subset M_k = M$$

be a chain of (graded) submodules of M with $M_i/M_{i-1} \cong L^\Bbbk(\lambda^i)\langle j_i \rangle$ with $\lambda^1, \ldots, \lambda^k \in \Lambda_\Bbbk^0$, $j_i \in \mathbb{Z}$ for $1 \leq i \leq k$. We have that

$$0 \subseteq M_1 e_\Pi \subseteq M_2 e_\Pi \subseteq \cdots \subseteq M_k e_\Pi = M e_\Pi$$

is a chain of submodules of $M e_\Pi$ with

$$M_i e_\Pi / M_{i-1} e_\Pi \cong \begin{cases} L_\Pi^\Bbbk(\lambda^i)\langle j_i \rangle & \text{for } \lambda^i \in \Pi \\ 0 & \text{for } \lambda^i \notin \Pi. \end{cases}$$

Proof. That truncating gives rise to a chain of strict submodules with the required simple subquotients follows directly from Theorem 6.12.8. □

Applying this corollary to the cell-modules of A, we realise that the decomposition matrix of A_Π is a submatrix of the decomposition matrix of A.

Corollary 6.12.10. *Let A be a (graded) weighted cellular $\Bbbk$-algebra with respect to Λ. Let $\Pi \subseteq \Lambda$ be a saturated subset and let $e_\Pi = \sum_{\pi \in \Pi} 1_\pi$. For $\lambda, \mu \in \Pi$, we have that*

$$[\Delta^\Bbbk(\lambda) : L^\Bbbk(\mu)] = [\Delta_\Pi^\Bbbk(\lambda) : L_\Pi^\Bbbk(\mu)]$$

and the graded version of this statement also holds:

$$\sum_{k \in \mathbb{Z}} [\Delta^{\Bbbk}(\lambda) : L^{\Bbbk}(\mu)\langle k \rangle] q^k = \sum_{k \in \mathbb{Z}} [\Delta_\Pi(\lambda) : L_\Pi(\mu)\langle k \rangle] q^k.$$

In other words, the (graded) decomposition matrix of A_Π is obtained from that of A by deleting the rows and columns labelled by $\lambda \notin \Pi$.

Definition 6.12.11. Let $\Bbbk$ be a field and A be a (graded) weighted cellular algebra with identity $1 = \sum_{\lambda \in \Lambda} 1_\lambda$. For any subset $\Lambda^+ \subseteq \Pi \subseteq \Lambda$ we say that the algebras A and $A_\Pi = e_\Pi A e_\Pi$ are **(graded) Morita equivalent**.

As a corollary of the above, and the definition of graded Morita equivalence, we have the following:

Corollary 6.12.12. *Let A be a (graded) weighted cellular $\Bbbk$-algebra with respect to Λ. Let $\Lambda^+ \subseteq \Pi \subseteq \Lambda$ and set $e_\Pi = \sum_{\pi \in \Pi} 1_\pi$. The $\Bbbk$-algebras A and $e_\Pi A e_\Pi$ have the same (graded) decomposition matrix.*

Example 6.12.13. Let $\Bbbk = \mathbb{F}_2$. The $\Bbbk$-algebras ZZ_2 and $S^{\Bbbk}(2)$ are graded Morita equivalent. We have already seen that these algebras have the same (graded) decomposition matrix. In fact, comparing Figure 6.7 and Exercise 6.11.17 we realise that the direct summands of ZZ_2 and $S^{\Bbbk}(2)$ have the same graded Alperin diagrams, this is part of a more general phenomenon, as we will soon see.

We claim that (graded) Morita equivalence preserves all submodule structures and hence "all representation theoretic information"; therefore instead of considering algebras "up to isomorphism" we will consider algebras "up to Morita equivalence".

Corollary 6.12.14. *Let A be a (graded) weighted cellular $\Bbbk$-algebra with respect to $\Lambda^+ \subseteq \Lambda$. Let $\Lambda^+ \subseteq \Pi \subseteq \Lambda$ so that A_Π and A are (graded) Morita equivalent. Let M be an A-module and*

$$0 \subset M_1 \subset M_2 \subset \cdots \subset M_k = M$$

be a chain of (graded) submodules of M with $M_i/M_{i-1} \cong L^{\Bbbk}(\lambda^i)\langle j_i \rangle$ with $\lambda^1, \ldots, \lambda^k \in \Lambda^0_{\Bbbk}$ and $j_i \in \mathbb{Z}$ for $1 \leq i \leq k$. We have that

$$0 \subseteq M_1 e_\Pi \subseteq M_2 e_\Pi \subseteq \cdots \subseteq M_k e_\Pi = M e_\Pi$$

is a chain of submodules of $M e_\Pi$ with

$$M_i e_\Pi / M_{i-1} e_\Pi \cong L^{\Bbbk}_\Pi(\lambda^i)\langle j_i \rangle$$

for $\lambda^1, \ldots, \lambda^k \in \Lambda$.

Proof. One direction was already proven in Corollary 6.12.9. For the converse, we let $V \subset W$ be an $e_\Pi A e_\Pi$-submodule with W/V some simple $e_\Pi A e_\Pi$-module, which we know is isomorphic to some $L^{\Bbbk}(\lambda) e_\Pi$ by Theorem 6.12.8. Then $V = V e_\Pi$ and $W = W e_\Pi$ and $V e_\Pi A \subseteq W e_\Pi A$ with $W e_\Pi A / V e_\Pi A \cong L^{\Bbbk}(\lambda) e_\Pi$ as required (since no simple module is killed by truncation by e_Π). □

In particular, if A and A_Π are (graded) Morita equivalent then truncation preserves (graded) Alperin diagrams.

6.12 Truncations, quotients, and saturation.

Finally, we make some (rather trivial) observations about quotients of cellular algebras. This idea was already used in the proof of Theorem 6.2.20 and so we do not linger too long over the details.

Proposition 6.12.15. *Let A be a (graded) weighted cellular $\Bbbk$-algebra with respect to $\Lambda^+ \subseteq \Lambda$. Let $\Pi \subseteq \Lambda$ be a saturated subset, then the algebra $A/Ae_\Pi A$ is a (graded) weighted cellular $\Bbbk$-algebra with respect to $\Lambda \setminus (\Lambda^+ \cap \Pi)$. The decomposition matrix of this algebra is obtained from that of A by deleting the rows and columns corresponding to Π.*

Proof. The proof is left as an exercise for the reader. □

Remark 13. Later in this book, we will often first truncate by a saturated idempotent and then quotient out by a saturated subalgebra in order to obtain a subquotient (cellular) algebra.

Further Reading on binary Schur algebras. This chapter focussed on the binary Schur algebra as a cellular algebra, using ideas from highest weight theory; we chose to skip aspects of the usual Lie theoretic story, for which we refer to [EW06, Hum78, FH91]. The beautiful Sierpinski-like combinatorics of the binary Schur algebra seen in this chapter was developed by Carter–Cline and Deriziotis [CC76, Der81] and subsequently generalised by James [Jam76] and Doty [Dot85, Dot89]; this has been visualised as part of Joel Gibson's *Lievis* project. The representation theoretic and cohomological structure of the binary Schur algebras has been richly developed by Cline [Cli79] and Erdmann and her collaborators [Erd93, Erd94, Erd95, CE00, EH02]; the definitive result in this area is Parker's calculation of all higher cohomological interactions between simple and cell modules [Par07]. Despite many years of study, the binary Schur algebras continue to be of interest as many questions concerning their structure admit complete solutions with rich combinatorial structure [DH05, Mar18b, DM20, PW21, TW21, MW22, TW22].

Further Reading on Cellularity. It is worth remarking that our definition of cellular is not precisely equivalent to the usual definition found in the literature [GL96]. The definition here is closer to the definition of cyclic cellular in [GG13] and was inspired by ideas of Enyang and Goodman [EG17, BEG18a]; these papers all emphasise the "factorisation" of the cellular basis through an ∗-fixed element of the cell-ideal. We believe that our definition is easier for newcomers to work with (and we are unaware of any interesting algebras which are cellular in the sense of [GL96] that are not cellular in our sense). Our "weighted cellular algebras" are a slimmed-down version of the "graded cellular algebras with a highest weight theory" from [BCS17] and are an attempt to incorporate idempotent-rich presentations into the cellular definition. Our "weighted cellular algebras" are all examples of quasi-hereditary algebras and we refer the reader to the excellent treatment of such algebras in [Don98, Appendix].

The notion of cellularity has also been generalised from algebras to categories [Wes09] and this has subsequently underdone a parallel development into the notion of (strongly) object-adapted cellular categories [EL16, EMTW20], incorporating the factorisation through an "object" (the analogue of an idempotent) into the cellular definition. For such categories (the Temperley–Lieb category and diagrammatic Soergel bimodules providing the motivating examples) this factorisation often has a particularly enjoyable visualisation in terms of "cinched" or "doubly trapezoidal" diagrams, which we will discuss later in the book.

The first prototypical examples of cellular bases to appear in the literature (pre-dating the definition of cellularity) were Murphy's basis of Hecke algebras of symmetric groups [Mur92, Mur95] and Green's codeterminant basis of the Schur algebra [Gre93]. The Kazhdan–Lusztig basis of the Hecke algebra of the symmetric group differs from Murphy's basis by uni-triangular transformation [Gec06] and therefore is also proven to be an example of a cellular basis; however this seems to be a particular quirk of the finite symmetric group (or at least, conjecturally, for finite Weyl groups [Gec07]) — Kazhdan–Lusztig bases are not, generally speaking, cellular and their axioms are of a completely different flavour [KL79]. However, the "Murphy-type" cellular bases *do generalise* to give many explicit (graded) cellular bases of all the Hecke algebras of the infinite families of complex reflection groups [DJM98, HM10, CGG12, Web17b, Bow22]; these bases arise as the shadows of richer (cellular) structures on (diagrammatic) Cherednik algebras.

There are an abundance of other cellular algebras which can be studied in a similar fashion to those considered here: there are (graded) Murphy-like constructions of cellular bases of the Brauer and walled Brauer algebras [GL96, Mar15, CDM09a, CDDM08, CD11], the endomorphism algebras of symplectic and orthogonal groups acting on tensor space [BEG18b], and blob algebras [MW00b]. There have been many generalisations of this notion of cellularity (each with particular motivations and examples in mind), we refer to [EG17, ET21, BS17, CMPX06, Tub24] for more details. For more on the general framework of cellular algebras in the spirit of this chapter, we refer to [Mat99]. In particular, a concrete way of constructing the radical layers of cell modules of binary Schur algebras is via their *Jantzen filtrations*, which are discussed in detail in [Mat99, Chapter 5].

Further Reading on Morita equivalences. In the final section of this chapter we sacrificed generality in favour of concreteness. There is a far more general notion of Morita equivalence, which applies to all finite dimensional $\Bbbk$-algebras (and beyond), we refer to [Ben98, Section 2.2] for more details.

Part III
Combinatorics

> "One starts out in life trying to do mathematics, and winds up doing combinatorics."
>
> Ian Macdonald

Combinatorics is the mathematics of "discrete phenomena", or "counting things". One can choose to read Ian Macdonald's famous quotation in two ways: either as a critique of combinatorics and its practitioners or as acknowledgement that, whatever your mathematical outlook, you will almost certainly encounter some form of discrete phenomena — whether you're a physicist interested in quantum behaviour or a biologist examining the twist and coils of DNA around itself — things are never as smooth as you think they are!

We are combinatorial representation theorists, and so the discrete objects that we count are simple and indecomposable representations, their dimensions, their simple extensions, their composition factor multiplicities, their tensor product decomposition multiplicities... These questions are often impossibly difficult to answer, or "wild". However, if you can answer an interesting question in non-semisimple representation theory, the answer will almost certainly be given in terms of the combinatorics of Kazhdan–Lusztig polynomials.

Kazhdan and Lusztig introduced their eponymous polynomials in order to deal with interesting problems across Lie theory, such as composition series of Verma modules, geometry of Schubert varieties, and primitive ideals in enveloping algebras [KL79, KL80, Lus80a]. Deodhar profoundly reimagined their theory in a much more combinatorial set-up [Deo87, Deo90], which we will introduce here. We begin by playing around with quantum binomials and graphs/posets and asking ourselves what we can determine about their structure. We then define the Kazhdan–Lusztig polynomials by counting paths in these graphs (with degree); we calculate these polynomials explicitly by borrowing ideas from the oriented Temperley–Lieb algebra. In Chapter 8 we reveal the extra complications encountered in the general definition of Kazhdan–Lusztig polynomials of arbitrary Bruhat graphs.

Deodhar's definition of Kazhdan–Lusztig polynomials is via two steps: first one counts certain paths in the Bruhat graph in an entirely combinatorial fashion, recording this information in a square uni-triangular matrix; then one factorises this matrix

Part III Combinatorics 159

in a natural fashion. The Kazhdan–Lusztig positivity conjecture states that the polynomials obtained in this fashion have non-negative coefficients.

Such positivity phenomena always surprise and intrigue combinatorists of all stripes: whilst "counting paths" must obviously produce polynomials with positive coefficients, there is no a priori reason why the matrix-factorisation step in this definition should honour this positivity. Proving such positivity phenomena usually involves finding a geometric or categorical interpretation for the coefficients in question. This case is no exception: geometric constructions explaining this positivity were found for the Kazhdan–Lusztig polynomials of Weyl groups in the 1980s, but generalising this to all Coxeter groups took another 30 years!

Much of the combinatorial structure of Kazhdan–Lusztig polynomials remains mysterious. Indeed, proving positivity of the coefficients is, in some sense, the most basic combinatorial question one can ask! Chief amongst these mysteries is the famous Lusztig–Dyer "combinatorial invariance" conjecture, which proposes that Kazhdan–Lusztig polynomials can be "determined locally" by the strong Bruhat graph. In the next two chapters we discuss this and other, smaller, mysteries and enjoyable combinatorial properties of the Kazhdan–Lusztig polynomials.

Chapter 7
Catalan combinatorics within Kazhdan–Lusztig theory

In the previous chapter, we studied the representation theory of binary Schur algebras in the combinatorial language of coloured Pascal triangles. The central problem of modular representation theory is to extend this analysis to more general algebraic objects, such as symmetric groups, (non-binary) Schur algebras, Hecke categories, as well as objects which are beyond the realms of this book (such as finite groups of Lie type, Lie algebras, Kac–Moody algebras, and quantum groups). Of course, this requires a suitably rich combinatorial language to replace our earlier use of coloured Pascal triangles; this new language is provided by Kazhdan–Lusztig theory.

Kazhdan–Lusztig theory is the study of walks in coset graphs for parabolic Coxeter systems. In this chapter, we will focus on the simplest parabolic Coxeter systems (those of $\mathfrak{S}_m \times \mathfrak{S}_n \leqslant \mathfrak{S}_{m+n}$) as a warm-up to the general case (which we consider in the next chapter). We will attempt to motivate the study of these parabolic Coxeter systems, their coset graphs, and their Kazhdan–Lusztig polynomials solely through the beautiful combinatorial properties they exhibit.

7.1 Quantum binomial coefficients

We can *quantise* the binomial coefficients of Pascal's triangle in order to obtain a family of combinatorially rich polynomials, which we now introduce. We define the quantum integers by setting $[0]_q = 1$ and for $n \in \mathbb{Z}_{\geqslant 0}$, we set

$$[n]_q = q^n + q^{n-1} + \cdots + q + 1 = \frac{1-q^n}{1-q}$$

and we define the quantum factorials in the natural manner

$$[n]_q! = [n]_q [n-1]_q \ldots [2]_q [1]_q.$$

Now we are able to define the quantum binomial coefficients as follows:

$$\begin{bmatrix} n \\ k \end{bmatrix}_q = \frac{[n]_q!}{[k]_q![n-k]_q!}.$$

While there are many interesting results concerning quantum binomial coefficients, we will only require the very simplest result regarding these coefficients: the quantised Pascal triangle rules.

Proposition 7.1.1 (The q-Pascal identities). *We have that*

$$q^k \begin{bmatrix} n-1 \\ k \end{bmatrix}_q + \begin{bmatrix} n-1 \\ k-1 \end{bmatrix}_q = \begin{bmatrix} n \\ k \end{bmatrix}_q = \begin{bmatrix} n-1 \\ k \end{bmatrix}_q + q^{n-k} \begin{bmatrix} n-1 \\ k-1 \end{bmatrix}_q$$

for all $0 \leqslant k \leqslant n$.

Proof. We prove the first equality, as the second is very similar. By the definitions above, we have that

$$\begin{bmatrix} n \\ k \end{bmatrix}_q = \frac{[n]_q!}{[k]_q![n-k]_q!} = \frac{[n]_q}{[n-k]_q} \cdot \frac{[n-1]_q!}{[k]_q![(n-1)-k]_q!} = \frac{1-q^n}{1-q^{n-k}} \begin{bmatrix} n-1 \\ k \end{bmatrix}_q$$

and similarly

$$\begin{bmatrix} n \\ k \end{bmatrix}_q = \frac{[n]_q!}{[k]_q![n-k]_q!} = \frac{[n]_q}{[k]_q} \cdot \frac{[n-1]_q!}{[k-1]_q![n-k]_q!} = \frac{1-q^n}{1-q^k} \begin{bmatrix} n-1 \\ k-1 \end{bmatrix}_q.$$

Putting these two statements together we deduce that

$$\frac{1-q^k}{1-q^{n-k}} \cdot \begin{bmatrix} n-1 \\ k \end{bmatrix}_q = \begin{bmatrix} n-1 \\ k-1 \end{bmatrix}_q. \tag{7.1.2}$$

Finally, we have that

$$\frac{1-q^n}{1-q^{n-k}} = \frac{(1-q^k) + (q^k - q^n)}{1-q^{n-k}} = \frac{1-q^k}{1-q^{n-k}} + q^k. \tag{7.1.3}$$

We can now substitute (7.1.3), followed by (7.1.2) into the first equation of the proof and hence obtain

$$\begin{bmatrix} n \\ k \end{bmatrix}_q = q^k \begin{bmatrix} n-1 \\ k \end{bmatrix}_q + \frac{1-q^k}{1-q^{n-k}} \begin{bmatrix} n-1 \\ k \end{bmatrix}_q = q^k \begin{bmatrix} n-1 \\ k \end{bmatrix}_q + \begin{bmatrix} n-1 \\ k-1 \end{bmatrix}_q$$

as required. □

Example 7.1.4. We have that

$$\begin{bmatrix} 4 \\ 1 \end{bmatrix}_q = 1 + q + q^2 + q^3 \qquad \begin{bmatrix} 4 \\ 2 \end{bmatrix}_q = q^4 + q^3 + 2q^2 + q + 1.$$

7.2 The poset of partitions in a rectangle

Given $m, n \in \mathbb{Z}_{>0}$ we consider an $(m \times n)$-rectangle tiled by (1×1)-tiles. We depict this rectangle so that its edges are at 135° and 45° to the horizontal axis. We label the tiles according to their x-coordinates as shown in Figure 7.1. We say that a pair of tiles within this $(m \times n)$-rectangle are **neighbouring** if they meet at an edge (which necessarily has an angle of 45° or 135°). Given a pair of neighbouring tiles X and Y, we say that Y **supports** X if X appears above Y. We say that a tile, X, is **fully supported** if every tile within the $(m \times n)$-rectangle which can support X does support X. We say that a collection of tiles, λ, is a **tile-partition** if every tile in λ is fully supported. We depict a tile-partition λ by colouring the tiles of λ, as in the rightmost diagram in Figure 7.1.

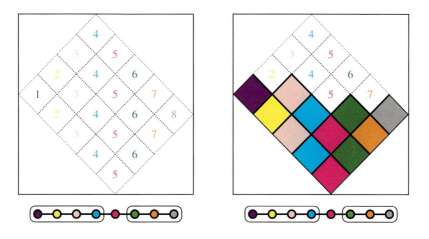

Fig. 7.1: On the left we depict the (5×4)-rectangle and within each tile, we depict the x-coordinate of said tile. On the right we draw a tile-partition for $\lambda \in \mathscr{P}_{5,4}$, we do this by colouring in the tiles belonging to λ. We associate to a tile with x-coordinate $i \in \mathbb{Z}_{>0}$, the Coxeter generator $s_i \in \mathfrak{S}_8$. (Note that the size of the rectangle determines the unique Coxeter generator not in the parabolic.)

Given two tile-partitions λ and μ we write $\lambda \subseteq \mu$ if every tile in λ also belongs to μ. We let $\mathscr{P}_{m,n}$ be the set of all tile-partitions in an $(m \times n)$-rectangle under the partial ordering $\subseteq$. If $\lambda \subseteq \mu$, we let $\mu \setminus \lambda$ be the set-difference between these collections of these tiles. We write $\square \in \mathrm{Add}(\lambda)$ if $\lambda \cup \{\square\}$ is a tile-partition. Similarly, we write $\square \in \mathrm{Rem}(\mu)$ if $\mu \setminus \{\square\}$ is a tile-partition. The Hasse graph of $\mathscr{P}_{m,n}$ is the graph with vertices on the kth level given by tile-partitions of k for $0 \leq k \leq mn$ and with coloured edges $\lambda \longrightarrow \mu$ for λ on the kth and μ on the $(k+1)$th level, if and only if $\mu = \lambda \cup \{\square\}$. An example of Hasse graph of the poset $\subseteq$ is depicted in Figure 7.2.

164   7 Catalan combinatorics within Kazhdan–Lusztig theory

We often wish to depict $\lambda \in \mathcal{P}_{m,n}$ in a non-pictorial fashion. To do this, we split λ into n distinct north-west-to-south-easterly diagonals and we record the number of boxes in each such diagonal. We hence obtain a weakly-decreasing sequences $(\lambda_1, \lambda_2, \ldots, \lambda_n)$ such that $0 \leq \lambda_1 + \lambda_2 + \cdots + \lambda_n \leq mn$. For example, the tile-partition λ in Figure 7.1 can be recorded as $\lambda = (5, 4, 2, 1)$.

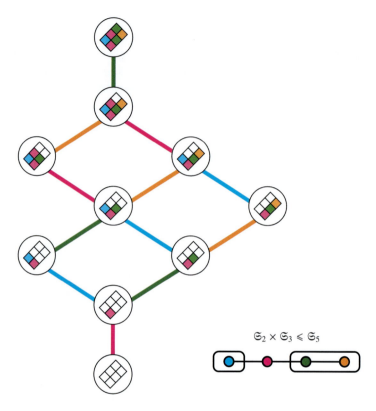

Fig. 7.2: The partial ordering $\subseteq$ on tile partitions with $m = 2$ and $n = 3$. Each coloured edge corresponds to an addable/removable node of the same colour.

One of the first things one might ask about a poset is exactly *how many* vertices appear on *each level* of the poset; we arrange this information into the *rank generating function* which counts these vertices from the top, mnth level, downwards:

Theorem 7.2.1. *The poset $\mathcal{P}_{m,n}$ has rank generating function*

$$\rho(\mathcal{P}_{m,n}) := \sum_{\lambda \in \mathcal{P}_{m,n}} q^{mn - \ell(\lambda)} = \begin{bmatrix} m + n \\ m \end{bmatrix}_q.$$

Proof. It will suffice to check, by induction on $m, n \geq 1$, that the rank-generating function satisfies the q-Pascal identity:

7.2 The poset of partitions in a rectangle

$$\rho(\mathscr{P}_{m,n}) = \rho(\mathscr{P}_{m-1,n}) + q^m \cdot \rho(\mathscr{P}_{m,n-1}) \tag{7.2.2}$$

along with the inductive base case where $m = n = 1$, $\rho(\mathscr{P}_{1,1}) = 1 + q$. To prove (7.2.2), we note that

$$\mathscr{P}_{m,n} = \{\lambda \in \mathscr{P}_{m,n} \mid \lambda_1 < m\} \sqcup \{\lambda \in \mathscr{P}_{m,n} \mid \lambda_1 = m\}$$

and we have bijections

$$\mathscr{P}_{m-1,n} \leftrightarrow \{\lambda \in \mathscr{P}_{m,n} \mid \lambda_1 < m\} \qquad \mathscr{P}_{m,n-1} \leftrightarrow \{\lambda \in \mathscr{P}_{m,n} \mid \lambda_1 = m\}$$

where the first bijection is trivial, the second is given by $(\lambda_1, \ldots, \lambda_{n-1}) \mapsto (m, \lambda_1, \ldots, \lambda_{n-1})$. (Note that the partition $(m, \lambda_1, \ldots, \lambda_{n-1})$ has first row equal to m and this accounts for the q^m in (7.2.2).) The result follows. $\square$

Example 7.2.3. The poset in Figure 7.2 has rank-generating function $\rho(\mathscr{P}_{2,3}) = 1 + q + 2q^2 + 2q^3 + q^4 + q^5$.

Having calculated the rank generating function, we can ask what nice properties this polynomial might satisfy. Given a polynomial

$$f(x) = a_0 + a_1 x + a_2 x^2 + \cdots + a_{n-2} x^{n-2} + a_{n-1} x^{n-1} + a_n x^n$$

we say that $f(x)$ is palindromic if its coefficients are symmetric in the sense that $a_i = a_{n-i}$ for $0 \leq i \leq n$; we say that $f(x)$ is unimodal if

$$a_0 \leq a_1 \leq \ldots \leq a_{k-1} \leq a_k \geq a_{k+1} \geq \ldots \geq a_{n-1} \geq a_n$$

for some $0 \leq k \leq n$.

Proposition 7.2.4. *The function $\rho(\mathscr{P}_{m,n})$ is palindromic and unimodal.*

Proof. The unimodality of quantum binomial coefficients was proven by Sylvester in 1878 (see [Sta80] for more details). The palindromic property follows from the map $(\lambda_1, \lambda_2, \ldots, \lambda_n) \mapsto (m - \lambda_n, \ldots, m - \lambda_2, m - \lambda_1)$ which is a bijection between the vertices on levels ℓ and $mn - \ell$ of the graph. $\square$

We are interested in the the subposets $\mathscr{P}_{m,n}^{\subseteq \mu} = \{\mu \in \mathscr{P}_{m,n} \mid \lambda \subseteq \mu\}$; whose rank generating functions we denote by $\rho(\mu) := \sum_{\lambda \subseteq \mu} q^{\ell(\mu) - \ell(\lambda)}$ (counting from the top downward, rather than the bottom upward). These polynomials are neither palindromic nor unimodal, for example, $\rho(8, 8, 4, 4)$ is equal to

$$1 + 2q + 5q^2 + 8q^3 + 14q^4 + 18q^5 + 24q^6 + 27q^7 + 31q^8 + 30q^9 + 31q^{10} + \cdots + q^{23} + q^{24}.$$

(The reader is invited to calculate the ... terms themselves!) We would say that the polynomials $\rho(\mu)$ are, in fact, not that interesting... putting aside our disappointment, we now go in search of a family of polynomials which provide a "correction factor" for this failure of unimodality and palindromy. These polynomials will also satisfy other beautiful combinatorial properties...

7.3 Kazhdan–Lusztig polynomials

We want to study combinatorial properties of subposets of $\mathscr{P}_{m,n}$. We have already seen in Section 7.2 that the rank generating functions of these posets are not entirely satisfying, and in particular they are not unimodal or palindromic. We now introduce a new family of polynomials, which can serve as a "correction factor" for this failure of the $\rho(\mu)$ for $\mu \in \mathscr{P}_{m,n}$ to be palindromic. Rather than counting the *points* of $\mathscr{P}_{m,n}^{\subseteq \mu}$, this new family of polynomials instead counts certain *paths* in $\mathscr{P}_{m,n}^{\subseteq \mu}$, as follows.

Definition 7.3.1. For each edge of $\lambda \longrightarrow \mu$ in the Hasse graph of $\mathscr{P}_{m,n}$ with $\lambda \subseteq \mu$, we allow this edge to be traversed in four possible ways. These paths/traversals are as follows,

- (U_i^1) The 'up' move, $\lambda \to \mu$, of degree 0;
- (U_i^0) The 'failed up', $\lambda \to \lambda$, of degree 1;
- (D_i^1) The 'down', $\mu \to \lambda$, of degree 0.
- (D_i^0) The 'failed down', $\mu \to \mu$, of degree -1.

We define a path, T, of shape $\lambda \in \mathscr{P}_{m,n}$ and length ℓ to be a sequence

$$\emptyset = \lambda^{(0)} \to \lambda^{(1)} \to \cdots \to \lambda^{(\ell)} = \lambda$$

composed of up, failed-up, down, and failed-down steps in the Bruhat graph.

Example 7.3.2. Four paths in the graph $\mathscr{P}_{2,3}$ are depicted in Figure 7.3.

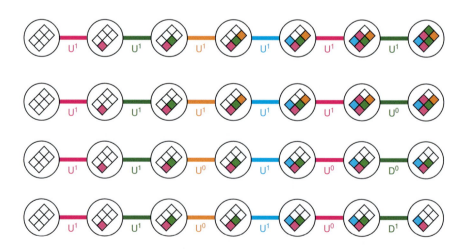

Fig. 7.3: We depict all possible paths in $\mathscr{P}_{2,3}$ of fixed colour sequence $s_2, s_3, s_4, s_1, s_2, s_3$. The degrees of these paths are 0, 1, 1, and 2 respectively. The first and last of these paths is depicted in Figure 7.4.

7.3 Kazhdan–Lusztig polynomials

Definition 7.3.3. We say that a path from $\emptyset$ to μ is reduced if it is of shortest possible length. We set $\ell(\mu)$ to be the length of a reduced path from $\emptyset$ to $\mu \in \Lambda_{m,n}$.

We let Path(λ) denote the set of all paths from the identity to the point λ in the Hasse graph of $\mathscr{P}_{m,n}$. Given $\mathsf{T} \in \text{Path}(\lambda)$, we write Shape($\mathsf{T}$) = λ and we say that the path has shape equal to λ. By picking a fixed reduced path, $\mathsf{T}_\mu \in \text{Path}(\mu)$, we are able to consider all possible paths with the same colour sequence as T_μ but which terminate at another point λ, as in Figure 7.3. For a fixed λ, we denote the set of all such paths by Path(λ, T_μ). We are now ready to define the Kazhdan–Lusztig polynomials of the poset $\mathscr{P}_{m,n}$ in terms of these paths.

Definition 7.3.4. The Kazhdan–Lusztig polynomials, $n_{\lambda,\mu}(q)$, of $(\mathscr{P}_{m,n}, \leq)$ are defined as follows

$$N := (n_{\lambda,\mu}(q))_{\lambda,\mu \in \mathscr{P}_{m,n}} \qquad n_{\lambda,\mu}(q) = \sum_{\mathsf{S} \in \text{Path}(\lambda, \mathsf{T}_\mu)} q^{\deg(\mathsf{S})}$$

for any choice of reduced path T_μ.

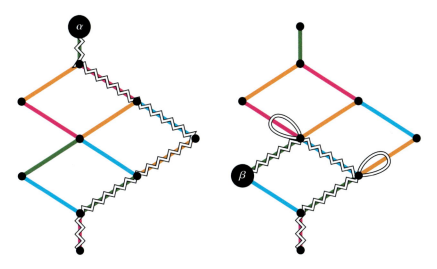

Fig. 7.4: We depict two paths in $\mathscr{P}_{2,3}$ with the same colouring sequence, $\mathsf{T}_\alpha = \mathsf{U}_2^1 \mathsf{U}_3^1 \mathsf{U}_4^1 \mathsf{U}_1^1 \mathsf{U}_2^1 \mathsf{U}_2^1 \in \text{Path}(\alpha, \mathsf{T}_\alpha)$ and $\mathsf{S} = \mathsf{U}_2^1 \mathsf{U}_3^1 \mathsf{U}_4^0 \mathsf{U}_1^1 \mathsf{U}_2^0 \mathsf{D}_3^1 \in \text{Path}(\beta, \mathsf{T}_\alpha)$.

We claim that the Kazhdan–Lusztig polynomials are of great importance in representation theory. In this chapter, we will motivate the study of these polynomials by proving that they satisfy many beautiful combinatorial properties. First, we will calculate the Kazhdan–Lusztig polynomials $n_{\lambda,\mu}(q)$ explicitly and prove that they are, indeed, independent of the choice of reduced path as claimed. Then we will show that the polynomials

$$\sum_{\lambda \subseteq \mu} q^{\ell(\mu) - \ell(\lambda)} n_{\lambda,\mu}(q)$$

are unimodal and palindromic (that is, the Kazhdan–Lusztig polynomials provide the desired "correction factor" to the failure of the rank-generating function to be unimodal and palindromic, as discussed at the end of Section 7.2).

Example 7.3.5. The Kazhdan–Lusztig polynomials for $\mathscr{P}_{2,2}$ are depicted in Table 7.1. For example, the first column can be calculated by counting the paths

$$U_2^1 U_3^1 U_1^1 U_2^1 \qquad U_2^1 U_3^1 U_1^1 U_2^0 \qquad U_2^1 U_3^0 U_1^0 D_2^0 \qquad U_2^1 U_3^0 U_1^0 D_2^1$$

of degrees 0, 1, 1, and 2 respectively.

Table 7.1: The Kazhdan–Lusztig polynomials $n_{\lambda,\mu}(q)$ for $\mathfrak{S}_2 \times \mathfrak{S}_2 \leqslant \mathfrak{S}_4$.

We notice that the polynomials $\sum_{\lambda \subseteq \mu} q^{\ell(\mu)-\ell(\lambda)} n_{\lambda,\mu}(q)$ for $\lambda, \mu \in \mathscr{P}_{2,2}$ are all palindromic. From the first two columns of Table 7.1, we see that

$$\sum_{\lambda \subseteq (2,1)} q^{3-\ell(\lambda)} n_{\lambda,\mu}(q) = 1 + 2q^2 + q^4$$
$$\sum_{\lambda \subseteq (2,2)} q^{4-\ell(\lambda)} n_{\lambda,\mu}(q) = 1 + q^2 + q^4 + q^6$$

thus the Kazhdan–Lusztig polynomials do provide the desired "correction factor" for the rank-generating functions to be palindromic, as claimed (in these examples).

Exercise 7.3.6. Calculate the complete (10×10)-table of all Kazhdan–Lusztig polynomials for $\mathscr{P}_{2,3}$. (Hint: for the first column, see Figure 7.3.)

Exercise 7.3.7. For $0 \leqslant m \leqslant n$, find every $\mu \in \mathscr{P}_{m,n}$ for which $n_{\emptyset,\mu} \neq 0$. (Hint: consider the colour of the final step in any path terminating at $\emptyset$.) Hence calculate the bottom row of the Kazhdan–Lusztig matrix.

7.4 Tile-partitions, weights, and Bruhat graphs

Those familiar with Kazhdan–Lusztig theory (and indeed those who read the introduction of this book!) will perhaps be confused as to why and how we have defined Kazhdan–Lusztig polynomials without mention of any (parabolic) Coxeter system. We now fix this oversight by recasting the ideas of the previous section in terms of coset graphs. Recall that our favourite (Coxeter) group is the symmetric group,

$$\mathfrak{S}_{m+n} = \langle s_i, 1 \leq i < m+n \mid s_i^2 = 1, (s_i s_{i+1})^3 = 1, (s_i s_j)^2 = 1 \text{ for } |i-j| > 1 \rangle.$$

We set $S_W = \{s_i \mid 1 \leq i < m+n\}$ to be the set of generators in the above presentation. This is recorded in the graph of Figure 7.5, which has a single line for each power 3 in the relations (we do not record the commuting cases, in which the powers are equal to 2). We refer to any subgroup, $P \leq W$, corresponding to a subset of the generators as a (standard) **parabolic subgroup**. Our favourite parabolics are the maximal ones $P = \mathfrak{S}_m \times \mathfrak{S}_n \leq \mathfrak{S}_{m+n} = W$ for any $m, n \in \mathbb{Z}_{>0}$. We set $S_P = \{s_i \mid 1 \leq i < m+n \text{ and } i \neq m\}$ to be the set of generators of this parabolic subgroup.

Fig. 7.5: The parabolic Coxeter graph for $\mathfrak{S}_m \times \mathfrak{S}_n \leq \mathfrak{S}_{m+n}$ for $m, n \in \mathbb{Z}_{>0}$.

Recall that an (m,n)-**weight** is a diagram obtained by labelling the points $1, \ldots, m+n$ by $\vee$ (down), $\wedge$ (up), and that we denote the set of all such (m,n)-weights by $\Lambda_{m,n}$. We draw the initial (m,n)-weight by putting m uparrows on the points $1, \ldots, m$ followed by n downarrows on the points $m+1, \ldots, m+n$. We denote the initial (m,n)-weight by $\emptyset$. For example, if $m=2$ and $n=3$ we have that

$$\emptyset = \boxed{ \wedge \wedge \vee \vee \vee } \, .$$

It is easy to see that $\mathfrak{S}_{m+n}$ permutes the set of all possible (m,n)-weights. The stabiliser of the initial (m,n)-weight is given by the parabolic $\mathfrak{S}_m \times \mathfrak{S}_n \leq \mathfrak{S}_{m+n}$ and so this gives us a diagrammatic way of constructing coset representatives.

Definition 7.4.1. The Bruhat graph for $\mathfrak{S}_m \times \mathfrak{S}_n \leq \mathfrak{S}_{m+n}$ is given as follows:

- Start by drawing the initial (m,n)-weight at the bottom.
- If applying a generator $\sigma \in S_W$ to an (m,n)-weight λ does result in a new weight μ, then record this in the next level up in the diagram and write $\sigma \in \mathrm{Add}(\lambda)$ and $\mu = \lambda + \sigma$, or $\sigma \in \mathrm{Rem}(\mu)$ and $\lambda = \mu - \sigma$, or simply $\lambda \leq \mu$.

Repeat the above until the process terminates with a graph $(\Lambda_{m,n}, \leq)$.

170   7 Catalan combinatorics within Kazhdan–Lusztig theory

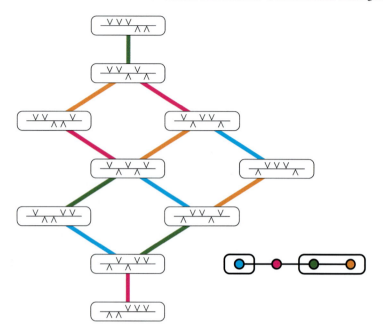

Fig. 7.6: The Bruhat graph $\Lambda_{2,3}$ for $\mathfrak{S}_2 \times \mathfrak{S}_3 \leqslant \mathfrak{S}_5$. Compare with Figure 7.2.

Our (m,n)-weights give a way of constructing cosets of $\mathfrak{S}_m \times \mathfrak{S}_n \leqslant \mathfrak{S}_{m+n}$ intuitively. Each coset is obtained by first drawing $\varnothing \in \Lambda_{m,n}$ along the southern edge of a rectangle and then drawing $\mu \in \Lambda_{m,n}$ along the northern edge of this rectangle. We then connect the ith $\vee$ along the bottom edge to the ith $\vee$ along the top edge for $1 \leqslant i \leqslant n$ and similarly for the $\wedge$ arrows. See Figure 7.7 for an example.

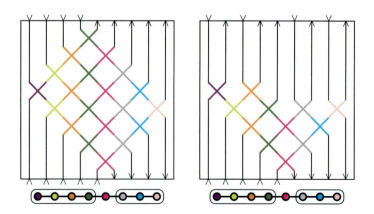

Fig. 7.7: Examples of $(5, 4)$-weights and cosets of $\mathfrak{S}_5 \times \mathfrak{S}_4 \leqslant \mathfrak{S}_9$.

7.4 Tile-partitions, weights, and Bruhat graphs

Proposition 7.4.2. *We define a map $\varphi : (\mathscr{P}_{m,n}, \subseteq) \to (\Lambda_{m,n}, \leqslant)$ by filling each tile $\square \in \lambda$ whose x-coordinate is $i \in \mathbb{Z}_{>0}$ with the Coxeter generator $s_i \in S_W$ and hence obtain a strand diagram of a permutation; apply this permutation to the initial (m, n)-weight to obtain an (m, n)-weight $\varphi(\lambda)$. The map φ is an isomorphism of graphs, that is φ is a bijection on vertices and $\lambda \subseteq \mu$ if and only if $\varphi(\lambda) \leqslant \varphi(\mu)$.*

Sketch of proof. We define a map $\psi : (\Lambda_{m,n}, \leqslant) \to (\mathscr{P}_{m,n}, \subseteq)$ as follows. We define $\psi(\wedge) = $ SW and $\psi(\vee) = $ NW and for $\lambda \in \Lambda_{m,n}$ we let $\psi(\lambda)$ denote the partition traced out by the resulting sequence of south-westerly and north-westerly moves. For example, the rightmost weight in Figure 7.7 has corresponding sequence

$$(\text{NW}, \text{SW}, \text{NW}, \text{SW}, \text{SW}, \text{NW}, \text{SW}, \text{NW}, \text{SW})$$

and we notice that the region underneath this path is the tile-partition in Figure 7.1. We leave it as an exercise for the reader to verify that φ and ψ are mutual inverses on the vertices of these graphs and respect the orderings. $\square$

In light of Proposition 7.4.2, we will refer to the polynomials from Definition 7.3.4 as the Kazhdan–Lusztig polynomials of type $\mathfrak{S}_m \times \mathfrak{S}_n \leqslant \mathfrak{S}_{m+n}$.

Exercise 7.4.3. The graph $(\Lambda_{2,2}, \leqslant)$ is depicted in Figure 7.8. Construct the isomorphic graph $(\mathscr{P}_{2,2}, \subseteq)$ and construct the coset representatives of $\mathfrak{S}_2 \times \mathfrak{S}_2 \leqslant \mathfrak{S}_4$.

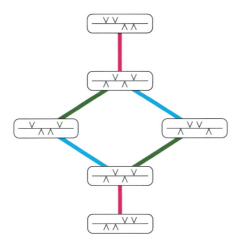

Fig. 7.8: The Bruhat graph $(\Lambda_{2,2}, \leqslant)$.

Exercise 7.4.4. Using the combinatorics of partitions, prove the following identity of quantum binomial coefficients

$$\begin{bmatrix} 2n \\ n \end{bmatrix}_q = \sum_{0 \leqslant k \leqslant n} q^{k^2} \left(\begin{bmatrix} n \\ k \end{bmatrix}_q \right)^2.$$

7.5 Oriented Temperley–Lieb combinatorics

Thus far, we have introduced the Kazhdan–Lusztig polynomials of $\mathscr{P}_{m,n}$; motivated their study via their beautiful combinatorial properties, which have been illustrated in examples; and described the connections to coset graphs and (m, n)-weight combinatorics. In this last section of this chapter, we will finally prove the claimed combinatorial properties. We do this by building on the (m, n)-weight combinatorics in the setting of the oriented Temperley–Lieb algebras of Section 5.5.

We begin by associating oriented Temperley–Lieb generators to single steps in the Bruhat graphs, as follows.

Definition 7.5.1. Let $\lambda \subseteq \mu$ and suppose that $\lambda \xrightarrow{s_i} \mu$ is a traversal of an edge in the Bruhat graph. We define the corresponding oriented Temperley–Lieb diagram, $E^i_{\lambda \to \mu}$ to be the diagram obtained by placing the weight λ along the bottom, the weight μ along the top, and then placing the corresponding the Temperley–Lieb generator E_i in between the two weights.

Example 7.5.2. Let's consider the edge $\lambda \xrightarrow{s_3} \mu$ in Figure 7.6 for

$$\lambda = \boxed{\begin{array}{c}\vee\;\;\vee\;\vee\\\wedge\;\;\wedge\end{array}} \qquad \mu = \boxed{\begin{array}{c}\vee\;\vee\;\;\vee\\\wedge\;\;\wedge\end{array}}$$

The four possible traversals of this edge correspond to the Temperley–Lieb-style elements depicted in Figure 7.9 and are of degrees $-1, 0, 0,$ and 1 respectively.

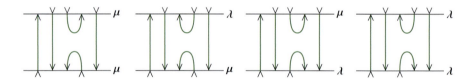

Fig. 7.9: Let λ and μ be as in Example 7.5.2. We picture the four Temperley–Lieb diagrams $E^3_{\mu \to \mu}, E^3_{\lambda \to \mu}, E^3_{\mu \to \lambda}, E^3_{\lambda \to \lambda}$ associated to the edge $\lambda \xrightarrow{s_3} \mu$.

We can now associate oriented Temperley–Lieb generators to entire paths in the Bruhat graphs, as follows.

Definition 7.5.3. Given $\mathsf{T} \in \mathrm{Path}(\lambda)$ a path of the form

$$\emptyset = \lambda^{(0)} \xrightarrow{i_1} \lambda^{(1)} \xrightarrow{i_2} \ldots \xrightarrow{i_\ell} \lambda^{(\ell)} = \lambda$$

we define

$$E_\mathsf{T} = E^{i_\ell}_{\lambda^{(\ell-1)} \to \lambda^{(\ell)}} \cdot \ldots \cdot E^{i_2}_{\lambda^{(1)} \to \lambda^{(2)}} E^{i_1}_{\lambda^{(0)} \to \lambda^{(1)}}.$$

An example is depicted in Figure 7.10.

7.5 Oriented Temperley–Lieb combinatorics

The following proposition is rather trivial, but it serves as a sign-post to where we are heading with all this Temperley–Lieb-malarky.

Proposition 7.5.4. *Given* $T \in \text{Path}(\lambda)$, *we have that* $\deg(E_S) = \deg(S)$. *Therefore the Kazhdan–Lusztig polynomials of* $\mathscr{P}_{m,n}$ *can be reinterpreted as follows:*

$$n_{\lambda,\mu}(q) = \sum_{S \in \text{Path}(\lambda, T_\mu)} q^{\deg(E_S)}$$

Proof. For paths of length 1, this follows immediately from Definitions 7.3.4 and 7.5.3. For paths of length greater than 1, the result follows by the fact that the oriented Temperley–Lieb algebra is a $\mathbb{Z}[q, q^{-1}]$-graded algebra. □

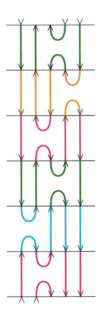

Fig. 7.10: An example of an element E_S for $S \in \text{Path}(\beta, T_\alpha)$ as in Figure 7.4. Notice that we have applied commutation rules to the blue and green generators in order to emphasise that this element is independent of the path chosen for T_α. This diagram is of degree 2 (which is equal to the degree of the underlying path!).

Exercise 7.5.5. Construct the elements E_S for each $S \in \text{Path}(\beta, T_\alpha)$ as in the four paths (of degree 0, 1, 1, and 2) listed in Figure 7.3.

Rather than repeating the arduous iterative construction of Definition 7.5.3 ad infinitum, we would like to have a simple combinatorial algorithm for constructing these elements, particularly in the case of reduced paths.

Definition 7.5.6. We let $e_\mu \in \mathrm{TL}^{\uparrow\downarrow}_{m,n}(q)$ be the element with northern weight μ, southern weight $\varnothing$, and degree 0. This can be constructed as follows:

- The northern half of the diagram, denoted $\overline{\mu}$, is obtained repeatedly connecting neighbouring northern vertices (in the sense that they are next to each other or only have vertices already connected by a cup between them) labelled by $\vee$ and $\wedge$ by northern anti-clockwise cups. We set $d(\mu) \in \mathbb{Z}_{\geqslant 0}$ to be the total number of such cups. Then attach vertical rays to the remaining $\wedge$ and $\vee$ arrows.
- The southern half of the diagram consists of precisely $d(\mu)$ concentric caps connecting neighbour vertices of the weight $\varnothing$ and all other strands propagating.

Examples of the diagrams produced by Definition 7.5.6 are depicted in Figure 7.11. We now reconcile our closed and iterative definitions as follows:

Proposition 7.5.7. *Given* T_μ *any reduced path for* $\mu \in \mathscr{P}_{m,n}$, *we have that* $E_{\mathsf{T}_\mu} = e_\mu$.

Proof. This follows from the correspondence between Temperley–Lieb diagrams and tilings in Theorem 5.2.3. □

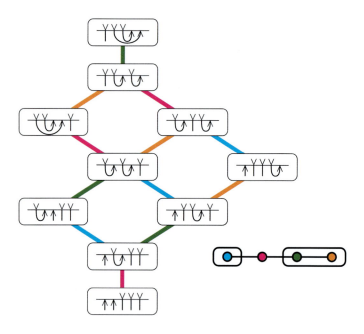

Fig. 7.11: The northern edges of the elements e_μ for $\mathfrak{S}_2 \times \mathfrak{S}_3 \leqslant \mathfrak{S}_5$. (The southern edges can easily be deduced using the second bullet point of Definition 7.5.6.)

We now wish to generalise the above to non-reduced paths. In other words, we wish to provide a non-iterative description of the diagrams E_T for $\mathsf{T} \in \mathrm{Path}(\lambda, \mathsf{T}_\mu)$ for all $\lambda \leqslant \mu$.

7.5 Oriented Temperley–Lieb combinatorics

Definition 7.5.8. Given $\lambda, \mu \in \mathscr{P}_{m,n}$ we let $\lambda\overline{\mu}$ be the diagram obtained from placing the weight λ on top of the diagram $\overline{\mu}$, if the resulting diagram is oriented. If the diagram is not oriented, we leave $\lambda\overline{\mu}$ undefined.

Example 7.5.9. The full matrix of diagrams $\lambda\overline{\mu}$ is provided in Figure 7.12 for $\mathfrak{S}_2 \times \mathfrak{S}_2 \leqslant \mathfrak{S}_4$. The columns are labelled by the cup diagrams $\overline{\mu}$; the rows are labelled by the weights λ. Calculating the degrees of these diagrams we obtain the Kazhdan–Lusztig polynomials in Table 7.1.

We now reconcile our iterative definition and closed combinatorial definitions of oriented Temperley–Lieb diagrams. In so doing, we obtain a closed combinatorial description of the Kazhdan–Lusztig polynomials of $\mathscr{P}_{m,n}$ as follows.

Theorem 7.5.10. Given $\lambda, \mu \in \Lambda_{m,n}$ we have that $\lambda\overline{\mu}$ is oriented if and only if there exists a (necessarily unique) path $\mathsf{T} \in \mathrm{Path}(\lambda, \mathsf{T}^\mu)$. If such a path exists, the diagram $\lambda\overline{\mu}$ is the top half of E_T. Therefore the Kazhdan–Lusztig polynomials of $\mathscr{P}_{m,n}$ can be calculated explicitly as follows:

$$n_{\lambda,\mu}(q) = \begin{cases} q^{\deg(\lambda\overline{\mu})} & \text{if } \lambda\overline{\mu} \text{ is defined} \\ 0 & \text{otherwise.} \end{cases}$$

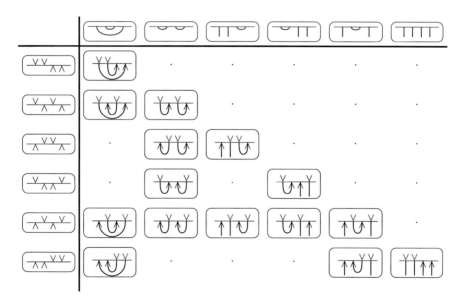

Fig. 7.12: The diagrams $\lambda\overline{\mu}$ which enumerate the Kazhdan–Lusztig polynomials $n_{\lambda,\mu}(q)$ for $\mathfrak{S}_2 \times \mathfrak{S}_2 \leqslant \mathfrak{S}_4$. Counting the number of clockwise arcs in a given diagram gives the degree of the corresponding polynomial in Table 7.1.

Proof of Theorem 7.5.10. In this proof, we freely identify the half diagram $\lambda\overline{\mu}$ with the oriented Temperley–Lieb diagram $\lambda e_\mu \varnothing$ obtained by adding $d(\mu)$ concentric arcs on the bottom, and joining up the propagating strands in the uniquely crossing-less manner as in Figures 7.13 and 7.14. Any oriented Temperley–Lieb diagram $\lambda e_\mu \varnothing$ can be constructed iteratively as a product of the generators (in general, in many possible ways) by way of its tiling, as in Figure 7.15.

Fig. 7.13: Half diagrams $\mu\overline{\mu}$ and $\lambda\overline{\mu}$ with $\lambda = (4, 3, 1^2)$, and $\mu = (5^2, 1^2)$.

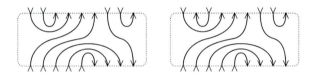

Fig. 7.14: Diagrams $\mu e_\mu \varnothing$ and $\lambda e_\mu \varnothing$ with $\lambda = (4, 3, 1^2)$, and $\mu = (5^2, 1^2)$.

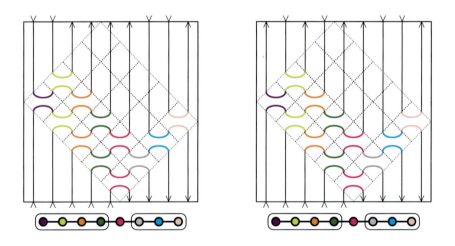

Fig. 7.15: The tilings of two oriented Temperley–Lieb diagrams. The former diagram is $E_{\mathsf{T}_{(5^2,1^2)}}$. The latter diagram is E_S for $\mathsf{S} \in \mathrm{Path}((4,3,1^2), \mathsf{T}_{(5^2,1^2)})$. The latter diagram is obtained by reorienting a single northern arc in the former.

7.5 Oriented Temperley–Lieb combinatorics

It remains to show that $\deg(\mathsf{S}) = \deg(\lambda\overline{\mu})$ for $\mathsf{S} \in \mathrm{Path}(\lambda, \mathsf{T}_\mu)$. This amounts to calculating the degree of the diagram $\lambda e_\mu \varnothing$ in two distinct ways. The first (iterative) way is to count the degree of each tile as we add it (this is $\deg(\mathsf{T})$ by Proposition 7.5.4); this summation has many cancellations and is not manifestly non-negative. This is illustrated on individual tiles in Figure 7.16.

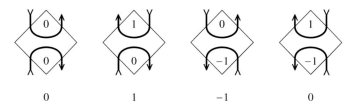

Fig. 7.16: We depict each of the four possible tiles. On each tile we depict the degree of the northern and southern arc. Below each tile we depict the total degree of the tile. The degree of T is equal to the sum over the total degrees of the tiles, by Proposition 7.5.4. In this way of counting it is not clear that the total degree of T is non-negative.

In the second way of counting, we calculate the degree of *each individual arc* in the diagram (this is $\deg(\lambda\overline{\mu})$). Any arc in a Temperley–Lieb diagram passes through an odd number of tiles and any propagating strand passes through an even number of tiles. Intersecting a strand with this set of tiles through which it passes, we obtain a "wiggle" or "ribbon" of the form depicted in Figure 7.17.

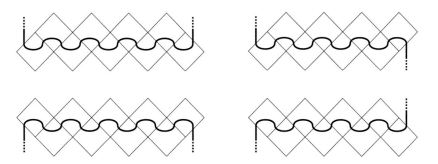

Fig. 7.17: Tilings of northern and southern arcs and (southeasterly and northeasterly) propagating strands. Reading from left to right, the strand oscillates between being either the "top" or "bottom" of a given Temperley–Lieb generator tile within the wider "ribbon" of tiles. Notice that the strands intersect an even number of tiles, whereas the arcs intersect an odd number of tiles.

We now count the degree contribution of each arc/strand in turn. We first consider clockwise-oriented northern arcs, whose ribbons are as in the top left diagram of Figure 7.17. Consider the intersection of this arc with any tile in this first ribbon. The degree contribution as it passes through the tiles in this ribbon is given by

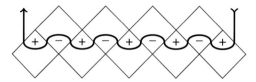

and summing over these local regions we obtain

$$+1 - 1 + 1 - 1 + 1 - 1 \cdots + 1 = 1$$

by (5.5.3) (notice that the strand is locally either a clockwise northern arc or an anticlockwise southern arc at each step, hence our alternating sum). This is the key observation of the proof: the "global" degree of the clockwise-oriented arc is +1 and this can be calculated much more quickly than considering all the "local" degrees in each tile. We will now repeat this trick with the other oriented arcs/strands.

If the arc is anti-clockwise oriented then the degree contribution as it passes through these tiles is given by

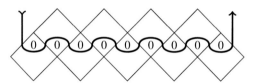

and summing over these local regions we obtain

$$0 + \cdots + 0 = 0.$$

Thus the total degree of an anti-clockwise oriented arc is indeed zero, as claimed. One can now repeat this exercise with southern arcs and the propagating strands and deduce the result. □

In our definition of the Kazhdan–Lusztig polynomials in terms of sums over degrees of paths from Path(λ, T^μ) it was not clear that these polynomials were (well-)defined independently of the choice of reduced path T^μ. Theorem 7.5.10 has the following immediate corollary.

Corollary 7.5.11. *Let $\lambda, \mu \in \mathcal{P}_{m,n}$. The Kazhdan–Lusztig polynomial $n_{\lambda,\mu}(q)$ is well-defined and independent of the choice of reduced path $\mathsf{T}_\mu \in \mathrm{Path}(\mu)$.*

Exercise 7.5.12. Repeat Exercise 7.3.7 using the interpretation of Kazhdan–Lusztig polynomials via the degree of $\lambda\overline{\mu}$. (You should find it much easier!)

7.5 Oriented Temperley–Lieb combinatorics

Having explicitly calculated the Kazhdan–Lusztig polynomials we are now able to check that they do satisfy the beautiful properties we discussed in Section 7.3. We first verify that the palindromic property does hold for the polynomials

$$\sum_{\lambda \subseteq \mu} q^{\ell(\mu)-\ell(\lambda)} n_{\lambda,\mu}(q),$$

and so the Kazhdan–Lusztig polynomials do indeed provide the desired "correction factor for the failure of the rank generating polynomial to be palindromic" as claimed. This requires a little more combinatorics...

Definition 7.5.13. Given $\mu \in \mathcal{P}_{m,n}$ we let $\overline{\mu}$ denote its cup diagram. For each cup, C, in $\overline{\mu}$ we let $w(C) \in \{2, 4, 6, 8, \dots\}$ denote the total number of symbols between its start and end points, counted inclusively (we think of this as the "width" of the cup).

Example 7.5.14. For $\mu \in \mathcal{P}_{2,2}$ the cup diagram

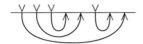

has 5 distinct cups, of widths 2, 2, 4, and 8.

Exercise 7.5.15. Let λ be obtained from $\mu\overline{\mu}$ by reorienting a single one of its cups, $C \in \overline{\mu}$. Prove that $\ell(\mu) - \ell(\lambda) = w(C) - 1$.

Proposition 7.5.16. Given $\mu \in \mathcal{P}_{m,n}$, the length-normalised μth column-sum of the Kazhdan–Lusztig matrix can be calculated as follows

$$\sum_{\lambda \subseteq \mu} q^{\ell(\mu)-\ell(\lambda)} n_{\lambda,\mu}(q) = \prod_{C \in \overline{\mu}} (1 + q^{w(C)}).$$

and in particular, it is both palindromic and unimodal.

Proof. The proof is left for an exercise for the reader. It amounts to generalising Exercise 7.5.15 to arbitrary reorientations of the arcs in $\mu\overline{\mu}$ in a natural fashion. □

Thus for $\mu \in \mathcal{P}_{m,n}$, the column-sum polynomials of the Kazhdan–Lusztig matrix are palindromic and unimodal and perhaps more aesthetically pleasing than the rank generating polynomials, let's see an example.

Example 7.5.17. If $\mu = (4, 3, 2, 1)$ then $d(\mu) = 4$ and the normalised $(4, 3, 2, 1)$-column sum of the Kazhdan–Lusztig matrix is given by

$$\sum_{\lambda \subseteq (4,3,2,1)} q^{\ell(4,3,2,1)-\ell(\lambda)} n_{\lambda\mu}(q) = (1 + q^2)^4 1 + 4q^2 + 6q^4 + 4q^6 + q^8.$$

Compare this with the rank generating function $\rho(4, 3, 2, 1)$ is given by

$$\sum_{\lambda \subseteq \mu} q^{\ell(\mu)-\ell(\lambda)} = 1 + 4q + 6q^2 + 7q^3 + 14q^4 + 10q^5 + 6q^6 + 3q^7 + 2q^8 + q^9 + q^{10}.$$

Example 7.5.18. For $\mu = (m, m - 1, \ldots, 2, 1) \in \mathcal{P}_{m,n}$, the μth length-normalised column-sum of the Kazhdan–Lusztig matrix is

$$\sum_{\lambda \in \mathcal{P}_{m,n}} q^{\ell(m,\ldots,2,1) - \ell(\lambda)} n_{\lambda,(m,\ldots,2,1)} = 1 + \binom{m}{1} q^2 + \binom{m}{2} q^4 + \cdots + \binom{m}{m-1} q^{2m-2} + q^{2m}$$

which quantises a row of Pascal's triangle, whereas the rank generating function for $\mu = (m^n)$ is given by

$$\rho(m^n) = \begin{bmatrix} m + n \\ m \end{bmatrix}_q$$

which quantises an entry in Pascal's triangle.

Exercise 7.5.19. Calculate the column labelled by the triangular partition $(k, k - 1, k - 2 \ldots, 2, 1)$. How many non-zero entries are there?

Exercise 7.5.20. Calculate the first column of the Kazhdan–Lusztig matrix of $\mathcal{P}_{m,n}$ for any $m, n \in \mathbb{Z}_{>0}$.

Further Reading. The Kazhdan–Lusztig polynomials of this section were first studied by Lascoux–Schützenberger in the 1980s [LS81], but were subsequently generalised and cast in different combinatorial settings by Boe, Brenti, and Brundan–Stroppel [Boe88, Bre07, BS11a]. The role of oriented Temperley–Lieb algebras was first hinted at in [BS11a] (where the diagrams first appear) and later developed in [BDF+] (where the algebra structure is defined).

The unimodality of rank generating functions of tile-partitions is studied in detail in [Sta90] and in, fact, the example given here (of $\rho(8, 8, 4, 4)$ of rank 24) is the smallest rank example for which unimodality fails..

Concerning the generalisations alluded to above: we say that a Bruhat graph is fully commutative if it contains no $2m$-gons for $m > 2$. Fully commutative Bruhat graphs were classified by Stembridge in [Ste96]. All of the material in this section was generalised to these Bruhat graphs, in same language as was used in this chapter, in [BDF+] (building on the ideas of [Boe88, Bre07, BS11a]).

There are two different "types" of Kazhdan–Lusztig polynomials associated to a parabolic Coxeter system: namely, the spherical and anti-spherical Kazhdan–Lusztig polynomials. The anti-spherical Kazhdan–Lusztig polynomials are the polynomials discussed in this chapter and, in fact, are the only ones we will consider explicitly in this book. Both families of Kazhdan–Lusztig polynomials can be seen as coming the shadow of "canonical bases" of induced modules of Hecke algebras, with the spherical polynomials corresponding to inducing the trivial representation and the anti-spherical polynomials corresponding to inducing the sign representation. If the parabolic is the trivial group then the spherical and anti-spherical Kazhdan–Lusztig polynomials coincide. We refer to [Deo87, KL79] for more details.

Chapter 8
General Kazhdan–Lusztig theory

In the previous chapter, we studied paths in the Bruhat graph of $\mathfrak{S}_m \times \mathfrak{S}_n \leqslant \mathfrak{S}_{m+n}$ and used this to define and calculate the associated Kazhdan–Lusztig polynomials. (Albeit in the language of tile-partitions and Temperley–Lieb combinatorics.) In this chapter, we will see that we secretly brushed quite a few details under the rug, and we will reveal a number of layers of added complexity in both the definition and the calculation of Kazhdan–Lusztig polynomials for arbitrary parabolic Coxeter systems.

As before, our motivation comes from representation theory. However, for the reader's benefit we will continue to emphasise the beautiful combinatorial properties that motivate the definition and study of Kazhdan–Lusztig polynomials. We will pay particular attention to the Kazhdan–Lusztig positivity conjecture and the Dyer–Lusztig combinatorial invariance conjecture for these polynomials.

8.1 Weak Bruhat graphs of parabolic Coxeter systems

Let (W, S_W) be a Coxeter system: W is the group generated by the finite set S_W subject to the relations $(\sigma\tau)^{m_{\sigma\tau}} = 1$ for $\sigma, \tau \in S_W$, $m_{\sigma\tau} \in \mathbb{Z}_{\geqslant 0} \cup \{\infty\}$ satisfying $m_{\sigma\tau} = m_{\tau\sigma}$, and $m_{\sigma\tau} = 1$ if and only if $\sigma = \tau$. Consider $S_P \subseteq S_W$ a subset and (P, S_P) its corresponding Coxeter system. We say that P is the **parabolic subgroup** corresponding to $S_P \subseteq S_W$.

Definition 8.1.1. The weak Bruhat graph for (W, P) is given as follows:

- start by writing the identity coset at the bottom.
- If applying a generator to some coset λ does result in a new coset μ, then record this in the next level up in the diagram and write $\sigma \in \text{Add}(\lambda)$ and $\mu = \lambda + \sigma$, or $\sigma \in \text{Rem}(\mu)$ and $\lambda = \mu - \sigma$ or simply $\lambda < \mu$. We record the colour of the reflection as an edge relating the two points in the graph.

Repeat the above until the process terminates with a graph $({}^P W, <)$.

Example 8.1.2. The weak Bruhat graph for $\mathfrak{S}_2 \times \mathfrak{S}_3 \leqslant \mathfrak{S}_5$ with labels given by cosets is depicted in Figure 8.1 below. Compare this with Figure 7.6

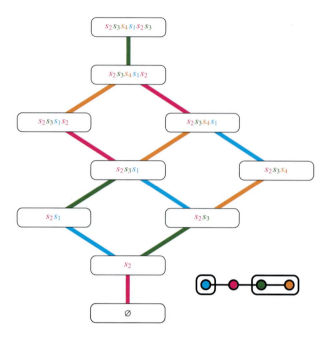

Fig. 8.1: The weak Bruhat graph $(^P W, <)$ for $(W, P) = (\mathfrak{S}_5, \mathfrak{S}_2 \times \mathfrak{S}_3)$.

Example 8.1.3. The weak Bruhat graph for $\mathfrak{S}_3$ is generated by the transpositions $\tau = (1, 2)$ and $\sigma = (2, 3)$ is given in Figure 8.2.

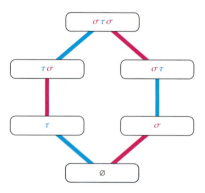

Fig. 8.2: The weak Bruhat graph for $(^P W, <)$ for $(W, P) = (\mathfrak{S}_3, \{1\})$.

8.1 Weak Bruhat graphs of parabolic Coxeter systems

Example 8.1.4. The symplectic affine Weyl group, $\widehat{C}_2$, has Coxeter presentation

$$\widehat{C}_2 = \langle \sigma, \tau, \rho \mid (\sigma\tau)^4 = 1, (\tau\rho)^4 = 1, (\sigma\rho)^2 = 1, \sigma^2 = \tau^2 = \rho^2 = 1 \rangle.$$

This group is infinite and its (infinite) weak Bruhat graph is recorded in Figure 8.3.

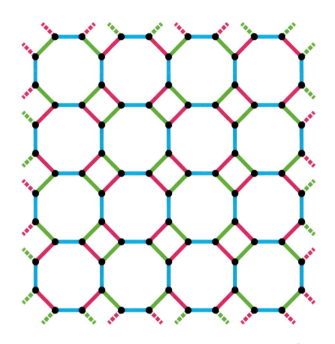

Fig. 8.3: The weak Bruhat graph $(^P W, \prec)$ for $(W, P) = (\widehat{C}_2, \{1\})$.

Example 8.1.5. The affine symmetric group $\widehat{\mathfrak{S}}_3$ has Coxeter presentation

$$\langle \sigma, \tau, \rho \mid (\sigma\tau)^3 = 1, (\tau\rho)^3 = 1, (\sigma\rho)^3 = 1, \sigma^2 = \tau^2 = \rho^2 = 1 \rangle.$$

This group is infinite and its weak Bruhat graph is given by a hexagonal tiling of the plane $\mathbb{R}^2$. We can take our parabolic subgroup to be the finite group $\mathfrak{S}_3 = \langle \tau, \rho \rangle \leqslant \widehat{\mathfrak{S}}_3$ of order 6. The weak Bruhat graph is a tiling of *one sixth* of the plane and is depicted in Figure 8.4.

Exercise 8.1.6. The dihedral group D_{2m} has Coxeter presentation $D_{2m} = \langle \sigma, \tau \mid (\sigma\tau)^m = 1, \sigma^2 = \tau^2 = 1 \rangle$. Notice that $D_6 \cong \mathfrak{S}_3$ and its weak Bruhat graph is the hexagon depicted in Figure 8.2. Construct the weak Bruhat graph of D_{2m} for $m > 3$.

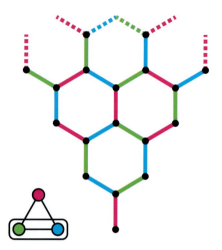

Fig. 8.4: The weak Bruhat graph $(^P W, <)$ for $(W, P) = (\widehat{\mathfrak{S}}_3, \mathfrak{S}_3)$. This graph tiles one sixth of the plane (because the parabolic $\mathfrak{S}_3$ has order 6).

8.2 Kazhdan–Lusztig polynomials in full generality

With our new and more general notion of weak Bruhat graphs in place, we wish to generalise the ideas of the previous chapter to this setting. We begin by repeating, verbatim, the the definition of the path-counting polynomials from the previous chapter. Our motivation for doing so remains the same: we seek palindromic, unimodal polynomials that encode the rich combinatorial structure of Bruhat graphs.

Definition 8.2.1. For each coloured edge of $\lambda \xrightarrow{\sigma} \mu$ the weak Bruhat graph with $\sigma \in W$, we allow it to be traversed in four possible ways. These paths are as follows,

(U^1_σ) The 'up' move, $\lambda \to \mu$, of degree 0;
(U^0_σ) The 'failed up', $\lambda \to \lambda$, of degree 1;
(D^1_σ) The 'down', $\mu \to \lambda$, of degree 0.
(D^0_σ) The 'failed down', $\mu \to \mu$, of degree -1.

We define a path, T, of shape $\lambda \in {}^P W$ and length ℓ to be a sequence

$$\emptyset = \lambda^{(0)} \to \lambda^{(1)} \to \cdots \to \lambda^{(\ell)} = \lambda$$

composed of up, failed-up, down, and failed-down steps in the weak Bruhat graph. We denote the set of all such paths by $\text{Path}_{(W,P)}(\lambda)$.

Definition 8.2.2. For a fixed reduced path, $\mathsf{T}_\mu \in \text{Path}_{(W,P)}(\mu)$, we are able to consider all possible paths with the same colour sequence as T_μ but which terminate at another point λ; we denote the set of all such paths by $\text{Path}_{(W,P)}(\lambda, \mathsf{T}_\mu)$.

8.2 Kazhdan–Lusztig polynomials in full generality

Definition 8.2.3. We say that a path from the identity to μ is **reduced** if it is of shortest possible length. We set $\ell(\mu)$ to be the length of a reduced path from the identity coset to $\mu \in {}^P W$.

We claim that the path-counting polynomials

$$\sum_{S \in \text{Path}(\lambda, \mathsf{T}_\mu)} q^{\deg(S)} \tag{8.2.4}$$

are undesirable as a verbatim replacement of Definition 7.3.4. These polynomials will depend on the *reduced path chosen*, or if you prefer, they cannot be well-defined for a pair of elements $\lambda, \mu \in {}^P W$. Let's see this first in an example. We let

$$W = \langle s_1, s_2 \mid (s_1 s_2)^3 = 1, s_1^2 = 1 = s_2^2 \rangle,$$

in which case we have two paths of shape $\mu = (1, 3)$, depicted in Figure 8.5.

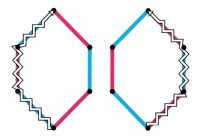

Fig. 8.5: The two choices of reduced path of shape $s_2 s_1 s_2 = (1, 3) = s_1 s_2 s_1$.

The 8 paths in Path$(\lambda, \mathsf{U}^1 \mathsf{U}^1 \mathsf{U}^1)$ for $\lambda \in \mathfrak{S}_3$ are depicted in Figure 8.7 and some of these paths are depicted on the weak Bruhat graph in Figure 8.6.

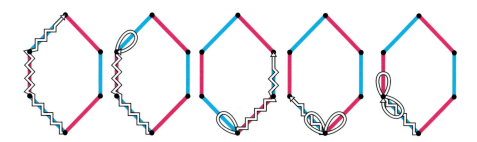

Fig. 8.6: Five of the eight paths of weight $s_1 s_2 s_1$ of degrees 0, 1, 1, 2, and 0 respectively.

186   8 General Kazhdan–Lusztig theory

The path-counting polynomials of shape $s_1 = (1,2)$ and shape $s_2 s_1 s_2 = (1,3) = s_1 s_2 s_1$ with respect to the two possible choices of reduced path are as follows

$$\sum_{S \in \text{Path}(s_1, \mathsf{U}^1 \mathsf{U}^1 \mathsf{U}^1)} q^{\deg(S)} = 1 + q^2 \neq q^2 = \sum_{S \in \text{Path}(s_1, \mathsf{U}^1 \mathsf{U}^1 \mathsf{U}^1)} q^{\deg(S)}. \qquad (8.2.5)$$

We enumerate all the path-counting polynomials for all choices of reduced paths in $W = \mathfrak{S}_3$ in Figure 8.8.

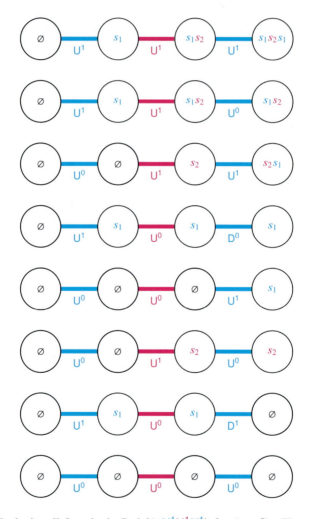

Fig. 8.7: We depict all 8 paths in $\text{Path}(\lambda, \mathsf{U}^1 \mathsf{U}^1 \mathsf{U}^1)$ for $\lambda \in \mathfrak{S}_3$. These paths have degree 0, 1, 1, 0, 2, 2, 1, and 3 respectively.

Exercise 8.2.6. Construct all 8 paths in $\text{Path}(\lambda, \mathsf{U}^1 \mathsf{U}^1 \mathsf{U}^1)$ for $\lambda \in \mathfrak{S}_3$.

8.2 Kazhdan–Lusztig polynomials in full generality

	$s_1s_2s_1$	s_2s_1	s_1s_2	s_1	s_2	$\emptyset$
$s_1s_2s_1$	1	·	·	·	·	·
s_2s_1	q	1	·	·	·	·
s_1s_2	q	·	1	·	·	·
s_1	$1+q^2$	q	q	1	·	·
s_2	q^2	q	q	·	1	·
$\emptyset$	$q+q^3$	q^2	q^2	q	q	1

	$s_2s_1s_2$	s_2s_1	s_1s_2	s_1	s_2	$\emptyset$
$s_2s_1s_2$	1	·	·	·	·	·
s_2s_1	q	1	·	·	·	·
s_1s_2	q	·	1	·	·	·
s_1	q^2	q	q	1	·	·
s_2	$1+q^2$	q	q	·	1	·
$\emptyset$	$q+q^3$	q^2	q^2	q	q	1

Fig. 8.8: The path-counting polynomials of (8.2.4) for $W = \mathfrak{S}_3$. In the first table we choose the reduced path $\mathsf{T}_{(1,3)} = \mathsf{U}_1^1\mathsf{U}_2^1\mathsf{U}_1^1$ and in the second we choose $\mathsf{T}_{(1,3)} = \mathsf{U}_2^1\mathsf{U}_1^1\mathsf{U}_2^1$. The first column of the first table enumerates the paths in Figure 8.7.

We want to modify the definition of (8.2.4) so that the polynomials are independent of the choice of reduced paths. So let's examine the points where the two matrices in Figure 8.8 differ from one another. This means considering the off-diagonal coefficient 1 in both matrices. The off-diagonal coefficient 1 in the former matrix is in the s_1-row and in the latter matrix is in the s_2-row. Simply subtracting these 1s seems a bit naive. Instead, let's try factorising/dividing the matrices in Figure 8.8 as a product of a pair of matrices as follows

$$\begin{bmatrix} 1 & \cdot & \cdot & \cdot & \cdot & \cdot \\ q & 1 & \cdot & \cdot & \cdot & \cdot \\ q & \cdot & 1 & \cdot & \cdot & \cdot \\ q^2 & q & q & 1 & \cdot & \cdot \\ q^2 & q & q & \cdot & 1 & \cdot \\ q^3 & q^2 & q^2 & q & q & 1 \end{bmatrix} \begin{bmatrix} 1 & \cdot & \cdot & \cdot & \cdot & \cdot \\ \cdot & 1 & \cdot & \cdot & \cdot & \cdot \\ \cdot & \cdot & 1 & \cdot & \cdot & \cdot \\ 1 & \cdot & \cdot & 1 & \cdot & \cdot \\ \cdot & \cdot & \cdot & \cdot & 1 & \cdot \\ \cdot & \cdot & \cdot & \cdot & \cdot & 1 \end{bmatrix} \quad (8.2.7)$$

for the former matrix, and

$$\begin{bmatrix} 1 & \cdot & \cdot & \cdot & \cdot & \cdot \\ q & 1 & \cdot & \cdot & \cdot & \cdot \\ q & \cdot & 1 & \cdot & \cdot & \cdot \\ q^2 & q & q & 1 & \cdot & \cdot \\ q^2 & q & q & \cdot & 1 & \cdot \\ q^3 & q^2 & q^2 & q & q & 1 \end{bmatrix} \begin{bmatrix} 1 & \cdot & \cdot & \cdot & \cdot & \cdot \\ \cdot & 1 & \cdot & \cdot & \cdot & \cdot \\ \cdot & \cdot & 1 & \cdot & \cdot & \cdot \\ \cdot & \cdot & \cdot & 1 & \cdot & \cdot \\ 1 & \cdot & \cdot & \cdot & 1 & \cdot \\ \cdot & \cdot & \cdot & \cdot & \cdot & 1 \end{bmatrix} \quad (8.2.8)$$

for the latter matrix. The leftmost matrices in (8.2.7) and (8.2.8) are equal (that is, independent of the choice of the reduced path) and the rightmost matrices consist only of 1s. (One can think of these unwanted path-choice-dependencies as being jettisoned into the righthand matrices in (8.2.7) and (8.2.8).) This sort of factorisation of matrices can be given representation theoretic meaning, as we will see in Chapter 10.

Taking what we have learnt from the $W = \mathfrak{S}_3$ case, we are now ready to define the Kazhdan–Lusztig polynomials for arbitrary parabolic Coxeter systems, as follows.

Definition 8.2.9. Let (W, P) be arbitrary. For each $\mu \in {}^P W$, fix a choice of reduced path T_μ. We define

$$\Delta^{(W,P)} := (\Delta_{\lambda,\mu}(q))_{\lambda,\mu \in {}^P W} \qquad \Delta_{\lambda,\mu}(q) = \sum_{S \in \mathrm{Path}(\lambda, \mathsf{T}_\mu)} q^{\deg(S)}$$

which is a (square) lower uni-triangular matrix. This matrix can be factorised *uniquely* as a product

$$\Delta^{(W,P)} = N^{(W,P)} B^{(W,P)}$$

of lower uni-triangular matrices

$$N^{(W,P)} := (n_{\lambda,\nu}(q))_{\lambda,\nu \in {}^P W} \qquad B^{(W,P)} := (b_{\nu,\mu}(q))_{\nu,\mu \in {}^P W}$$

such that $n_{\lambda,\nu}(q) \in q\mathbb{Z}[q]$ for $\lambda \neq \nu$ and $b_{\nu,\mu}(q) \in \mathbb{Z}[q+q^{-1}]$. A recursive algorithm for this matrix factorisation is given by setting $b_{\lambda,\lambda}(q) = 1 = n_{\lambda,\lambda}(q)$ and defining the polynomials

$$b_{\lambda,\mu}(q) \in \mathbb{Z}[q + q^{-1}] \qquad n_{\lambda,\mu}(q) \in q\mathbb{Z}[q]$$

by induction as follows

$$b_{\lambda,\mu}(q) + n_{\lambda,\mu}(q) = \sum_{S \in \mathrm{Path}(\lambda, \mathsf{T}_\mu)} q^{\deg(S)} - \sum_{\substack{\mathrm{Path}(\nu, \mathsf{T}^\mu) \neq \emptyset \\ \mathrm{Path}(\lambda, \mathsf{T}^\nu) \neq \emptyset}} n_{\lambda,\nu}(q) b_{\nu,\mu}(q). \qquad (8.2.10)$$

The polynomial $n_{\lambda,\mu}(q)$ is called the **Kazhdan–Lusztig polynomial** for $\lambda, \mu \in {}^P W$.

The following theorem is incredibly important, but can only be proven using the theory of canonical bases which is beyond the topics of this book.

Theorem 8.2.11 ([Hum90, Chapter 7]). *The polynomials $n_{\lambda,\mu}(q)$ for $\lambda, \mu \in {}^P W$ are well-defined and independent of the choice of the reduced paths T^μ for $\mu \in {}^P W$.*

Whilst the path-counting polynomials have non-negative coefficients, it is not at all clear from Definition 8.2.9 that the Kazhdan–Lusztig polynomials should too. We thus arrive at the first of many beautiful conjectures concerning these polynomials...

Conjecture 8.2.12 (The Kazhdan–Lusztig positivity conjecture). *For (W, P) arbitrary, the polynomials $n_{\lambda,\mu}(q)$ for $\lambda, \mu \in {}^P W$ have non-negative coefficients.*

The Kazhdan–Lusztig positivity conjecture was beguiling to combinatorists, geometers, and representation theorists alike. Positivity phenomena in mathematics are often shadows of interesting algebraic or geometric structures. We will see that this is indeed the case for Kazhdan–Lusztig polynomials in Part IV of this book.

Example 8.2.13. For $(W, P) = (\mathfrak{S}_2 \wr \mathfrak{S}_3, \mathfrak{S}_3)$ the Bruhat graph is depicted in Figure 8.9 and the factorisation $\Delta^{(W,P)} = N^{(W,P)} \times B^{(W,P)}$ in in Figure 8.10.

8.2 Kazhdan–Lusztig polynomials in full generality

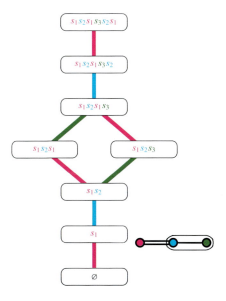

Fig. 8.9: The Bruhat graph for $(W, P) = (\mathfrak{S}_2 \wr \mathfrak{S}_3, \mathfrak{S}_3)$.

$$\begin{bmatrix} 1 & \cdot & \cdot & \cdot & \cdot & \cdot & \cdot & \cdot \\ q & 1 & \cdot & \cdot & \cdot & \cdot & \cdot & \cdot \\ 1 & q & 1 & \cdot & \cdot & \cdot & \cdot & \cdot \\ q & \cdot & q & 1 & \cdot & \cdot & \cdot & \cdot \\ q & \cdot & q & \cdot & 1 & \cdot & \cdot & \cdot \\ q^2 & q & q^2 & q & q & 1 & \cdot & \cdot \\ q & q^2 & \cdot & \cdot & 1 & q & 1 & \cdot \\ q^2 & \cdot & \cdot & \cdot & q & \cdot & q & 1 \end{bmatrix} = \begin{bmatrix} 1 & \cdot & \cdot & \cdot & \cdot & \cdot & \cdot & \cdot \\ q & 1 & \cdot & \cdot & \cdot & \cdot & \cdot & \cdot \\ \cdot & q & 1 & \cdot & \cdot & \cdot & \cdot & \cdot \\ \cdot & \cdot & q & 1 & \cdot & \cdot & \cdot & \cdot \\ \cdot & \cdot & q & \cdot & 1 & \cdot & \cdot & \cdot \\ \cdot & q & q^2 & q & q & 1 & \cdot & \cdot \\ q & q^2 & \cdot & \cdot & \cdot & q & 1 & \cdot \\ q^2 & \cdot & \cdot & \cdot & \cdot & \cdot & q & 1 \end{bmatrix} \begin{bmatrix} 1 & \cdot & \cdot & \cdot & \cdot & \cdot & \cdot & \cdot \\ \cdot & 1 & \cdot & \cdot & \cdot & \cdot & \cdot & \cdot \\ 1 & \cdot & 1 & \cdot & \cdot & \cdot & \cdot & \cdot \\ \cdot & \cdot & \cdot & 1 & \cdot & \cdot & \cdot & \cdot \\ \cdot & \cdot & \cdot & \cdot & 1 & \cdot & \cdot & \cdot \\ \cdot & \cdot & \cdot & \cdot & \cdot & 1 & \cdot & \cdot \\ \cdot & \cdot & \cdot & \cdot & \cdot & 1 & \cdot & 1 & \cdot \\ \cdot & \cdot & \cdot & \cdot & \cdot & \cdot & \cdot & 1 \end{bmatrix}$$

Fig. 8.10: The product $\Delta^{(W,P)} = N^{(W,P)} \times B^{(W,P)}$ for $(W, P) = (\mathfrak{S}_2 \wr \mathfrak{S}_3, \mathfrak{S}_3)$.

Exercise 8.2.14. Explicitly construct $\mathrm{Path}_{(W,P)}(\lambda, \mathsf{T}_\mu)$ for $\lambda, \mu \in {}^P W$ for $(W, P) = (\mathfrak{S}_2 \wr \mathfrak{S}_3, \mathfrak{S}_3)$ and hence verify that the $n_{\lambda,\mu}(q)$ for $\lambda, \mu \in {}^P W$ are as in Figure 8.10.

Example 8.2.15. The subtracted term is zero for $\lambda = \mu$ or λ and μ adjacent in the weak Bruhat order (that is, $|\ell(\mu) - \ell(\lambda)| \leq 1$). There is a single element of $\mathrm{Path}(\lambda, \mathsf{T}_\mu)$ of degree 0 or 1 respectively in these cases. Therefore

$$b_{\lambda,\lambda}(q) = 1 \qquad n_{\lambda,\lambda}(q) = 1 \qquad b_{\lambda,\mu}(q) = 0 \qquad n_{\lambda,\mu}(q) = q$$

for $\lambda \lessdot \mu$ with $\ell(\mu) = \ell(\lambda) + 1$.

Exercise 8.2.16. The parabolic Coxeter graph of $(W, P) = (E_6, D_5)$ and its weak Bruhat graph is depicted in Figure 8.11. Calculate the corresponding Kazhdan–Lusztig polynomials.

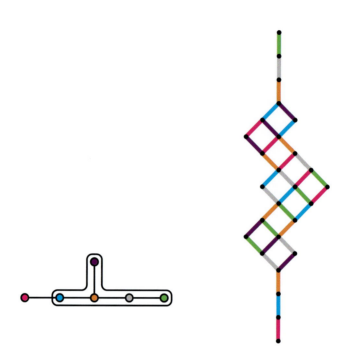

Fig. 8.11: The Coxeter graph and weak Bruhat graph of $(W, P) = (E_6, D_5)$.

Example 8.2.17. Let $W = \mathfrak{S}_4$ and set $\lambda = (2,3)$ and $\mu = (2,3)(1,2)(3,4)(2,3)$. We have that
$$n_{\lambda,\mu} = q + q^3$$
and this is the smallest example of a Kazhdan–Lusztig polynomial that evaluates (at $q = 1$) to be (greater than or) equal to 2. Hint: the points λ and μ are highlighted within the weak Bruhat of $\mathfrak{S}_4$, depicted in Figure 8.12.

Exercise 8.2.18. Calculate the Kazhdan–Lusztig polynomials for $\mathfrak{S}_4$.

Example 8.2.19. For $\mathfrak{S}_m \times \mathfrak{S}_n \leq \mathfrak{S}_{m+n}$ we have that $b_{\lambda,\mu}(q) = 0$ for all $\lambda \neq \mu$ and this is why we were able to artificially simplify Definition 7.3.4.

8.3 The strong Bruhat order and combinatorial-invariance

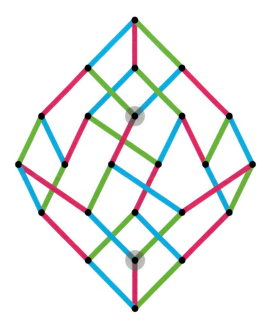

Fig. 8.12: The weak Bruhat graph of $\mathfrak{S}_4$ with $\lambda = (2,3)$ and $\mu = (2,3)(1,2)(3,4)(2,3)$ highlighted.

8.3 The strong Bruhat order and combinatorial-invariance

We now define a new*ish* partial ordering, $\leqslant$, on ^PW and use this to formulate a surprising conjecture, which was proposed independently by both Lusztig and Dyer. This conjecture is widely regarded as geometrically counterintuitive, and yet remarkable evidence for this conjecture has been provided by combinatorists [BCM06, BLP23] and Google Deepmind's artificial intelligence [BBD+22, DVB+21].

Definition 8.3.1. For $\lambda, \mu \in {}^PW$ we write $\lambda \leqslant \mu$ if $\mathrm{Path}_{(W,P)}(\lambda, \mathsf{T}_\mu) \neq \emptyset$ for some reduced path $\mathsf{T}_\mu \in \mathrm{Path}_{(W,P)}(\mu)$. We extend this to a partial order, called the strong Bruhat order $\leqslant$, by transitivity.

We note that the weak and strong Bruhat orders, $\prec$ and $\leqslant$, are distinct. That is, the strong Bruhat order has more edges in its Hasse graph than the weak Bruhat order does (we colour these edges in grey, as they cannot be assigned to the colours of the generators of the Coxeter system). This is because the strong order is defined in terms of the combinatorics of *paths* in the Hasse graph of the weak order. An example is depicted in Figures 8.13 and 8.14 below.

We now discuss how one can "truncate" Kazhdan–Lusztig-theoretic combinatorics. This is mostly a formal exercise which follows immediately from the definitions. This process is entirely analogous to the algebraic truncations of Section 6.12. We will bring these (apparently different) notions of truncation together later in this book, where we will apply the algebraic truncations of Section 6.12 to certain diagrammatic algebras and hence recover the combinatorial truncations of this section as a consequence. We will consider specific subposets of $(^PW, \leqslant)$ the set of coset representatives of $P \leqslant W$ under the strong Bruhat order.

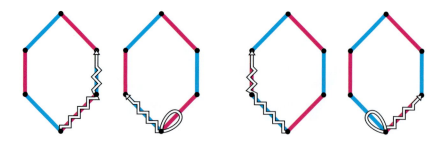

Fig. 8.13: The pairs $\lambda, \mu \in \mathfrak{S}_3$ such that $\lambda < \mu$ in the strong Bruhat order, but which are not relatable in the weak Bruhat order.

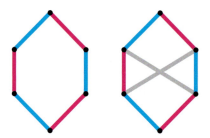

Fig. 8.14: The weak and strong Bruhat orders for $W = \mathfrak{S}_3$. Notice that we cannot colour the extra edges of the strong Bruhat graph as these edges do not correspond to right multiplication by Coxeter generators.

Definition 8.3.2. A subset $\Pi \subseteq {}^PW$ is saturated if $\mu \in \Pi$ and $\lambda \in {}^PW$ with $\lambda \leqslant \mu$ implies that $\lambda \in \Pi$.

Example 8.3.3. Let $(W, P) = (\widehat{\mathfrak{S}}_3, \mathfrak{S}_3)$ where

$$\mathfrak{S}_3 = \langle \rho, \tau \mid (\tau\rho)^3 = 1, \tau^2 = 1 = \rho^2 \rangle$$
$$\widehat{\mathfrak{S}}_3 = \langle \sigma, \rho, \tau \mid (\sigma\rho)^3 = (\sigma\tau)^3 = (\rho\tau)^3 = 1, \sigma^2 = \tau^2 = \rho^2 = 1 \rangle.$$

The set $\Pi = \{\lambda \mid \lambda \leqslant \sigma\rho\tau\sigma\rho\tau\} \subseteq {}^PW$ is saturated and is depicted in Figure 8.15.

Exercise 8.3.4. Let $W = \widehat{\mathfrak{S}}_3 = \langle \sigma, \rho, \tau \mid (\sigma\rho)^3 = (\sigma\tau)^3 = (\rho\tau)^3 = 1, \sigma^2 = \tau^2 = \rho^2 = 1 \rangle$ and let P be the trivial parabolic. Construct the paths Path$(\lambda, \mathsf{T}^\mu)$ for $\mu \in \Pi$. (Hint: these paths are enumerated in Table 8.1.)

8.3 The strong Bruhat order and combinatorial-invariance

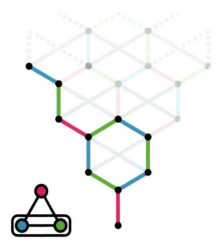

Fig. 8.15: The strong Bruhat graph of a saturated subset of $(^PW, \leqslant)$ for $(W, P) = (\widehat{\mathfrak{S}}_3, \mathfrak{S}_3)$.

Definition 8.3.5. A subset $\Pi \subseteq {}^PW$ is co-saturated if its complement in ${}^PW$ is saturated. A subset $\Pi \subseteq {}^PW$ is closed if it is the intersection of a saturated set and a co-saturated set.

Let $\Pi \subseteq {}^PW$ be a closed subset. We define

$$\Delta^{\Pi} := (\Delta_{\lambda,\mu}(q))_{\lambda,\mu \in \Pi} \qquad \Delta_{\lambda,\mu}(q) = \sum_{S \in \text{Path}(\lambda, \mathsf{T}_\mu)} q^{\deg(S)}$$

which is a submatrix of $\Delta^{(W,P)}$. This matrix factorises as a product

$$\Delta^{\Pi} = N^{\Pi} B^{\Pi}$$

of lower uni-triangular matrices

$$N^{\Pi} := (n_{\lambda,\nu}(q))_{\lambda,\mu \in \Pi} \qquad B^{\Pi} := (b_{\nu,\mu}(q))_{\nu,\mu \in \Pi}$$

such that $n_{\lambda,\nu}(q) \in q\mathbb{Z}[q]$ for $\lambda \neq \nu$ and $b_{\nu,\mu}(q) \in \mathbb{Z}[q + q^{-1}]$. We note that restricting our attention to submatrices labelled by closed subsets $\Pi \subseteq {}^PW$ commutes with this matrix factorisation. (One way to see this is by noting that the algorithm in (8.2.10) respects saturated and co-saturated sets, by its definition in terms counting paths in the weak Bruhat graph.)

Exercise 8.3.6. Let $(W, P) = (\widehat{\mathfrak{S}}_3, \mathfrak{S}_3)$ and Π be as in Example 8.3.3. Factorise the matrix in Table 8.1 and hence compute the associated Kazhdan–Lusztig polynomials for the set Π of Example 8.3.3.

	στρσρτ	σρτσρ	σρτσ	στρσ	στρ	σρτ	σρ	στ	σ	∅
στρσρτ	1	·	·	·	·	·	·	·	·	·
στρσρ	q	1	·	·	·	·	·	·	·	·
στρσ	·	q	1	·	·	·	·	·	·	·
στρτ	1	q	·	1	·	·	·	·	·	·
σρτ	2q	q^2	q	q	1	·	·	·	·	·
στρ	q	·	·	q	·	1	·	·	·	·
σρ	$2q^2$	q	·	$1+q^2$	q	q	1	·	·	·
στ	q^2	·	·	q^2	q	q	·	1	·	·
σ	q^3	q^3	q	$q+q^3$	q^2	q^2	q	q	1	·
∅	·	·	q^2	·	·	·	·	·	q	1

Table 8.1: The Δ^Π matrix for Π as in Example 8.3.3. We have highlighted a 6×6 submatrix of Π which is closed (under the strong Bruhat order).

For the purposes of disproving Lusztig's conjecture, the most important closed subsets in the Bruhat order come from *affinization* of the finite symmetric group. The finite symmetric group $\mathfrak{S}_n$ is generated by the reflections

$$(1,2), (2,3), (3,4), \ldots, (n-2, n-1), (n-1, n)$$

and that these are *precisely the permutations* which can be pictured as string diagrams drawn in a rectangular frame so that they have a *single pair of crossing strands*. We record these generators in the Coxeter graph

Now, if we picture string diagrams *on a cylinder* then we can draw one new generator using a single pair of crossing strands, namely we can draw $s_0 := (1, n)$ so that these two strands go around the edge of the cylinder *without crossing the other strands*. An example for $n = 6$ is depicted in Figure 8.16.

Picturing our string diagrams on a cylinder, we see that $s_{n-1} s_0 s_{n-1} = s_0 s_{n-1} s_0$ and $s_1 s_0 s_1 = s_0 s_1 s_0$ and indeed we record this in the Coxeter graph in Figure 8.17. We hence obtain n distinct embeddings : $\mathfrak{S}_n \to \widehat{\mathfrak{S}}_n$. This allows us to see infinitely many copies of the finite Bruhat graph of $\mathfrak{S}_n$ within the infinite Bruhat graph of $\widehat{\mathfrak{S}}_n$. An example is depicted in Figure 8.18. These embeddings of Bruhat graphs are all quite intuitive, however we will need to pay a bit of extra care when dealing with non-trivial parabolic subgroups of $\widehat{\mathfrak{S}}_n$. This motivates the following definition.

8.3 The strong Bruhat order and combinatorial-invariance

Fig. 8.16: The generators s_1, s_2, s_0 of $\widehat{\mathfrak{S}}_6$ pictured from the bottom of the cylinder on which they are drawn (with some artistic license).

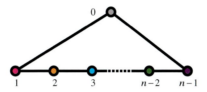

Fig. 8.17: The Coxeter graph of the affine symmetric group $\widehat{\mathfrak{S}}_n$.

Definition 8.3.7. Set $(W, P) = (\widehat{\mathfrak{S}}_n, \mathfrak{S}_n)$. We say that a vertex $\pi \in {}^P W$ generates a σ-Steinberg region for $\sigma \in S_W$ if $\pi\sigma < \pi$ and $\pi < \pi\tau$ for all $\sigma \neq \tau \in S_W$.

Example 8.3.8. For $(W, P) = (\widehat{\mathfrak{S}}_3, \mathfrak{S}_3)$, the element $\pi = \sigma \in {}^P W$ generates a σ-Steinberg. To see this, note that $\sigma\sigma = 1 < \sigma$ whereas the elements

$$\sigma, \quad \sigma\rho, \quad \sigma\tau, \quad \sigma\rho\tau, \quad \sigma\tau\rho, \quad \sigma\rho\tau\rho$$

are all greater than σ in the strong Bruhat order. These six elements form a copy of the finite $\mathfrak{S}_3$ which is highlighted in Figure 8.18.

Example 8.3.9. For $(W, P) = (\widehat{\mathfrak{S}}_3, \mathfrak{S}_3)$, the element $\pi = \sigma\rho\tau\sigma \in {}^P W$ is not a σ-Steinberg (or indeed a τ-Steinberg or a ρ-Steinberg). To see this, note that $\pi < \pi\sigma, \pi\tau$ and indeed the latter element belongs to the parabolic subgroup, see Figure 8.18.

Theorem 8.3.10. Set $(W, P) = (\widehat{\mathfrak{S}}_n, \mathfrak{S}_n)$. Suppose that $\pi \in {}^P W$ is below a σ-Steinberg. We let $\mathrm{St}(\pi) = \{\pi w \mid w \in \langle \tau \mid \sigma \neq \tau \in S_W \rangle\}$. We have a bijective map, $\mathrm{st} : \mathfrak{S}_n \to \mathrm{St}(\pi)$ which respects the Bruhat ordering. Under this map, the finite and affine Kazhdan–Lusztig polynomials coincide, that is

$$n_{\lambda\mu} = n_{\mathrm{st}(\lambda), \mathrm{st}(\mu)}$$

for $\lambda, \mu \in \mathfrak{S}_n = \langle \tau \mid \sigma \neq \tau \in S_W \rangle \leqslant \widehat{\mathfrak{S}}_n$.

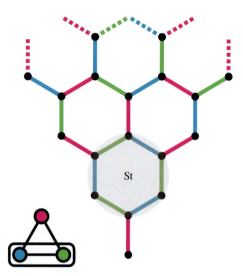

Fig. 8.18: The embedding of $\mathfrak{S}_3$ to the "Steinberg" region of the weak Bruhat graph for $\mathfrak{S}_3 \leqslant \widehat{\mathfrak{S}}_3$. Notice the circled region is the Bruhat graph for $\mathfrak{S}_3$.

Proof. Our vertex $\pi \in {}^P W$ has been chosen so that it is the minimal element in $\mathrm{St}(\pi)$ and so that the elements of $\mathrm{St}(\pi)$ are of the form πw for some $w \in \mathfrak{S}_n = \langle \tau \mid \sigma \neq \tau \in S_W \rangle \leqslant \widehat{\mathfrak{S}}_n$. Thus the map $\mathrm{st} : \mathfrak{S}_n \to \mathrm{St}(\pi)$ given by $\mathrm{st}(w) = \pi w$ is a bijection and it respects the Bruhat ordering. Furthermore, we can choose our reduced paths $\mathsf{T}_{\pi w}$ for $\pi w \in \mathrm{St}(\pi)$ so that they are all of the form

$$\mathsf{T}_{\pi w} = \mathsf{T}_\pi \otimes \mathsf{T}_w$$

for any choice of reduced path T_π terminating at π and any $w \in \mathfrak{S}_n$. Moreover

$$\sum_{\mathsf{T} \in \mathrm{Path}_{(W,P)}(\lambda, \mathsf{T}^\mu)} q^{\deg(\mathsf{T})} = \sum_{\mathsf{S} \in \mathrm{Path}_{(W,P)}(\mathrm{st}(\lambda), \mathsf{T}_{\pi\mu})} q^{\deg(\mathsf{S})}$$

and the result follows. □

Example 8.3.11. Continuing with Example 8.3.8, compare the highlighted (6×6)-submatrix of Table 8.1 with the matrix in Figure 8.8 and hence deduce that Theorem 8.3.10 holds in this case.

Example 8.3.12. For $(W, P) = (\widehat{\mathfrak{S}}_3, \mathfrak{S}_3)$ and continue with Example 8.3.9. We notice that $n_{\pi, \pi \rho \tau} = 0$ and, in fact, the conditions of Theorem 8.3.10 are necessary if we wish to ignore the parabolic subgroup.

We have seen that certain natural isomorphisms between closed subsets of Bruhat graphs can be lifted to the level of Kazhdan–Lusztig polynomials. The following conjecture generalises this idea as far as possible and justifies our claim that *Kazhdan–Lusztig polynomials encode a great deal of information about Bruhat graphs.*

8.3 The strong Bruhat order and combinatorial-invariance

Conjecture 8.3.13 (The combinatorial invariance conjecture). *Let W and W' be Coxeter groups and let $\lambda, \mu \in W$ and $\alpha, \beta \in W'$. Let $\Pi \subseteq W$ and $\Pi' \subseteq W'$ be closed subsets and suppose that $(\Pi, \leqslant)$ and $(\Pi', \leqslant)$ are isomorphic as posets. We have that $n_{\lambda,\mu}(q) = n_{\alpha,\beta}(q)$ for $\lambda, \mu \in \Pi$ and $\alpha, \beta \in \Pi'$.*

Example 8.3.14. For $\mathfrak{S}_4$, the strong Bruhat order is depicted in Figure 8.19 and

$$n_{(2,3),(1,3)(2,4)} = q + q^3 = n_{(1,2)(3,4),(1,4)}.$$

The isomorphic sub-posets $\{\lambda \mid (2,3) \leqslant \lambda \leqslant (1,3)(2,4)\}$ and $\{\alpha \mid (1,2)(3,4) \leqslant \lambda \leqslant (1,4)\}$ are both known as "the 4-crown" and are depicted in Figure 8.20.

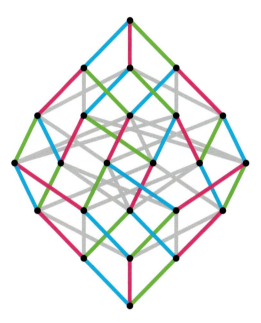

Fig. 8.19: The strong Bruhat graph of $\mathfrak{S}_4$.

Fig. 8.20: The 4-crown subposet of the strong Bruhat order on $\mathfrak{S}_4$.

8.4 The infinite dihedral group

We now consider the infinite parabolic Coxeter systems for which we can provide complete answers: namely, the infinite dihedral group

$$W = \widehat{\mathfrak{S}}_2 = \langle \sigma, \tau \mid \sigma^2 = 1 = \tau^2 \rangle$$

and its trivial and maximal parabolics, whose weak Bruhat graphs are depicted in Figure 8.21. For any $\mu \in W$ we have that T_μ is unique and that

$$\sum_{\{\lambda \mid \lambda \leq \mu\}} |\mathrm{Path}_W(\lambda, \mathsf{T}_\mu)| = 2^{\ell(\mu)}.$$

Exercise 8.4.1. Show that the $|\mathrm{Path}_W(\lambda, \mathsf{T}_\mu)|$ for $\lambda, \mu \in W$ are equal to binomial coefficients of the form $\binom{\ell(\mu)-1}{k}$ for some $0 \leq k < \ell(\mu)$ dependent on $\ell(\lambda)$ and whether or not λ and μ have the same leftmost reflection.

Exercise 8.4.2. Prove that the entries of the Δ-matrix of $\widehat{\mathfrak{S}}_2$ belong to $\mathbb{Z}_{\geq 0}[q]$.

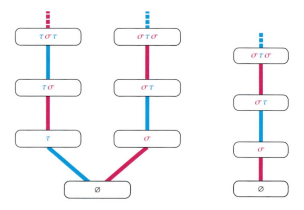

Fig. 8.21: The weak Bruhat graphs of $(W, P) = (\widehat{\mathfrak{S}}_2, 1)$ and $(\widehat{\mathfrak{S}}_2, \mathfrak{S}_2)$, for $\mathfrak{S}_2 = \langle \tau \rangle$.

By Exercise 8.4.2, the matrix $(B_{\lambda,\mu})_{\lambda,\mu \in W}$ is concentrated entirely in degree zero and, indeed, counts precisely the paths of degree zero. The matrix $(B_{\lambda,\mu})_{\lambda,\mu \in W}$ can be calculated using the "half Pascal triangle" (depicted in Figure 8.23) in an intuitive, but fiddly, manner. We refer to Figure 8.22 for an example which illustrates the idea. The Kazhdan–Lusztig polynomials are equal to $n_{\lambda,\mu} = q^{\ell(\mu)-\ell(\lambda)}$ for $\lambda \leq \mu$, see for example Figure 8.22.

8.4 The infinite dihedral group

	στστστσ	τστστστ	στστστ	τστστσ	στστσ	τστστ	στστ	τστσ	στσ	τστ	στ	τσ	σ	τ	∅
στστστσ	1	·	·	·	·	·	·	·	·	·	·	·	·	·	·
τστστστ	·	1	·	·	·	·	·	·	·	·	·	·	·	·	·
στστστ	q	q	1	·	·	·	·	·	·	·	·	·	·	·	·
τστστσ	q	q	·	1	·	·	·	·	·	·	·	·	·	·	·
στστσ	q^2	q^2	q	q	1	·	·	·	·	·	·	·	·	·	·
τστστ	q^2	q^2	q	q	·	1	·	·	·	·	·	·	·	·	·
στστ	q^3	q^3	q^2	q^2	q	q	1	·	·	·	·	·	·	·	·
τστσ	q^3	q^3	q^2	q^2	q	q	·	1	·	·	·	·	·	·	·
στσ	q^4	q^4	q^3	q^3	q^2	q^2	q	q	1	·	·	·	·	·	·
τστ	q^4	q^4	q^3	q^3	q^2	q^2	q	q	·	1	·	·	·	·	·
στ	q^5	q^5	q^4	q^4	q^3	q^3	q^2	q^2	q	q	1	·	·	·	·
τσ	q^5	q^5	q^4	q^4	q^3	q^3	q^2	q^2	q	q	·	1	·	·	·
σ	q^6	q^6	q^5	q^5	q^4	q^4	q^3	q^3	q^2	q^2	q	q	1	·	·
τ	q^6	q^6	q^5	q^5	q^4	q^4	q^3	q^3	q^2	q^2	q	q	·	1	·
∅	q^7	q^7	q^6	q^6	q^5	q^5	q^4	q^4	q^3	q^3	q^2	q^2	q	q	1

	στστστσ	τστστστ	στστστ	τστστσ	στστσ	τστστ	στστ	τστσ	στσ	τστ	στ	τσ	σ	τ	∅
στστστσ	1	·	·	·	·	·	·	·	·	·	·	·	·	·	·
τστστστ	·	1	·	·	·	·	·	·	·	·	·	·	·	·	·
στστστ	·	·	1	·	·	·	·	·	·	·	·	·	·	·	·
τστστσ	·	·	·	1	·	·	·	·	·	·	·	·	·	·	·
στστσ	1	·	·	·	1	·	·	·	·	·	·	·	·	·	·
τστστ	·	1	·	·	·	1	·	·	·	·	·	·	·	·	·
στστ	·	·	2	·	·	·	1	·	·	·	·	·	·	·	·
τστσ	·	·	·	2	·	·	·	1	·	·	·	·	·	·	·
στσ	2	·	·	·	3	·	·	·	1	·	·	·	·	·	·
τστ	·	2	·	·	·	3	·	·	·	1	·	·	·	·	·
στ	·	·	5	·	·	·	4	·	·	·	1	·	·	·	·
τσ	·	·	·	5	·	·	·	4	·	·	·	1	·	·	·
σ	5	·	·	·	9	·	·	·	5	·	·	·	1	·	·
τ	·	5	·	·	·	9	·	·	·	5	·	·	·	1	·
∅	·	·	14	14	·	·	14	14	·	·	6	6	·	·	1

Fig. 8.22: The matrices $(N_{\lambda,\mu})_{\lambda,\mu\in\Pi}$ and $(B_{\lambda,\mu})_{\lambda,\mu\in\Pi}$ for $\Pi = \{v \in \widehat{\mathfrak{S}}_2 \mid \ell(v) \leqslant 7\}$.

200   8 General Kazhdan–Lusztig theory

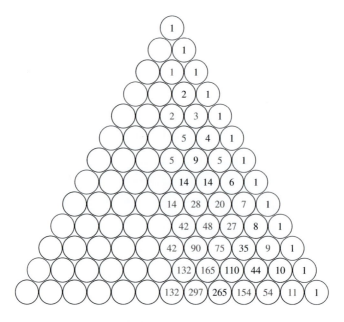

Fig. 8.23: The first few layers of the "half-Pascal triangle". Compare with the second matrix in Figure 8.22.

Example 8.4.3. The strong Bruhat graph of $\widehat{\mathfrak{S}}_2$ has infinitely many copies of the strong Bruhat graph of $\mathfrak{S}_3$, we identify one such copy in Figure 8.24.

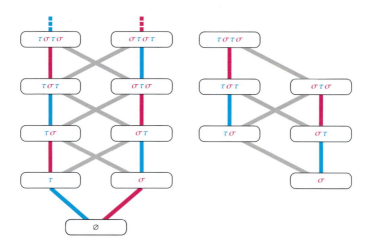

Fig. 8.24: The strong Bruhat graph of $\widehat{\mathfrak{S}}_2$ and a subposet which is isomorphic to the strong Bruhat graph of $\mathfrak{S}_3$.

We emphasise that the Kazhdan–Lusztig polynomials do match-up under these poset isomorphisms, but the entries of the B-matrix (of self dual polynomials) do not. In more detail, the B-matrix for $\mathfrak{S}_3$ has a single off-diagonal non-zero entry given by 1, whereas varying over the corresponding B-matrices under the poset isomorphisms we obtain off-diagonal entries given by all $n \in \mathbb{Z}_{\geqslant 0}$.

Exercise 8.4.4. Calculate the path-counting polynomials for $W = \widehat{\mathfrak{S}}_2$ and $P = \mathfrak{S}_2 = \langle \tau \rangle$. Factorise this matrix and hence compute the corresponding Kazhdan–Lusztig polynomials.

8.5 Kazhdan–Lusztig combinatorics?

We begin this section with a sort of "negative" result which basically says *"every polynomial is a Kazhdan–Lusztig polynomial"* (up to renormalisation). In light of this result, one might surrender ones hopes that the Kazhdan–Lusztig polynomials could ever be nice combinatorial objects (and indeed ask why we ever bothered naming these polynomials in the first place!). We will then attempt to rebuild the reader's faith by taking a tour through some of the beautiful combinatorial results which have been proven for these polynomials.

Theorem 8.5.1. *Let $\ell \in \mathbb{Z}_{\geqslant 0}$ and coefficients $a_1, a_2, \ldots, a_\ell \in \mathbb{Z}_{\geqslant 0}$ be arbitrary. There exists $\lambda, \mu \in \mathfrak{S}_n$ for some $n \in \mathbb{Z}_{>0}$ with $2\ell = \ell(\mu) - \ell(\lambda) - 2$ such that*

$$n_{\lambda,\mu}(q) = q^{2\ell+2} + a_\ell q^{2\ell} + \cdots + a_2 q^4 + a_1 q^2 \in \mathbb{Z}_{\geqslant 0}[q^2].$$

In other words, every monic polynomial in $\mathbb{Z}_{\geqslant 0}[q^2]$ with constant term equal to zero is equal to a Kazhdan–Lusztig polynomial of the finite symmetric group.

Recall that one of the enjoyable properties of quantum binomial coefficients was that their coefficients are unimodal Proposition 7.2.4 (they increase and then decrease). The columns of the Kazhdan–Lusztig matrix satisfy a similarly beautiful property:

Theorem 8.5.2 (Monotonicity [Irv88, BM01, Pla17]). *Let W be a Coxeter group. For $\lambda < \nu \leqslant \mu \in W$ we have that*

$$n_{\lambda,\mu}(q) - q^{\ell(\lambda)-\ell(\nu)} n_{\nu,\mu}(q) \in \mathbb{Z}_{\geqslant 0}[q].$$

In other words, reading down the μth column of the Kazhdan–Lusztig matrix the polynomials obtained are coefficient-wise weakly increasing (after appropriate renormalisation).

Another of our motivations for studying Kazhdan–Lusztig polynomials was our desire for "correct factors" for the failure of the rank generating functions to be palindromic. We now see that this is indeed a general property of these polynomials.

Theorem 8.5.3 (Palindromy [KL80]). *For $\mu \in W$ we have that the μth (length normalised) column-sum of Kazhdan–Lusztig polynomials* palindromic *in the sense that*

$$\sum_{\lambda \in W} q^{\ell(\mu)-\ell(\lambda)} n_{\lambda,\mu}(q) = a_0 + a_1 q^2 + \cdots + a_{\ell(\mu)-1} q^{2\ell(\mu)-2} + a_{\ell(\mu)} q^{2\ell(\mu)}$$

satisfies $a_k = a_{2\ell(\mu)-k}$ for all $0 \leq k \leq \ell(\mu)$.

One of the most exciting open questions in Kazhdan–Lusztig theory (certainly from a combinatorial perspective) is the famous combinatorial invariance conjecture (Conjecture 8.3.13). Remarkable new advances in our understanding of combinatorial invariance for parabolic Coxeter systems have come from both mathematicians [Mar18a, BLP23] and even Google Deepmind's artificial intelligence [BBD+22, DVB+21]. We now recall the most strikingly simple and provocative of these results.

Theorem 8.5.4 (Combinatorial Invariance for saturated sets [BCM06]). *Let W and W' be Coxeter groups. For $\alpha \in W$ and $\mu \in W'$ we suppose that the posets $\{\lambda \mid \lambda \leq \mu\}$ and $\{\beta \mid \beta \leq \alpha\}$ are isomorphic via a map φ. Then*

$$n_{\lambda,\nu}(q) = n_{\varphi(\lambda),\varphi(\nu)}(q)$$

for all $\lambda, \nu \leq \mu$.

Remark 14 (Quantised symmetric functions). We say that function in n variables is **symmetric** if it is invariant under permuting these variables. For example the functions

$$X_1^3 + X_1^2 X_2 + X_1 X_2^2 + X_2^3 \qquad X_1^2 X_2 + X_1 X_2^2 \qquad (8.5.5)$$

are both symmetric functions in 2 variables. Indeed every symmetric function of degree 3 in two variables can be written as a linear combination of the polynomials in (8.5.5) and so we say that they provide a basis for the symmetric polynomials (of degree 3 in two variables). The "LLT polynomials" were introduced in order to provide a basis for the quantised symmetric functions, and as such are some of the most important objects in algebraic combinatorics (there is an extensive literature on this subject, but recent highlights include the proof of the shuffle conjecture [CM18] and the Loehr–Warrington conjecture [JBS]). These LLT polynomials are, in fact, parabolic Kazhdan–Lusztig polynomials [LT00]. Therefore it is fair to say that Kazhdan–Lusztig polynomials are of central interest in algebraic combinatorics.

Remark 15. We recall from Example 6.1.6 that the trivial $TL_3(\delta)$ is spanned by the linear combination of diagrams:

8.5 Kazhdan–Lusztig combinatorics?

The coefficients of these diagrams in the trivial representation can be calculated recursively. The search closed (non-iterative) combinatorial formula for these coefficients has been of great interest to mathematicians and physicists (in particular, those working in quantum field theory). In recent work of Joe Baine, such a formula was given in terms of Kazhdan–Lusztig polynomials [Bai].

Remark 16. In [BBD+22, DVB+21] they remark that their supervised learning model was able to predict the Kazhdan–Lusztig polynomial from a given subposet of the Bruhat interval with reasonably high accuracy. They were unable to divine much mathematical intuition as to how the the model was doing this, and yet it is still encouraging to think that AI can intuit some direct connection between the shape of a closed region of the strong Bruhat graph and the forms of the associated Kazhdan–Lusztig polynomials. This gives further hope that mathematicians might one day find a direct way of combinatorially understanding this relationship.

Further Reading. We emphasise that the matrices

$$(\Delta_{\lambda,\mu}(q))_{\lambda,\mu \in {}^P W} \qquad (B_{\lambda,\mu}(q))_{\lambda,\mu \in {}^P W}$$

depend heavily on the choice of reduced paths. Thus it remains to show that the Kazhdan–Lusztig polynomials are independent of these choices, that is, that the polynomials $n_{\lambda,\mu}(q)$ are *well defined* for $\lambda, \mu \in {}^P W$. We do not prove this here (except in the case $(W, P) = \mathfrak{S}_{m+n}, \mathfrak{S}_m \times \mathfrak{S}_n$ of the previous chapter) but instead we refer to [Hum90, Chapter 7]. We remark that this proof requires that one lift the combinatorics of this section to that of *canonical bases* of Hecke algebras, which are deformations of the corresponding Coxeter groups by a quantum parameter. This really is the highlight of the theory of canonical bases and illustrates the power and importance of the Hecke algebra — therefore we recommend Humphrey's book [Hum90] as a must-read for anyone wanting to delve further into representation theory

Recall that for the *non-parabolic* Kazhdan–Lusztig polynomials, the *highest order* term is in degree $\ell(\mu) - \ell(\lambda)$ and the coefficient of this term is 1 (that is, $n_{\lambda,\mu} = q^{\ell(\mu)-\ell(\lambda)} + \ldots$ where the $\ldots$ are lower order terms). One can renormalise these polynomials so that the *lowest order* term is in degree 0 and the coefficient of this term is 1 — this convention was common in the early literature on the subject (and was adopted by the SAGE programming community). The convention that we have chosen is superior because it lifts to a structural level (in terms of radical and grading filtrations of cell modules) as we will soon see in Part IV of this book (the convention we have used here is that preferred by Soergel [Soe97]).

Finally, we remark that the polynomials we consider in this book are called the *anti-spherical Kazhdan–Lusztig polynomials* in the literature. There is a dual notion of *spherical Kazhdan–Lusztig polynomials* which we do not discuss in this book (except to say in the vaguest terms that they are "dual"). For those who wish to learn more about spherical and anti-spherical Kazhdan–Lusztig polynomials, my personal favourite reference is Soergel's excellent synopsis [Soe97] (which is where I learnt the majority of this material myself).

Part IV
Categorification

> *"I don't remember statements that are proved, but rather I try to keep a collection of pictures in my mind"*
>
> Pierre Deligne

Categorical Lie theory is the search for richer structures, or the "meta-symmetries" lying behind classical mathematical symmetries. These meta-symmetries secretly underpin classical mathematical problems: highlights include Khovanov's homological knot theory [Kho00], Crane's conjectural quantum gravity [Cra95], and the recent resolution of the Jantzen, Lusztig, and Kazhdan–Lusztig positivity conjectures [Wil16, EW14, Wil17]. In the spirit of Deligne's emphasis on visual understanding in mathematics, Part 4 of this book provides an entirely pictorial formulation of the disproof of Lusztig's conjecture — we hope you find it memorable!

Much of the categorification programme (and research mathematics in general!) is devoted to the construction of elaborate algebraic and geometric objects which explain the existence of natural dualities and physical phenomena. As a consequence, many of the most celebrated mathematical results have proofs which are understood by only a few living experts. However, all of this is ultimately just one mode of analytic thought. Diagrammatic calculus provides a new, wide-access, conduit to algebro-geometric ideas. It allows us to encode the "symmetries between symmetries" into an elementary visual game, which in Part 4 of this book is introduced as *the Hecke category* $\mathcal{H}_{(W,P)}$. Sophisticated geometric ideas can be encapsulated in "Tetris-style" rules of "game play". One does not require decades of mathematical experience in order to employ this diagrammatic reasoning. It is visual, intuitive, and can be used to prove powerful results in a self-contained manner — without ever appealing to the geometric and algebraic ideas lurking in the background. Over the following two chapters, we promise to guide the reader from *categorical Lie theory novice* all the way to *Hecke category expert*, arming the reader with a complete understanding of Williamson's explosive examples of p-torsion. We do this all within the diagrammatic incarnation of the Hecke category, explicitly constructed by Elias–Williamson [EW16, EW23] and Libedinsky–Williamson [LW22], building on the foundational work of Elias–Khovanov [EK10] and Elias [Eli16].

Whilst we have employed a diagrammatic approach, it is important to acknowledge that this is a modern lens. The following historical discussion of these ideas is primarily intended as grounding for the reader who is already versed in the geometric or algebraic foundations. Those new to the subject may safely skip this discussion.

The diagrammatic theory presented here was the culmination of the *algebraic* categorification programme, initiated by Soergel [Soe92, Soe00, Soe07]. Soergel provided an algebraic theory of "Soergel bimodules" and he conjectured the existence of concrete indecomposable bimodules, whose characters are expressible in terms of Kazhdan–Lusztig polynomials (thus providing an innately positive interpretation of the coefficients in the Kazhdan–Lusztig polynomials and resolving Conjecture 8.2.12). Soergel's algebraic theory extends to Coxeter groups, generalising earlier geometric work by Soergel, Kazhdan–Lusztig, Beilinson–Bernstein, and Brylinski–Kashiwara for Weyl groups [KL79, KL80, BK81, BB81, Soe92, Soe00]. Elias–Williamson finally proved Soergel's conjecture in [EW14], accrediting the diagrammatic calculus as the key to their breakthrough (see also [EW13]). We will discuss neither the geometric, nor the bimodule-theoretic incarnation of the Hecke category here, instead working entirely in the elementary (but powerful!) diagrammatic setting.

Lusztig's conjecture eclipses all other conjectures discussed here in terms of recognition and impact within the mathematical community. In the language of this book, this conjecture stated that the Hecke category of $(W, P) = (\widetilde{\mathfrak{S}}_n, \mathfrak{S}_n)$ should be torsion-free for $p > 2n - 4$. Lusztig's conjecture was a cornerstone of representation theory for decades, and has been tackled with increasing sophistication over that time. Early attempts by Cline, Parshall, and Scott explored the conjecture's connection to Koszul algebras. Lusztig himself outlined a geometric programme for large primes, later realised by Kashiwara–Tanisaki [KT95, KT96], Kazhdan–Lusztig [KL93, KL94a, KL94b], and Andersen–Jantzen–Soergel [AJS94]. A geometric approach championed by Bezrukavnikov, Mirkovic, Rumynin, Ginzburg, and Arkhipov, leveraging Springer fibres, has led to a subsequent reproof for $p \to \infty$ [ABG04, BBM04, BM13, BMR08]. Fiebig's groundbreaking work used sheaves on moment graphs in order to establish an explicit, if astronomically large, bound guaranteeing the conjecture's validity for all primes beyond this threshold [Fie12]. Although not explicitly articulated in the literature, the work of Elias–Williamson offers an additional proof of the Lusztig conjecture for large primes in the context of the diagrammatic Hecke category — we will discuss this later in Chapter 10.

The sheer energy invested in proving (and then reproving!) the asymptotic version of Lusztig's conjecture (as $p \to \infty$) by the world's most famous geometers and representation theorists does speak volumes about the community's unwavering belief in its correctness. The community's confidence in Lusztig's conjecture was further evidenced by the adoption of Soergel's reframing of a key part of this conjecture — "the Hecke category of $W = \mathfrak{S}_n$ should be torsion-free for $p > 2n - 4$" [Soe00] — simply as the "toy model".

The "toy model" (otherwise known as "the region around the Steinberg weight") is easier for computation and it was expected that this would provide a warm-up to a general proof of Lusztig's conjecture — in actuality this was where Lusztig's conjecture would later be *disproved!* In 2002, Tom Braden provided the first glimpse of p-torsion in the toy model for $W = \mathfrak{S}_8$, for $p = 2$ and in 2011 Patrick Polo found the first infinite families of p-torsion for $x \in \mathfrak{S}_{4n}$ and p dividing $n - 1$ (although neither of these calculations were published at the time). Finally, in 2013 Williamson provided his diagrammatic counterexamples within the "toy model" upending over 30 years of research and forcing modular representation theorists to reconsider the realms of what is possible and impossible within the field. We present Geordie's counterexamples as the main result of Chapter 10 of this book.

At the beginning of this book we chose to be optimistic about the future, marvelling at the dichotomy between the behaviour encountered for small primes, versus the limit $p \to \infty$. In the final section of Chapter 10 we provide evidence for our optimistic viewpoint. We present possible explicit bounds for a new "toy model" for modular Lie theory, these strikingly simple bounds were suggested by He–Williamson. Taking the limit as $p \to \infty$, we explore the possibility of developing a combinatorial theory for all p-Kazhdan-Lusztig polynomials, potentially exceeding Lusztig's and Andersen's initial hopes and expectations!

Chapter 9
The diagrammatic algebra for $\mathfrak{S}_m \times \mathfrak{S}_n \leqslant \mathfrak{S}_{m+n}$

We will now proceed to "categorify" the Kazhdan–Lusztig polynomials of Part 3 of this book. What does this mean? Recall that the Kazhdan–Lusztig polynomials of (W, P) were defined in Part 3 of this book via two steps:

Step 1: We first formed a uni-triangular matrix $(\Delta_{\lambda\mu})_{\lambda,\mu \in {}^P W}$ whose λth row and μth column records the paths of shape λ and weight μ as a polynomial. In this chapter, we "categorify" this matrix by constructing a diagrammatic algebra $\mathcal{H}_{(W,P)}$ with idempotents 1_μ for $\mu \in {}^P W$ and with weighted cellular basis such that the graded dimension of $\Delta^\Bbbk(\lambda)1_\mu$ is equal to this polynomial. The key to this construction is Libedinsky and Williamson's proof that the "light leaves" diagrams form a basis of $\mathcal{H}_{(W,P)}$. This proof is well beyond the scope of this book, however we do provide a proof in the case that $(W, P) = (\mathfrak{S}_{m+n}, \mathfrak{S}_m \times \mathfrak{S}_n)$.

Step 2: We then factorised the matrix $(\Delta_{\lambda\mu})_{\lambda,\mu \in {}^P W}$ as a product of uni-triangular matrices

$$(\Delta_{\lambda\mu})_{\lambda,\mu \in {}^P W} = (N_{\lambda\nu})_{\lambda,\nu \in {}^P W}(B_{\nu\mu})_{\nu,\mu \in {}^P W} \qquad (9.0.1)$$

where the off-diagonal entries of N belonged to $q\mathbb{Z}[q]$ and the entries of B belonged to $\mathbb{Z}[q+q^{-1}]$. The Kazhdan–Lusztig polynomial $n_{\lambda\mu}(q)$ was defined to be the entry in the λth row and μth column of the matrix N. We will "categorify" this matrix factorisation by observing that any basis element in $\Delta^{\mathbb{C}}(\lambda)1_\mu$ belongs to some simple composition factor $L^{\mathbb{C}}(\nu)\langle k \rangle$ of $\Delta^{\mathbb{C}}(\lambda)$ for $k \geq 0$ (with $k = 0$ only if $\lambda = \nu$). This allows us to interpret the entry in the λth row and νth column of $(N_{\lambda\nu})_{\lambda,\nu \in {}^P W}$ as the graded composition factor multiplicity

$$n_{\lambda,\nu}(q) = \sum_{k \in \mathbb{Z}}[\Delta^{\mathbb{C}}(\lambda) : L^{\mathbb{C}}(\nu)\langle k \rangle]q^k \qquad (9.0.2)$$

and the entry in the νth row and μth column of $(B_{\nu\mu})_{\nu\mu \in {}^P W}$ as the graded dimension

$$b_{\nu\mu}(q) = \dim_q(L^\Bbbk(\nu)1_\mu). \qquad (9.0.3)$$

9 The diagrammatic algebra for $\mathfrak{S}_m \times \mathfrak{S}_n \leqslant \mathfrak{S}_{m+n}$

We note that composition factor multiplicities in (9.0.2) are innately non-negative and so this proves that the coefficients of Kazhdan–Lusztig polynomials are non-negative: this was the subject of the famous Kazhdan–Lusztig positivity conjecture. The key to the categorification of (9.0.1) is Elias–Williamson's "Hodge theoretic" proof that $[\Delta^{\mathbb{C}}(\lambda) : L^{\mathbb{C}}(\nu)\langle k \rangle] \in q\mathbb{Z}[q]$ for $\lambda \neq \nu$. This proof is well beyond the scope of this book, however we do prove the result in the case that $(W, P) = (\mathfrak{S}_{m+n}, \mathfrak{S}_m \times \mathfrak{S}_n)$, where we have already seen (in Part 3 of this book) that the matrix factorisation $\Delta = NB$ is given trivially by $N = \Delta$ and $B = \mathrm{id}$.

This chapter focusses on the diagrammatic algebra $\mathcal{H}_{(W,P)}$ for $(W, P) = (\mathfrak{S}_{m+n}, \mathfrak{S}_m \times \mathfrak{S}_n)$, the parabolic Coxeter system that formed the focus of Chapter 7. Recall that the Bruhat graph of $(W, P) = (\mathfrak{S}_{m+n}, \mathfrak{S}_m \times \mathfrak{S}_n)$ does not contain any $2m$-gons for $m > 2$ and that this simplification allowed us to obtain a concrete understanding of the associated Kazhdan–Lusztig polynomials. In this chapter, we will be able to lift this to a concrete understanding of the important structural results for the diagram algebras $\mathcal{H}_{(W,P)}$ for $(W, P) = (\mathfrak{S}_{m+n}, \mathfrak{S}_m \times \mathfrak{S}_n)$ (complete with proofs!). It is also worth emphasising that for general Coxeter systems *even the definition of $\mathcal{H}_{(W,P)}$ is difficult* (as we will see in the next chapter) and that by first focussing on the case of $(W, P) = (\mathfrak{S}_{m+n}, \mathfrak{S}_m \times \mathfrak{S}_n)$ the reader will have the opportunity to play with the diagrammatics before encountering the full complexity of the presentation. In this chapter we will continue to utilise our combinatorial interpretation of cosets for $(W, P) = (\mathfrak{S}_{m+n}, \mathfrak{S}_m \times \mathfrak{S}_n)$ in terms of tile-partitions $\lambda \in \mathcal{P}_{m,n}$ from Chapter 7.

9.1 From paths to diagrammatic algebras

We wish to lift the path combinatorics to the level of (diagrammatic) algebra. Each σ-edge in the Bruhat graph has an identity map to itself and we have a corresponding idempotent 1_σ. Each reduced path T has an associated diagram (denoted by 1_T) which we obtain by horizontally concatenating (denoted $\otimes$) the identity generator for each edge traversed by the path as depicted in Figure 9.1. We also associate a generator, $1_\emptyset$, to the empty path and we represent this by an empty frame.

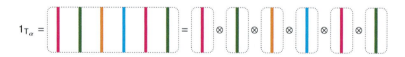

Fig. 9.1: The idempotent for the path T_α depicted in Figure 7.4.

Let $(W, P) = (\mathfrak{S}_{m+n}, \mathfrak{S}_m \times \mathfrak{S}_n)$. Fix S, T two reduced paths of shape $\lambda, \mu \in {}^P W$. We define an (S, T)-**Soergel diagram** to be any diagram whose top and bottom edges are given by the idempotents 1_S and 1_T and any local region is one of the following

9.1 From paths to diagrammatic algebras

monochrome diagrams:

$$1_\emptyset = \quad\quad 1_\sigma = \quad\quad \mathsf{spot}^\emptyset_\sigma = \quad\quad \mathsf{fork}^\sigma_{\sigma\sigma} = \qquad\qquad (9.1.1)$$

which we refer to as the empty, idempotent, spot and fork diagrams associated to any $\sigma \in S_W$; or is the two-coloured braid diagram

$$\mathsf{braid}^{\tau\sigma}_{\sigma\tau} = \qquad\qquad (9.1.2)$$

for $\sigma, \tau \in S_W$ with $m_{\sigma,\tau} = 2$; or is obtained by flipping such a diagram through the horizontal axis.

Alternatively, an (S, T)-Soergel diagram is any diagram formed from the horizontal and vertical concatenation of the diagrams in (9.1.1) and (9.1.2) (so that the top and bottom of the resulting diagram have the correct colour sequences). Here the vertical concatenation of an (S, T)-Soergel diagram on top of a (U, V)-Soergel diagram is zero if T ≠ U. We say that an (S, T)-Soergel diagram has northern colour sequence S and southern colour sequence T. Examples are given in Figure 9.2 and Example 9.1.5.

The spot and fork diagrams will correspond to shrinking and growing paths; the braid diagram will correspond to bending a path around the perimeter of a $2m$-gon in the Bruhat graph. The only $2m$-gons in the Bruhat graph for $(W, P) = (\mathfrak{S}_{m+n}, \mathfrak{S}_m \times \mathfrak{S}_n)$ are squares (whereas we have already seen that general Coxeter groups have hexagons and $2m$-gons in their Bruhat graphs for $m \geqslant 2$) — this makes the situation of this chapter simpler and hence a manageable introduction to the general case (which we will consider in the next chapter).

Given S, T any two reduced paths of shape λ we have an associated braid diagram $\mathsf{braid}^\mathsf{T}_\mathsf{S}$. In the case of $(W, P) = (\mathfrak{S}_{m+n}, \mathfrak{S}_m \times \mathfrak{S}_n)$ all our cosets are fully commutative and so the only polygons in the Bruhat graph are squares corresponding to $m_{\sigma\tau} = 2$ in the Coxeter presentation. These *commuting* braids are particularly easy to describe: we simply define $\mathsf{braid}^\mathsf{T}_\mathsf{S}$ to be the colour-preserving bijection between the pair of paths S, T drawn with the minimal number of crossings.

Fig. 9.2: Braid diagrams for $\mathsf{S} = \mathsf{U}^1_2\mathsf{U}^1_3\mathsf{U}^1_1\mathsf{U}^1_4$, $\mathsf{T} = \mathsf{U}^1_2\mathsf{U}^1_1\mathsf{U}^1_3\mathsf{U}^1_4$ and $\mathsf{U} = \mathsf{U}^1_2\mathsf{U}^1_1\mathsf{U}^1_4\mathsf{U}^1_3$ paths in the Bruhat graph pictured in Figure 7.2.

Definition 9.1.3. We define the graded degree of a spot generator to be +1 and that of the fork generator to be −1. All other regions of a diagram have degree 0.

Let T_μ be a fixed choice of reduced path to $\mu \in \mathscr{P}_{m,n}$. Given $S \in \text{Path}(\lambda, T_\mu)$, we construct an associated diagram d_S as follows. We proceed by induction, assuming that the next step in the path has colour σ. There are four cases (corresponding to the four possible traversals of an edge) to consider.

Definition 9.1.4. We define up and down operators on diagrams as follows:

- Suppose that D has northern colour sequence $S \in \text{Path}(\lambda)$ and that $\sigma \in \text{Add}(\lambda)$. Fix a preferred choice of $T \in \text{Path}(\lambda + \sigma)$ and $U \in \text{Path}(\lambda)$. We define

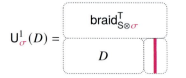

- Suppose that $\sigma \in \text{Rem}(\lambda)$ and that D has northern colour sequence $S \otimes \sigma \in \text{Path}(\lambda)$. Fix a preferred choice of $T \in \text{Path}(\lambda)$ and $U \in \text{Path}(\lambda - \sigma)$. We define

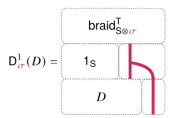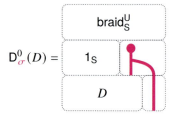

For $S \in \text{Path}(\lambda, T_\mu)$ written as a concatenation of a sequence of U_σ^0, U_σ^1 and D_σ^0, D_σ^1 steps in the Bruhat graph we define d_S to be the corresponding sequence of up U_σ^0, U_σ^1 and D_σ^0, D_σ^1 operators applied to the empty diagram.

This procedure is best illustrated via an example.

Example 9.1.5. We now construct the element d_S for $S = U_2^1 U_3^1 U_4^0 U_1^1 U_2^0 D_3^1$ from Figure 7.4, step-by-step. The first two steps are very easy

Now, we we take our first "failed up" U_4^0 step (followed by another, boring step):

9.1 From paths to diagrammatic algebras

Now, we perform another "failed up" step to obtain

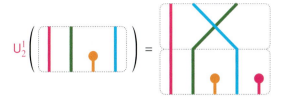

where here we have braided the top of the diagram in such a manner as to prepare us for the next step D_3^1 step. Having braided our diagram in the previous step so that we have a green strand at the far right of the diagram, we are now able to take a D_3^1 step,

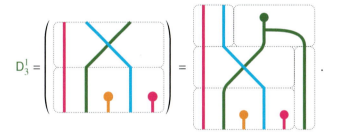

Example 9.1.6. The diagrams d_S for $\mathsf{S} \in \operatorname{Path}(\lambda, \mathsf{T}_\mu)$ for a fixed choice of reduced path for each $\lambda, \mu \in \mathscr{P}_{2,2}$ are depicted in Figure 9.3. The degrees of these diagrams should be compared with the entries in Table 7.1.

Exercise 9.1.7. Fix a preferred choice of reduced path T_μ for each $\mu \in \mathscr{P}_{2,3}$. Construct d_S indexed by each of the elements $\mathsf{S} \in \operatorname{Path}(\lambda, \mathsf{T}_\mu)$ for $\lambda \in \mathscr{P}_{2,3}$ and T_μ the leftmost path in Figure 7.4.

Example 9.1.8. Continuing with Example 9.1.5 with $\mathsf{S} = \mathsf{U}_2^1 \mathsf{U}_3^1 \mathsf{U}_4^0 \mathsf{U}_1^1 \mathsf{U}_2^0 \mathsf{D}_3^1$ we consider the diagram d_S. This diagram has northern colour sequence $s_2 s_1$ and southern colour sequence $s_2 s_3 s_4 s_1 s_2 s_3$.

We let $\ast$ denotes the map which flips a diagram through its horizontal axis.

Definition 9.1.9. Let $(W, P) = (\mathfrak{S}_{m+n}, \mathfrak{S}_m \times \mathfrak{S}_n)$ and $\Bbbk$ a field. For each $\mu \in \mathscr{P}_{m,n}$ we fix a preferred choice of reduced path T_μ to μ. We define the diagrammatic algebra $\mathscr{H}_{(W,P)}$ to be the $\Bbbk$-algebra generated by the diagrams d_S and their flips $d_\mathsf{S}^\ast$ for $\mathsf{S} \in \operatorname{Path}(-, \mathsf{T}_\mu)$ with multiplication given by $\circ$ modulo isotopy and the monochrome relations (M1) and (M2') and the multi-colour relations (S1) to (S3) together with the cyclotomic relations (C) and (P), which we will introduce throughout this chapter.

Example 9.1.10. For each choice of reduced path T_μ for $\mu \in \mathscr{P}_{m,n}$ the associated element 1_{T_μ} is an idempotent in $\mathscr{H}_{(W,P)}$. To see this, simply note that $1_{\mathsf{T}_\mu} \circ 1_{\mathsf{T}_\mu}$ is merely "twice as tall" as 1_{T_μ}, and so these diagrams are equal by isotopy.

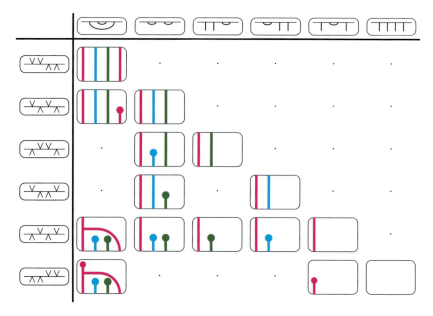

Fig. 9.3: The Soergel diagrams d_S for $S \in \text{Path}(\lambda, T_\mu)$ for $\mathfrak{S}_2 \times \mathfrak{S}_2 \leq \mathfrak{S}_4$. Here we have chosen T_μ so that any blue steps occur before any green steps. Compare this with Table 7.1 in the light of the the discussion at the beginning of this chapter (regarding "what is categorification").

9.2 The one colour relations

We begin with the monochrome relations for diagrammatic algebras. To any single reflection $\sigma \in S_W$, the associated one-colour local relations are as follows

$$\text{[diagram]} = \text{[diagram]} \qquad \text{[diagram]} = \text{[diagram]} \qquad \text{[diagram]} = 0 \quad \text{(M1)}$$

which we refer to as the "double-fork", "fork-spot contraction" and "circle annihilation" relations, together with the following "cinching" relation

$$\text{[diagram]} = \text{[diagram]} + \text{[diagram]} - \text{[diagram]} \quad \text{(M2)}$$

and their vertical and horizontal flips.

9.2 The one colour relations

Finally, we have the non-local (one colour) cyclotomic relation

$$\begin{array}{c}\rule{0pt}{2ex}\end{array} \otimes D = 0 \qquad (C)$$

for $\sigma \in S_W$ and D any diagram.

Remark 17. We emphasise that any of the "local" relations can be applied to an arbitrary region of the diagram. On the other hand, the non-local relation (C) says that if a barbell gets *all the way to the left* of the diagram, then it is zero. This is *not* a local relation (of course!!). If it were local, it would mean that *any diagram* with a barbell is zero, which would beg the question as to why we had bothered with these barbells in the first place.

We now consider the diagram algebra $\mathcal{H}_W$ associated to $W = \mathfrak{S}_2$, the finite symmetric group on 2 letters. The Bruhat graph for $W = \mathfrak{S}_2$ consists of two vertices labelled by the identity and the transposition $\sigma = (1,2) \in \mathfrak{S}_2$ (equivalently, the corresponding weight labels from the previous section) which are connected by a single edge.

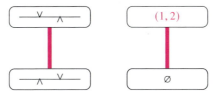

Fig. 9.4: The Bruhat graph for $W = \mathfrak{S}_2$ (via weights and permutations).

There are 2 reduced paths: the empty path $\mathsf{T}_\emptyset$ and $\mathsf{T}_\sigma = \mathsf{U}_\sigma^1$; there is also one non-reduced path $\mathsf{U}_\sigma^0 \in \text{Path}(\emptyset, \mathsf{T}_\sigma)$. The corresponding diagrams are

and taking naive products in $\mathcal{H}_{\mathfrak{S}_2}$ we obtain the five elements

(9.2.1)

and, taking further products, we also obtain diagrams with (powers of) the barbell; however these diagrams with barbells are all zero by (C). Therefore the elements of

(9.2.1) form a spanning set. By the fact that $1_\emptyset 1_\sigma = 0$ and degree considerations, we deduce that these diagrams are also linearly independent.

The multiplication table for $\mathscr{H}_{\mathfrak{S}_2}$ is given in Figure 9.5. These calculations are quite easy: many of the zeroes come from the idempotent relations; the other zeroes follow immediately from (C); the non-zero products are simply given by concatenation.

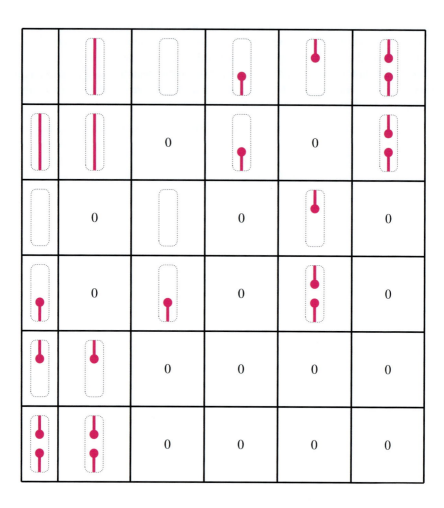

Fig. 9.5: The multiplication table for $\mathscr{H}_{\mathfrak{S}_2}$.

As a right $\mathscr{H}_{\mathfrak{S}_2}$-module, we have that $\mathscr{H}_{\mathfrak{S}_2}$ decomposes as a direct sum of two indecomposable modules

$$1_{T_\emptyset} \mathscr{H}_{\mathfrak{S}_2} \oplus 1_{T_\sigma} \mathscr{H}_{\mathfrak{S}_2}$$

9.2 The one colour relations

these two indecomposable modules are depicted in Figures 9.6 and 9.7. We note that $\mathcal{H}_{\mathfrak{S}_2}$ is generated by the degree 0 elements (the idempotents) together with the degree 1 elements (the spot and its dual) and we further record the action of these degree 1 generators in Figures 9.6 and 9.7 (the action of the idempotents is trivial).

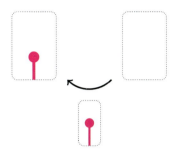

Fig. 9.6: The action of non-idempotent generators of $\mathcal{H}_{\mathfrak{S}_2}$ on $1_\varnothing \mathcal{H}_{\mathfrak{S}_2}$. (Any non-idempotent generator not depicted acts as zero.)

Fig. 9.7: The action of non-idempotent generators of $\mathcal{H}_{\mathfrak{S}_2}$ on $1_\sigma \mathcal{H}_{\mathfrak{S}_2}$. (Any non-idempotent generator not depicted acts as zero.)

The algebra $\mathcal{H}_{\mathfrak{S}_2}$ has two non-isomorphic simple modules: these are the 1-dimensional spaces spanned by the idempotents (modulo the spot generators acting as zero).

Exercise 9.2.2. Check that this algebra is graded weighted cellular with idempotents 1_σ and $1_\varnothing$. Construct the right cell modules $\Delta^{\Bbbk}(\varnothing)$ and $\Delta^{\Bbbk}(\sigma)$ and identify the filtrations of $1_\varnothing \mathcal{H}_{\mathfrak{S}_2}$ and $1_\sigma \mathcal{H}_{\mathfrak{S}_2}$ by cell modules.

Exercise 9.2.3. Construct the graded Alperin diagrams of the indecomposable modules $1_\varnothing \mathcal{H}_{\mathfrak{S}_2}$ and $1_\sigma \mathcal{H}_{\mathfrak{S}_2}$.

We now observe that we have actually seen the algebra $\mathcal{H}_{\mathfrak{S}_2}$ in a different context, earlier in this book...

Proposition 9.2.4. *Let $\Bbbk = \mathbb{F}_2$. Recall our construction of the zig-zag algebra, ZZ_2, which was Morita equivalent to the Schur algebra $S^{\Bbbk}(2)$ from Example 6.12.2. The map : $\mathcal{H}_{\mathfrak{S}_2} \to ZZ_2$ given by mapping the ordered 5 elements*

to the ordered 5 elements

$$\begin{bmatrix} 0 & 0 & 0 & 0 \\ 0 & 1 & 0 & 0 \\ 0 & 0 & 1 & 0 \\ 0 & 0 & 0 & 0 \end{bmatrix} \begin{bmatrix} 1 & 0 & 0 & 0 \\ 0 & 0 & 0 & 0 \\ 0 & 0 & 0 & 0 \\ 0 & 0 & 0 & 0 \end{bmatrix} \begin{bmatrix} 0 & 0 & 0 & 0 \\ 1 & 0 & 0 & 0 \\ 1 & 0 & 0 & 0 \\ 0 & 0 & 0 & 0 \end{bmatrix} \begin{bmatrix} 0 & 1 & 1 & 0 \\ 0 & 0 & 0 & 0 \\ 0 & 0 & 0 & 0 \\ 0 & 0 & 0 & 0 \end{bmatrix} \begin{bmatrix} 0 & 0 & 0 & 0 \\ 0 & 1 & 1 & 0 \\ 0 & 1 & 1 & 0 \\ 0 & 0 & 0 & 0 \end{bmatrix}$$

is a graded $\Bbbk$-algebra isomorphism.

Proof. This can be checked by calculating the products of the matrices and comparing the answer with Figure 9.5. This is left as an exercise for the reader. $\square$

Relation (M2) will allow us to move barbells leftwards through the diagram (and this will be important later on in this chapter). To see this, we pre and post multiply (M2) with spot generators in order to obtain

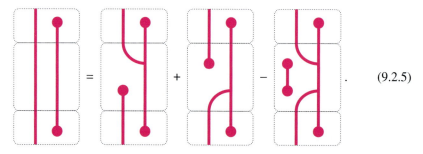 (9.2.5)

Now, we apply the fork-spot contraction relations to the above to tidy this up, and we hence obtain the so-called "Demazure" relation

$$\left| \begin{array}{c} \bullet \\ \bullet \end{array} \right| = 2 \left| \begin{array}{c} \bullet \\ \bullet \end{array} \right| - \left| \begin{array}{c} \bullet \\ \bullet \end{array} \right| \qquad (\text{M2'})$$

This Demazure relation is commonly listed as one of the defining relations of the Hecke category, but we prefer to deduce it from (M2).

Exercise 9.2.6. Prove that relation (M2') implies relation (M2) (providing $p \neq 2$).

9.3 The multi-colour relations for $\mathfrak{S}_m \times \mathfrak{S}_n \leqslant \mathfrak{S}_{m+n}$

We now consider the two-colour relations for $(W, P) = (\mathfrak{S}_{m+n}, \mathfrak{S}_m \times \mathfrak{S}_n)$. We suppose that $\sigma, \tau \in W = \mathfrak{S}_{m+n}$ are two *non-commuting Coxeter reflections* (that is, $m_{\sigma,\tau} = 3$) and $\rho, \tau, \pi \in W$ are *commuting Coxeter reflections* (that is, $m_{\rho,\tau} = m_{\rho,\pi} = m_{\pi,\tau} = 2$). We have the local two-colour barbell relation which allows us to move a barbell through a strand at the expense of two error terms, as follows:

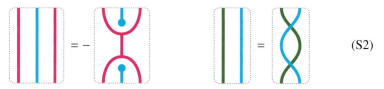
(S1)

and the local Temperley–Lieb relations

(S2)

and the local commutativity relations

(S3)

together with their flips through the horizontal and vertical axes. Finally, we have the non-local "parabolic annihilation" relation

$$\bigg| \otimes D = 0 \tag{P}$$

for any $\tau \in S_P$, and D any diagram. For the purposes of exposition we now (re)define the diagrammatic algebra for $(W, P) = (\mathfrak{S}_{m+n}, \mathfrak{S}_m \times \mathfrak{S}_n)$.

Definition 9.3.1. Let $(W, P) = (\mathfrak{S}_{m+n}, \mathfrak{S}_m \times \mathfrak{S}_n)$ and $\Bbbk$ be a field. We define $\mathscr{H}_{(W,P)}$ to be the $\Bbbk$-algebra generated by all d_S and their duals with multiplication given by vertical concatenation of diagrams modulo isotopy and the monochrome relations (M1) and (M2), the multi-colour relations (S1) to (S3), together with the cyclotomic relations (C) and (P).

Remark 18. We label these relation by (S1) to (S3) as these are the relations in the case that the Bruhat graphs contains only squares.

Exercise 9.3.2. Let S, T be two reduced paths of the same shape. Prove that $\text{braid}_T^S \text{braid}_S^T = 1_S$.

9.4 How to manipulate diagrams

We now discuss how one uses this diagrammatics in practice. When manipulating diagrams, a hugely useful idea is that of *factoring through* an idempotent. That is, we consider an arbitrary horizontal cut through a diagram; we take this cut in such a fashion as to avoid any spots, forks or braids (by isotopy, this is easy to do); we hence obtain a diagram of the form

$$1_\sigma \otimes 1_\tau \otimes \ldots$$

for some sequence of colours which *need not correspond to a path* in the weak Bruhat graph. For example, we depict several such cuts and the corresponding colour sequences in Figure 9.8.

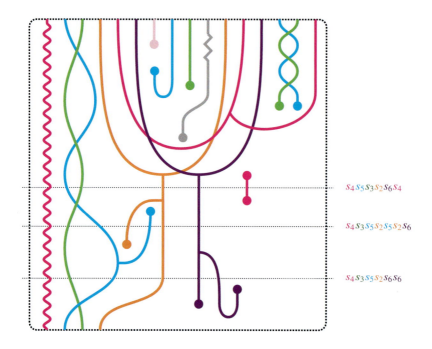

Fig. 9.8: Let $(W, P) = (\mathfrak{S}_8, \mathfrak{S}_4 \times \mathfrak{S}_4)$. We depict a complicated Soergel diagram with horizontal cuts corresponding to colour sequences $(4,5)(5,6)(3,4)(2,3)(6,7)(4,5)$, and $(4,5)(3,4)(5,6)(2,3)(5,6)(2,3)(6,7)$ and $(4,5)(3,4)(5,6)(2,3)(6,7)(6,7)$ respectively. See Figure 9.9 for the colouring of the Coxeter graph.

The whole point of categorification is to lift the combinatorics of paths in Bruhat graphs (and therefore Kazhdan–Lusztig polynomials) to the level of diagrammatic

9.4 How to manipulate diagrams

algebras. Indeed, we defined the generators of $\mathcal{H}_{(W,P)}$ to be the diagrams obtained from such paths. Later in this chapter we will prove that pairs of such paths index a basis of $\mathcal{H}_{(W,P)}$ via a "double trapezoid" shape illustrated in Figure 9.10. Notice that the top, bottom, and the cut through the centre of the diagram in Figure 9.10 are all colour sequences which correspond to reduced paths. Moreover, the centre of the diagram is "cinched" so as to form a "waist" of the diagram (in other words, the centre is skinnier than the top and bottom).

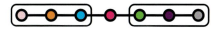

Fig. 9.9: The colouring of the reflections for $(W, P) = (\mathfrak{S}_8, \mathfrak{S}_4 \times \mathfrak{S}_4)$.

When manipulating diagrams, we should always try to bear in mind that we are trying to use cuts and paths as our organising principle and we should try to obtain the "cinched" double-trapezoid shape highlighted in Figure 9.10. There are two obvious ways in which a diagram might have a "fatter" middle than its top and bottom: the first is if there are barbells in the diagram (as these do not touch the top and bottom of the diagram) and the second is through the use of fork-branching. We now illustrate how to deal with each of these phenomena via small examples.

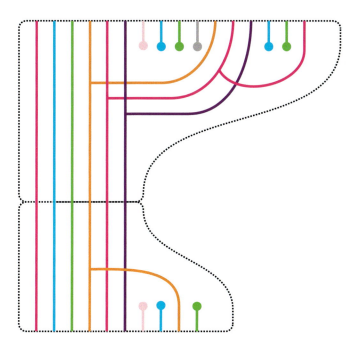

Fig. 9.10: An example of a cinched diagram with double trapezoidal shape.

Example 9.4.1. Let $(W, P) = (\mathfrak{S}_8, \mathfrak{S}_4 \times \mathfrak{S}_4)$ be coloured as in Figure 9.9. We will now illustrate how to remove barbells from diagrams. We do this by first pushing them all the way to the left of the diagram. We have that

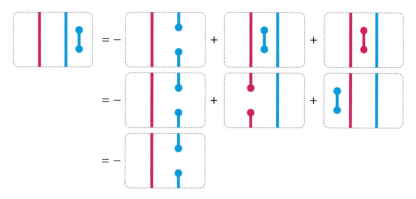

where the first two equalities move barbells to the left (using relations (M2') and (S1)); the third equality follows from (C) and (P).

Exercise 9.4.2. Continue with $(W, P) = (\mathfrak{S}_8, \mathfrak{S}_4 \times \mathfrak{S}_4)$ coloured as in Figure 9.9. Building on the ideas in Example 9.4.1, verify the following relation:

We note that in terms of paths in the graph in Figure 7.6, the above equation can be rewritten as

$$d_\mathsf{S} d_\mathsf{S}^* = -d_\mathsf{T}^* d_\mathsf{T} - d_\mathsf{U}^* d_\mathsf{U} \qquad (9.4.3)$$

where $\mathsf{S} \in \mathrm{Path}((2,1),(2,2))$, $\mathsf{T} \in \mathrm{Path}((2),(2,1))$ and $\mathsf{U} \in \mathrm{Path}((1^2),(2,1))$

Exercise 9.4.4. Continue with $(W, P) = (\mathfrak{S}_8, \mathfrak{S}_4 \times \mathfrak{S}_4)$ coloured as in Figure 9.9. Rewrite the following diagram as a linear combination of cinched/trapezoidal diagrams:

We remark that this case is slightly more complicated than that of Exercise 9.4.2. In particular, you will need to apply the non-commuting Temperley–Lieb relation (S2) to simplify some of the diagrams.

Exercise 9.4.5. Rewrite your answer to Exercise 9.4.4 in the path-theoretic form used in (9.4.3).

Exercise 9.4.6. For $1 \leq a \leq m, n$ consider the square partition $(a^a) \in \mathscr{P}_{m,n}$ and the partition $\lambda = (a^{a-1}, a) \in \mathscr{P}_{m,n}$ obtained from $\mu = (a^a)$ by removing the unique

9.4 How to manipulate diagrams

removable tile. For $S \in \text{Path}((a^{a-1}, a), (a^a))$, provide a rule for writing $d_S d_S^*$ as a linear combination of cinched diagrams of the form $d_T^* d_T$ for $T \in \text{Path}(\nu, T_\lambda)$ for $\nu \leqslant \lambda$. Hint: this is a difficult exercise, but your answers to Exercises 9.4.2 and 9.4.4 give the $a = 2$ and $a = 3$ cases.

So, we now understand the basic notion of "getting rid of barbells" by moving them all the way to the left (more to follow!). We now provide a few examples of how to rewrite other "fat in the middle" diagrams. We first consider the case in which our diagram looks like a "lollipop".

Example 9.4.7. Let $(W, P) = (\mathfrak{S}_8, \mathfrak{S}_4 \times \mathfrak{S}_4)$, coloured as in Figure 9.9. Let us consider an example of a "lollipop" product of diagrams whose "fat middle" comes from a combination of branching forks and barbells:

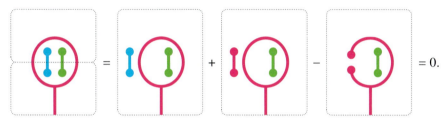

The first equality follows by applying the $\sigma\tau$-barbell relation to move the τ-barbell through the σ-strand immediately to its left. We hence obtain 3 diagrams each of which has a barbell which can be istoped all the way to the lefthand edge of the diagram and therefore each diagram is zero by relation (C).

Example 9.4.8. Let $(W, P) = (\mathfrak{S}_8, \mathfrak{S}_4 \times \mathfrak{S}_4)$. We now consider another product of two (non-zero) elements $d_S d_T^*$ for $S \in \text{Path}(\emptyset, T_{\sigma\tau\rho\sigma})$ and $T \in \text{Path}(\sigma\tau\rho, T_{\sigma\tau\rho\sigma})$. We have that

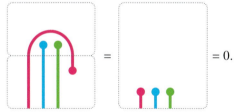

The first equality follows from fork-spot contraction and the second equality follows from relation (P) (since the blue or green strand can now be isotoped to touch the lefthand-side and both blue/green coloured reflections belong the parabolic $\mathfrak{S}_2 \times \mathfrak{S}_2$).

Having illustrated some of the ways to cinch a diagram, we now go into more detail on the kind of colour sequences one will see when making a horizontal cut as in Figure 9.8. We claim that the "waist" of our diagram (the skinniest idempotent through which the diagram factors) can always be rewritten as the colour sequence of a reduced path as in Figure 9.10. To illustrate how we might go about this inductive cinching process, let's first consider an example where the waist of the diagram is not the colour sequence of a reduced path.

Example 9.4.9. Applying the relation (S2), we have that

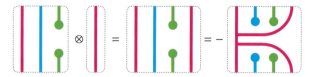

where the righthand-side is equal to $-d_T^* 1_\sigma d_T$ for $T \in \text{Path}(\sigma, T_{\sigma\tau\rho\sigma})$. What we have done here is taken a diagram whose middle *does not correspond to a reduced path* and rewritten it in terms of a diagram *with a skinnier middle* which *does correspond to a reduced path*. We refer to the skinnier diagram on the righthand-side of the equality as "cinched".

Example 9.4.10. Let $S \in \text{Path}(\sigma, T_{\sigma\tau\rho})$. We have that

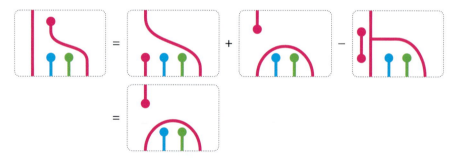

We now illustrate how the (inductively constructed) diagram $\mathsf{U}_\sigma^0(d_S)$ can be rewritten in terms of the elements $d_S^* 1_\lambda d_T$. We first apply isotopy to stretch the pink spot upwards, this allows us to easily see that we can apply (M2) to the two adjacent pink strands in the middle of the diagram to obtain

where the second equality follows from the relation (C) and (P). Again, we emphasise that what we have done here is taken a diagram whose middle *does not correspond to a reduced path* and rewritten it in terms of a diagram *with a skinnier middle* which *does correspond to a reduced path* (in this case, $T_\emptyset$ the empty path). We refer to the skinnier diagram on the righthand-side of the equality as "cinched".

What have we learnt thus far? For $m_{\sigma\tau} = 3 = m_{\sigma\rho}$, any idempotent of the form

$$1_\sigma \otimes 1_\tau \otimes 1_\sigma \quad \text{or} \quad 1_\sigma \otimes 1_\rho \otimes 1_\sigma \quad \text{or} \quad 1_\sigma \otimes 1_\sigma$$

can be cinched to factor through 1_σ using relation (S2), (S2) or (M2), respectively. Recall that for $(W, P) = (\mathfrak{S}_{m+n}, \mathfrak{S}_m \times \mathfrak{S}_n)$ the elements of $^P W$ can be though of

9.4 How to manipulate diagrams

as partition-tilings in $\mathscr{P}_{m,n}$. In this language, our observation can be rephrased as follows: given an idempotent corresponding to a tiling of the form

(9.4.11)

we can "collapse the tower" under gravity and hence obtain an idempotent corresponding to a single pink box. This leads us to a very intuitive way of cinching our idempotents:

- an arbitrary colour sequence can be recorded by dropping coloured-tiles (in a Tetris-like fashion) in the order in which they occur in the sequence (we record the order in which the tiles are dropped as in (9.4.11));
- if the resulting picture is a tile-partition $\lambda \in \mathscr{P}_{m,n}$ and the labels increase up rows and columns, then we are done;
- otherwise, collapse the components of the tower of the form (9.4.11) until one obtains a tile-partition with increasing row/column labels.

Since $\lambda \in \mathscr{P}_{m,n}$ is a tile-partition if and only if it contains no region of one of the forms depicted in (9.4.11), this implies that the procedure above terminates with a "cinched" diagram corresponding to a reduced path. This is best illustrated via an example.

Example 9.4.12. We let $(W, P) = (\mathfrak{S}_8, \mathfrak{S}_4 \times \mathfrak{S}_4)$. We colour the reflections as in Figure 9.9. We now illustrate how to cinch an idempotent (from a non-reduced path) to obtain a skinnier idempotent (from a reduced path) using the procedure above, using the tiling combinatorics to keep track of the steps.

We begin with an element of $\mathscr{H}_{(W,P)}$ and we take a horizontal cut in order to obtain a diagram which perhaps takes the following form:

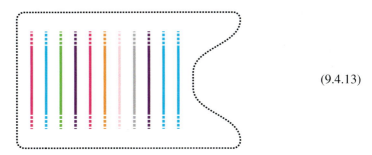

(9.4.13)

which factors through an idempotent that does not correspond to a reduced path. The colour sequence of this idempotent is recorded in a tiling as follows:

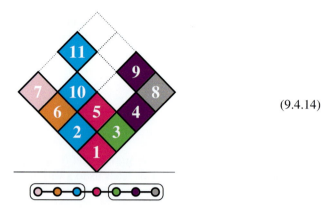

(9.4.14)

where the numbers in each tile correspond to the order in which these colours appear in the idempotent in (9.4.13). The first point at which a tile is missing from the tiling in (9.4.14) is after 9 steps, where we see a missing $(5,6)$-tile. We can therefore collapse the $(6,7)(7,8)(6,7)$ shape to obtain the following new tiling:

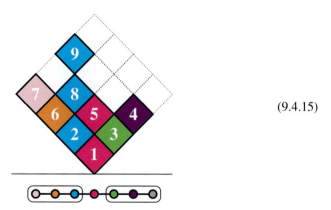

(9.4.15)

and we perform the corresponding cinching to the Soergel diagram in order to obtain the following diagram:

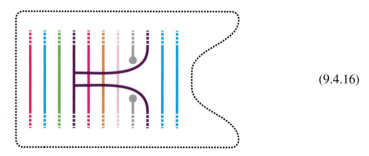

(9.4.16)

which has been "cinched" and now factors through an idempotent with 2 fewer vertical strands. We now consider the colour sequence of the skinniest idempotent-

9.4 How to manipulate diagrams

cut in the new diagram (9.4.16), (recorded as a tiling in (9.4.15)). The first point at which a tile is missing from the tiling in (9.4.15) is after 9 steps, where we see a pair of missing $(2, 3)$ and $(4, 5)$-tiles. We can therefore collapse the $(3, 4)(3, 4)$ shape in (9.4.15) to obtain the following new tiling:

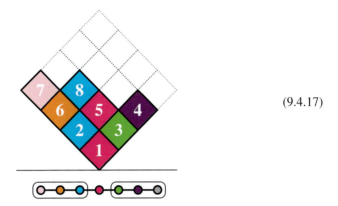

(9.4.17)

We perform the corresponding cinching to the diagram in (9.4.16) to obtain a new linear combination of diagrams:

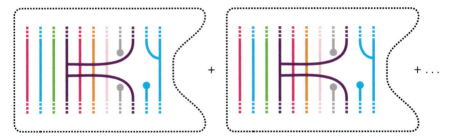

where the ... indicates a term with a barbell (see Exercise 9.4.18). We can now stop as we have a tile-partition (in (9.4.17)) with increasing row/column labels, as required.

Exercise 9.4.18. The following diagram is the barbell term obtained when we cinch the diagram in (9.4.16); Arguing as in Example 9.4.1, cinch this diagram

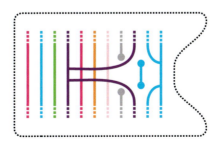

to obtain a *single diagram* which factors through a skinnier idempotent.

9.5 Return of the zig-zag algebras

In this section we take a short detour in order to establish that we have already seen some of these (apparently new!) diagram algebras before. This is not required for anything later in the book, our motivation is simply to help ground the reader in familiar material.

Example 9.5.1. We set $(W, P) = (\mathfrak{S}_3, \mathfrak{S}_2)$ where, to fix notation, we set $\mathfrak{S}_3 = \langle (1,2), (2,3) \rangle$ and $\mathfrak{S}_2 = \langle (2,3) \rangle$. The map

$$\varphi : \mathsf{ZZ}_3 \to \mathcal{H}_{(W,P)}$$

defined as follows:

$\varphi(e_0) = $ [diagram] $\varphi(e_1) = $ [diagram] $\varphi(e_2) = $ [diagram]

$\varphi(a_1^0) = $ [diagram] $\varphi(a_0^1) = $ [diagram] $\varphi(a_2^1) = $ [diagram]

$\varphi(a_1^2) = $ [diagram] $\varphi(a_0^1 a_1^0) = $ [diagram] $\varphi(a_1^2 a_2^1) = $ [diagram]

is a $\mathbb{Z}$-graded isomorphism of $\Bbbk$-algebras. That φ respects the idempotent relations of ZZ_2 is obvious; we now verify the degree 2 relations. We have that

$$\varphi(a_1^0)\varphi(a_2^1) = \text{[diagram]} = 0 = \varphi(0) \tag{9.5.2}$$

where the second equality follows from isotopy and relation (P) (the dual relation follows by flipping the diagrams). We have that

$$\varphi(a_1^0)\varphi(a_0^1) = \text{[diagram]} = 0 = \varphi(0) \tag{9.5.3}$$

9.6 Cellularity of $\mathcal{H}_{(W,P)}$ for $(W,P) = (\mathfrak{S}_{m+n}, \mathfrak{S}_m \times \mathfrak{S}_n)$

where the second equality follows from (C) (as the τ-strand can be isotoped all the way to the left edge of the rectangle). Finally, we have that

$$\varphi(a_2^1)\varphi(a_1^2) = \quad\cdots\quad = -\quad\cdots\quad = -\varphi(a_0^1)\varphi(a_1^0) \qquad (9.5.4)$$

as required. We have checked all the relations of Definition 5.4.1, and so φ is a surjective $\Bbbk$-algebra homomorphism (the reader is invited to check bijectivity).

In fact, the above example generalises completely. We have a $\mathbb{Z}$-graded $\Bbbk$-algebra isomorphism between the zig-zag algebra ZZ_n and the diagrammatic algebras $\mathcal{H}_{(W,P)}$ for $(W,P) = (\mathfrak{S}_n, \mathfrak{S}_{n-1})$ for all $n \geq 2$. Moreover, the proof is no more difficult than checking the degree 2 relations above.

Theorem 9.5.5. *For $n \geq 2$, we set $(W,P) = (\mathfrak{S}_n, \mathfrak{S}_{n-1})$ where, to fix notation, we set $\mathfrak{S}_n = \langle (k, k+1) \mid 1 \leq k < n \rangle$ and $\mathfrak{S}_{n-1} = \langle (k, k+1) \mid 2 \leq k < n \rangle$. The map $\varphi : ZZ_n \to \mathcal{H}_{(W,P)}$ defined on the generators by*

$$\varphi(e_k) = 1_{(1,2)} \otimes 1_{(2,3)} \otimes \cdots \otimes 1_{(k-1,k)} \otimes 1_{(k,k+1)}$$
$$\varphi(a_k^{k-1}) = 1_{(1,2)} \otimes 1_{(2,3)} \otimes \cdots \otimes 1_{(k-1,k)} \otimes \text{spot}^\varnothing_{(k,k+1)}$$

for $(k, k+1) \in S_W$ (and extending to linearly and taking duals) is an isomorphism of $\mathbb{Z}$-graded $\Bbbk$-algebras.

Exercise 9.5.6. Prove that $\varphi : ZZ_n \to \mathcal{H}_{(W,P)}$ is an isomorphism of $\mathbb{Z}$-graded $\Bbbk$-algebras for $n \geq 2$. Hint: the general case is essentially the same, but one will need to also apply commutativity relations when checking the analogues of (9.5.2) and (9.5.4). (The analogue of (9.5.4) further requires *repeated* applications of 2-colour barbell relations to move the barbells all the way to the left.)

9.6 Cellularity of $\mathcal{H}_{(W,P)}$ for $(W,P) = (\mathfrak{S}_{m+n}, \mathfrak{S}_m \times \mathfrak{S}_n)$

In this section we prove the cellularity of our diagrammatic algebras $\mathcal{H}_{(W,P)}$ for $\mathfrak{S}_m \times \mathfrak{S}_n \leq \mathfrak{S}_{m+n}$. While this proof is quite simple, it is worth paying attention: the cellularity formalism for $\mathcal{H}_{(W,P)}$ is the key ingredient of the definition of the p-Kazhdan–Lusztig polynomials and of Elias–Williamson's proof of the Kazhdan–Lusztig positivity conjecture. This section should give some insight into how the proof works in the general case, but avoiding technicalities.

Remark 19. We will fix a choice of reduced path T_μ to each $\mu \in \mathscr{P}_{m,n}$. By Exercise 9.3.2 it does not matter what choice we make. To simplify notation, we will often write 1_μ for the idempotent 1_{T_μ}.

Theorem 9.6.1. *Let $(W, P) = (\mathfrak{S}_{m+n}, \mathfrak{S}_m \times \mathfrak{S}_n)$. The algebra $\mathcal{H}_{(W,P)}$ is a graded weighted cellular algebra with idempotents 1_μ for $\mu \in \mathscr{P}_{m,n}$, the the anti-involution given by flipping a diagram through its horizontal axis, and the basis*

$$\{d_S^* 1_\lambda d_T \mid S \in \text{Path}(\lambda, T_\mu), T \in \text{Path}(\lambda, T_\nu), \lambda, \mu, \nu \in \mathscr{P}_{m,n}\} \quad (9.6.2)$$

and with respect to any total refinement of the Bruhat order on $\mathscr{P}_{m,n}$.

Before embarking on the proof, we first consider an illustrative example.

Example 9.6.3. Let $(W, P) = (\mathfrak{S}_4, \mathfrak{S}_2 \times \mathfrak{S}_2)$. We depict the basis of the 2-sided ideal $\mathcal{H}_{(W,P)} 1_\varnothing \mathcal{H}_{(W,P)}$ in Figure 9.11.

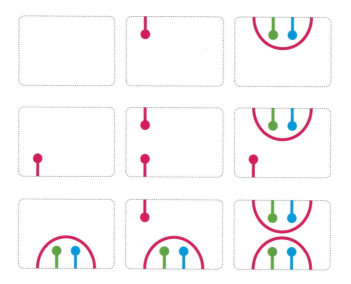

Fig. 9.11: The basis of the 9-dimensional ideal of $\mathcal{H}_{(W,P)}$ generated by $1_\varnothing$ for $(W, P) = (\mathfrak{S}_4, \mathfrak{S}_2 \times \mathfrak{S}_2)$. Compare with Figure 9.3.

Proof of Theorem 9.6.1. The idempotent, anti-involution, and 2-sided ideal conditions all follow trivially once we prove that (9.6.2) is indeed a basis. We will rewrite any choice of diagram $D \in 1_\mu \mathcal{H}_{(W,P)} 1_\lambda \mathcal{H}_{(W,P)} 1_\nu$ in the form

$$D = \pm d_S^* 1_\lambda d_T + \sum_{\substack{\alpha < \lambda \\ U \in \text{Path}(\alpha, T_\mu) \\ V \in \text{Path}(\alpha, T_\nu)}} d_U^* 1_\alpha d_V$$

for some $S \in \text{Path}(\lambda, T_\mu), T \in \text{Path}(\lambda, T_\nu)$. We can cinch the diagram D into the form $D = D' 1_\lambda D''$ as in Section 9.4 and hence proceed by induction on $\ell(\lambda) + \ell(\mu) + \ell(\nu)$, assuming that 1_λ is the minimal idempotent through which D factors.

9.6 Cellularity of $\mathcal{H}_{(W,P)}$ for $(W, P) = (\mathfrak{S}_{m+n}, \mathfrak{S}_m \times \mathfrak{S}_n)$

We consider each half diagram in the factorisation $D = D'1_\lambda D''$ in turn, thus working by induction on $\ell = \ell(\lambda) + \ell(\nu)$. The $\ell = 0$ case being trivial and we assume the result is true for all $\lambda', \nu' \in {}^P W$ such that $\ell(\lambda') + \ell(\nu') < \ell$. We have that $\nu = \nu' + \sigma$ for some $\sigma \in S_W$. By induction, we have four cases to consider:

$$D'' = \mathsf{U}^1_\sigma(d_T) \quad D'' = \mathsf{U}^0_\sigma(d_T) \quad D'' = \mathsf{D}^1_\sigma(d_T) \quad D'' = \mathsf{D}^0_\sigma(d_T)$$

for $T \in \text{Path}(\lambda', T_{\nu'})$ for $\lambda' \in \{\lambda, \lambda - \sigma\}$. By Exercise 9.3.2 we will always be able to (implicitly) make a choice of T_λ which makes drawing our diagrams as easy as possible.

Case 1. We first consider the case that $D'' = \mathsf{U}^1_\sigma(d_T)$ for $T \in \text{Path}(\lambda', T_{\nu'})$ with $\nu' + \sigma = \nu$. By assumption $T \in \text{Path}(\lambda', T_{\nu'})$ and $\sigma \in \text{Rem}(\lambda)$, $\sigma \in \text{Rem}(\nu)$. Therefore $\mathsf{U}^1_\sigma(T) \in \text{Path}(\lambda, T_\nu)$ and $D'' = d_{\mathsf{U}^1_\sigma(T)}$, as required. Pictorially, we have that

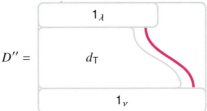

and this is an element of (9.6.2), as required.

Case 2. We now consider the case that $D'' = \mathsf{U}^0_\sigma(d_T)$ for $T \in \text{Path}(\lambda, T_{\nu'})$ with $\nu' + \sigma = \nu$ and $\lambda' = \lambda$. There are three cases to consider: $\sigma \in \text{Add}(\lambda)$, $\sigma \in \text{Rem}(\lambda)$, or $\sigma \notin \text{Add}(\lambda) \cup \text{Rem}(\lambda)$. In the first case, $\sigma \in \text{Add}(\lambda)$ and we have that $\mathsf{U}^0_\sigma(T) \in \text{Path}(\lambda, T_\nu)$ and $D'' = d_{\mathsf{U}^0_\sigma(T)}$, as required.

We now suppose that $\sigma \notin \text{Add}(\lambda) \cup \text{Rem}(\lambda)$. We have that $\lambda + \sigma$ is not a partition and so D'' contains a region as in (9.4.11) which can be simplified using (S2) and the commutativity relations (as in Example 9.4.9). Diagrammatically, we can picture this as follows

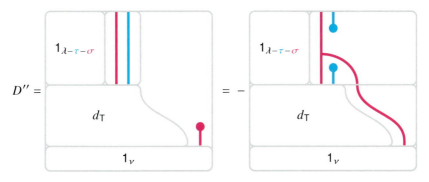

we hence contradict our assumption that T_λ is the minimal idempotent through which our diagram D'' factors. (We have taken some artistic license here, by assuming

$\tau \in \text{Rem}(\lambda)$ — this is not the case in general, but the strands to the right of the rightmost τ-strand will all trivially commute with the σ-strand.)

Finally, if $\sigma \in \text{Rem}(\lambda)$, then (by isotopy) our diagram has a local region of the form $1_\sigma \otimes \text{spot}_\sigma^\varnothing$. We apply (M2) to this region and hence obtain a sum of three diagrams: this first of which is of the desired form; the second of which immediately factors through the skinnier idempotent $1_{\lambda'-\sigma}$; the third contains a barbell diagram and so can be rewritten as factoring through the ideal $\mathscr{H}_{(W,P)}1_{<\lambda}\mathscr{H}_{(W,P)}$. (For an example, see Example 9.4.10.) Diagrammatically, we have that

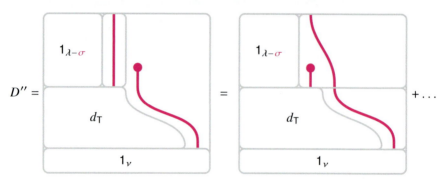

where the first equality is simply the definition and some isotopy; the second equality follows from (M2). Notice that $d_T = U_\sigma^1 U_\sigma^0(d_S)$ for $S \in \text{Path}(\lambda' - \sigma, T_{\nu'})$ and so is of the of the required form by first applying Case 1 and then the $\sigma \in \text{Add}(\lambda)$ subcase of Case 2.

Case 3. We now consider the case that $D'' = \mathsf{D}_\sigma^0(d_T)$ for $T \in \text{Path}(\lambda, T_{\nu'})$ with $\nu' + \sigma = \nu$. By definition of D_σ^0 we have that $\lambda' = \lambda$ and $\sigma \in \text{Rem}(\lambda)$. In which case, $\mathsf{D}_\sigma^0(T) \in \text{Path}(\lambda, T_\nu)$ and $D'' = d_{\mathsf{D}_\sigma^0(T)}$ is of the required form. Pictorially, we have that

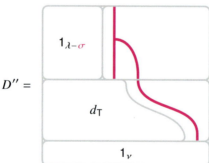

and this is an element of (9.6.2), as required.

Case 4. We now consider the case that $D'' = \mathsf{D}_\sigma^1(d_T)$ for $T \in \text{Path}(\lambda', T_{\nu'})$ with $\nu' + \sigma = \nu$ and $\lambda' - \sigma = \lambda$. As in Case 2, we have three cases to consider: $\sigma \in \text{Add}(\lambda)$, $\sigma \in \text{Rem}(\lambda)$, $\sigma \notin \text{Add}(\lambda) \cup \text{Rem}(\lambda)$. If $\sigma \in \text{Add}(\lambda)$, then $\mathsf{D}_\sigma^1(T) \in \text{Path}(\lambda, T_\nu)$ and $D'' = d_{\mathsf{D}_\sigma^1(T)}$, as required.

9.6 Cellularity of $\mathcal{H}_{(W,P)}$ for $(W, P) = (\mathfrak{S}_{m+n}, \mathfrak{S}_m \times \mathfrak{S}_n)$

Now suppose that $\sigma \notin \mathsf{Add}(\lambda) \cup \mathsf{Rem}(\lambda)$; we have that $\lambda + \sigma$ is not a partition and so d_T contains a region of the form $1_\sigma \otimes 1_\tau \otimes 1_\sigma$ and such diagrams contradict our inductive assumption that $\mathsf{T}_{\lambda'}$ is the minimal idempotent through which our diagram d_T factors (we should already have applied (S2) and obtained a diagram which factored through a smaller idempotent). Pictorially we have that

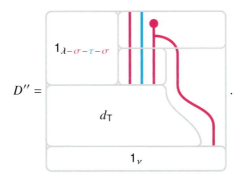

(We have again taken some artistic license here, by assuming $\tau \in \mathsf{Rem}(\lambda - \sigma)$ — this is not the case in general, but the strands to the right of the rightmost τ-strand will all trivially commute with the σ-strand.)

Now suppose that $\sigma \in \mathsf{Rem}(\lambda)$; we have that λ' is not a partition and d_T contains a region of the form $1_\sigma \otimes 1_\sigma$ and such diagrams contradict our inductive assumption that $\mathsf{T}_{\lambda'}$ is the minimal idempotent through which our diagram d_T factors (we should already have applied (M2) and obtained a diagram which factored through a smaller idempotent). Pictorially we have that

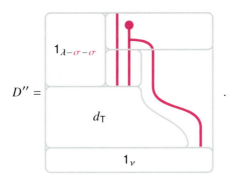

The result follows. □

Exercise 9.6.4. Let $(W, P) = (\mathfrak{S}_4, \mathfrak{S}_2 \times \mathfrak{S}_2)$. Construct the basis of the 2-sided ideal
$$\mathcal{H}_{(W,P)} 1_{\leqslant \sigma} \mathcal{H}_{(W,P)} \qquad 1_{\leqslant \sigma} = 1_\sigma + 1_\varnothing$$
for $\sigma = (2, 3) \in \mathfrak{S}_4$. You will need 25 diagrams in addition to the 9 diagrams already constructed in Figure 9.11.

Example 9.6.5. Let $(W, P) = (\mathfrak{S}_4, \mathfrak{S}_2 \times \mathfrak{S}_2)$. Calculate the graded formal character of the $\mathcal{H}_{(W,P)}$-module $\Delta^{\Bbbk}((2,3))$.

9.7 The categorification theorem

We are now in a position to provide an answer the question *"why should I care about the diagram algebras $\mathcal{H}_{(W,P)}$?"*. The following theorem can be thought of as saying *"we care about these diagram algebras because they lift the combinatorially defined Kazhdan–Lusztig polynomials to a richer structural level"*. This idea is often referred to as "categorification".

Theorem 9.7.1. *Let $(W, P) = (\mathfrak{S}_{m+n}, \mathfrak{S}_m \times \mathfrak{S}_n)$ and let $\Bbbk$ be a field. The $\mathcal{H}_{(W,P)}$-modules are constructible as the 1-dimensional quotients*

$$L^{\Bbbk}(\lambda) = \Delta^{\Bbbk}(\lambda)/\mathrm{rad}^{\Bbbk}\Delta^{\Bbbk}(\lambda) = \{1_{\mathsf{T}_\lambda} + \mathrm{rad}^{\Bbbk}(\Delta^{\Bbbk}(\lambda))\}$$

for $\lambda \in \mathcal{P}_{m,n}$, up to isomorphism and grading shift.

Proof. We have that $\mathsf{S} \in \mathrm{Path}(\lambda, \mu)$ for $\lambda \neq \mu$ implies $\deg(\mathsf{T}) > 0$ by Theorem 7.5.10. The grading trick implies that $\mathrm{rad}\langle -, \rangle^{\Bbbk}_\lambda = \{d_{\mathsf{S}} \mid \mathsf{T}_\lambda \neq \mathsf{S} \in \mathrm{Path}(\lambda, -)\}$ as required. □

Example 9.7.2. The $\mathcal{H}_{(\mathfrak{S}_4, \mathfrak{S}_2 \times \mathfrak{S}_2)}$ cell-module $1_{\varnothing}\mathcal{H}_{(\mathfrak{S}_4, \mathfrak{S}_2 \times \mathfrak{S}_2)} = \Delta^{\Bbbk}(\varnothing)$ is 3-dimensional with basis given by the diagrams in the final row of the table in Figure 9.3. By Theorem 9.7.1, each of these three diagrams spans a (grading shifted) 1-dimensional simple module which appears as a composition factor of $\Delta^{\Bbbk}(\varnothing)$. The Alperin diagram of $\Delta^{\Bbbk}(\varnothing)$ is given in Figure 9.12. We notice that this module "lifts" (or "categorifies") the Kazhdan–Lusztig polynomials in the final row of Table 7.1 by interpreting an entry q^k in the μth column as a a copy of $L^{\Bbbk}(\mu)\langle k \rangle$ in the kth radical layer of $\Delta^{\Bbbk}(\varnothing)$.

More generally, we have the following theorem:

Theorem 9.7.3. *Let $(W, P) = (\mathfrak{S}_{m+n}, \mathfrak{S}_m \times \mathfrak{S}_n)$ and $\Bbbk$ be an arbitrary field. The Kazhdan–Lusztig polynomials $n_{\lambda,\mu}(q)$ can be lifted to the structural level of graded composition factor multiplicities*

$$n_{\lambda,\mu}(q) = \sum_{k \in \mathbb{Z}} [\Delta^{\Bbbk}(\lambda) : L^{\Bbbk}(\mu)\langle k \rangle] q^k$$

of simple $\mathcal{H}_{(W,P)}$-modules within the $\mathcal{H}_{(W,P)}$-cell modules.

Proof. This follows immediately from the fact that the algebra is graded cellular and positively graded, as in Section 6.7. □

9.7 The categorification theorem

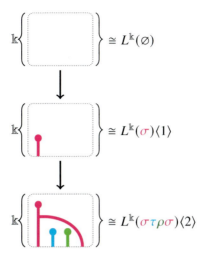

Fig. 9.12: The Alperin diagram of the $\mathscr{H}_{(\mathfrak{S}_4,\mathfrak{S}_2\times\mathfrak{S}_2)}$-module $\Delta^{\Bbbk}(\emptyset)$

Example 9.7.4. In Figure 9.13 we provide a vector-by-vector example of the submodule structure of the $\mathscr{H}_{(W,P)}$-module $\Delta^{\Bbbk}(s_2s_3s_1)$ for $(W,P) = (\mathfrak{S}_5, \mathfrak{S}_2 \times \mathfrak{S}_3)$.

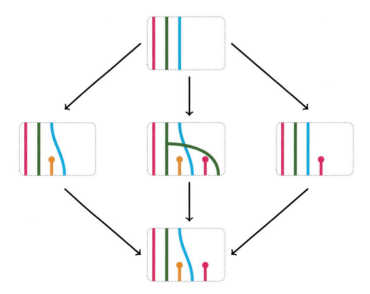

Fig. 9.13: The Alperin diagram of the 5-dimensional $\mathscr{H}_{(W,P)}$-module $\Delta^{\Bbbk}(s_2s_3s_1)$ for $(W,P) = (\mathfrak{S}_5, \mathfrak{S}_2 \times \mathfrak{S}_3)$.

236   9 The diagrammatic algebra for $\mathfrak{S}_m \times \mathfrak{S}_n \leqslant \mathfrak{S}_{m+n}$

Example 9.7.5. In Figure 9.14 we provide the full submodule structure of the $\mathcal{H}_{(W,P)}$-module $\Delta^{\Bbbk}(2,1)$ for $(W,P) = (\mathfrak{S}_6, \mathfrak{S}_3 \times \mathfrak{S}_3)$. We have indicated the simple composition factors $L^{\Bbbk}(\lambda)$ within $\Delta^{\Bbbk}(2,1)$ and leave it as an exercise for the reader to draw the corresponding Soergel diagrams. As a further exercise for the enthusiastic reader, we ask if you can that each edge does correspond to an inclusion of submodules within $\Delta^{\Bbbk}(2,1)$: one can do this by providing the diagrams which correspond to these edges as we did in Figures 9.6 and 9.7.

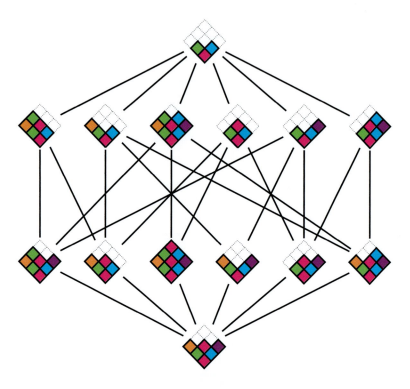

Fig. 9.14: The full submodule lattice of the cell module $\Delta(2,1)$ of $\mathcal{H}_{(W,P)}$ for $(W,P) = (\mathfrak{S}_6, \mathfrak{S}_3 \times \mathfrak{S}_3)$. We represent each simple module by the corresponding partition and highlight the 3×3 rectangle in which the partition exists.

The observant reader might have made the following observation from the above examples...

Theorem 9.7.6. *The radical and grading layers of $\mathcal{H}_{(\mathfrak{S}_{m+n}, \mathfrak{S}_m \times \mathfrak{S}_n)}$-cell modules coincide. In other words,*

$$\mathrm{rad}_k(\Delta^{\Bbbk}(\lambda)) = \bigoplus_{\{\mu \mid n_{\lambda,\mu}(q) = q^k\}} L(\mu)\langle k \rangle.$$

9.7 The categorification theorem

Proof. It is enough to show that the algebra $\mathscr{H}_{(\mathfrak{S}_{m+n},\mathfrak{S}_m\times\mathfrak{S}_n)}$ is generated by elements of degree 0 and 1 (see Exercise 6.11.18). An example of how one establishes this fact is given in Figure 9.15, but formalising this into a proof is left as an exercise for the ambitious reader who enjoys combinatorics. □

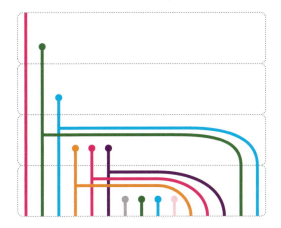

Fig. 9.15: A light leaves basis element for $\mathscr{H}_{(W,P)}$ for $(W, P) = (\mathfrak{S}_8, \mathfrak{S}_4 \times \mathfrak{S}_4)$ drawn according to the colouring of Figure 9.9. We have drawn this basis element in such a way as to highlight its construction as a product of degree 1 (light leaves) elements. (That is, we emphasise vertical concatenation rather than horizontal concatenation.)

Remark 20. The upshot of Theorems 9.7.3 and 9.7.6 is that the Kazhdan–Lusztig polynomials carry a stupendous amount of information on the structure of $\mathscr{H}_{(W,P)}$. Not only do they tell us composition factor multiplicities, but they tell us *where these composition factors appear* within modules!!

Exercise 9.7.7. Calculate the Alperin diagrams of the $\mathscr{H}_{(\mathfrak{S}_5,\mathfrak{S}_2\times\mathfrak{S}_3)}$ -modules $\Delta^{\Bbbk}(\lambda)$ for $\lambda \in \mathscr{P}_{2,3}$.

Exercise 9.7.8. Calculate the Alperin diagram of the $\mathscr{H}_{(\mathfrak{S}_8,\mathfrak{S}_4\times\mathfrak{S}_4)}$-module $\Delta^{\Bbbk}(2, 2)$.

Remark 21. It is quite striking that the structure of the diagram algebras $\mathscr{H}_{(W,P)}$ considered in this chapter (and the associated p-Kazhdan–Lusztig polynomials) are entirely independent of the field. This almost never happens! We will see in the next chapter that the composition factor multiplicities

$$\sum_{k\in\mathbb{Z}}[\Delta^{\Bbbk}(\lambda) : L^{\Bbbk}(\mu)\langle k\rangle]q^k$$

will always be Kazhdan–Lusztig polynomials when $\Bbbk = \mathbb{C}$, but will be wildly unmanageable for finite fields in full generality.

Further Reading. Most of the ideas of this chapter will be generalised in the next chapter, so we do not provide a very in depth list of further reading here. However, we do emphasise that the diagram algebras of this chapter are isomorphic to the *extended Khovanov arc algebras* (see [BHDS] for a statement and proof of this isomorphism). These (extended) Khovanov arc algebras were defined via topological quantum field theory and can be viewed as a "categorification of the Temperley–Lieb algebra"; their first applications were in categorical knot theory [Kho00, Str09]. Further work "categorifying the Temperley–Lieb algebra" was carried out by Bar-Natan [BN05], Bernstein–Frenkel–Khovanov [BFK99], and Elias [Eli10].

The (extended) Khovanov arc algebras have subsequently been studied from the point of view of their cohomological and representation theoretic structure [BS10, BS11b, BS12a, BS12b, BW, BDD$^+$a, BDD$^+$b], and symplectic geometry [MS22] and have further inspired much generalisation: from the Temperley–Lieb setting to web diagrams [MPT14, Mac14, Tub20, Tub14] and also from the "even" setting to "super" [Sar16] and "odd" settings [NV18], as well as to the orthosymplectic case [ES16b, ES16a, ES17]. In summary, these algebras form the prototype for knot-theoretic categorification, we refer to Stroppel's 2022 ICM address for more details [Str23]. So, even though these algebras are the simplest examples of diagram algebras we shall see in this book, they are also incredible rich in structure and touch-on many different areas of mathematics.

Finally, we wish to discuss the coincidence between grading and radical filtrations on cell modules of Theorem 9.7.6. This is a particularly beautiful property which is characteristic of complex Lie theory. This property is typically proven as a consequence of the strong cohomological property of **standard Koszulity**, which is beyond the remit of this book (the definitive works regarding Koszulity in representation theory are [BGS96, ADL03]). Here we simply note that many well-loved objects in Lie theory are Koszul, for example the quantum Schur algebras [Sha12], extended Khovanov arc algebras [BS10], and the (diagrammatic) Cherednik algebras [RSVV16, Los16, Web17b].

Chapter 10
Lusztig's conjecture in the diagrammatic algebra $\mathcal{H}_{(W,P)}$

In this chapter we make the considerable leap in difficulty that comes with studying the Hecke categories of more general Coxeter systems (W, P) such that $m_{\sigma\tau} \in \{2, 3, \infty\}$ for all $\sigma, \tau \in S_W$ (in other words, those whose Bruhat graphs contain only squares and hexagons).

Our diagrammatic algebras are much richer than their underlying Coxeter groups and their presentations are correspondingly more complicated. For instance, the Coxeter generators of the symmetric group are of rank 2 and the most complicated relation for the symmetric group is the *braid relation* which tells us that the most complicated products of these generators in rank 3 are "the same", that is

On the other hand, the diagrammatic algebra $\mathcal{H}_{(W,P)}$ does not have a braid relation, but rather a braid *generator* of rank 3. The most complicated relation in $\mathcal{H}_{(W,P)}$ is the rank 4 *permutohedron relation* depicted in Figure 10.1 below.

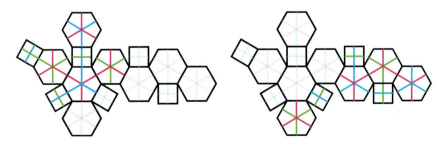

Fig. 10.1: The two sides of the permutohedron relation of $\mathcal{H}_{(W,P)}$, viewed as decorations of "two halves" of a permutahedron.

The permutohedron relation tells us that the most complicated products of braid-generators in rank 4 are "the same" in $\mathcal{H}_{(W,P)}$. We have already seen that $\mathfrak{S}_4$ acts faithfully on the permutohedron, and this is why we can realise this relation on the decorated net for the permutahedron. We shall make precise momentarily...

Once we have a firm grip on the diagrammatic algebras $\mathcal{H}_{(W,P)}$, we delve into the calculation of their intersection forms. We then conclude the chapter by examining the counterexamples to the expected bounds of Lusztig's conjecture within $\mathcal{H}_{(W,P)}$.

10.1 From paths to diagrammatic algebras (again)

In the previous section, we defined $\mathcal{H}_{(W,P)}$ for $(W, P) = (\mathfrak{S}_{m+n}, \mathfrak{S}_m \times \mathfrak{S}_n)$ in terms of maps between paths in the associated weak Bruhat graph. The only polygons in the weak Bruhat graph were squares, which corresponded to *commuting* pairs of $\sigma, \tau \in S_W$ with $m_{\sigma,\tau} = 2$. For the purposes of disproving Lusztig's conjecture, we must consider the weak Bruhat graphs for more general parabolic Coxeter systems.

The polygons in the weak Bruhat graphs of the Coxeter groups we consider here are either squares (which we have already seen for $m_{\sigma\tau} = 2$) or hexagons (for $m_{\sigma\tau} = 3$), we also allow for $m_{\sigma\tau} = \infty$ (but there is no such thing as an ∞-gon). Thus we require a new "hexagonal" braid generator which will allow us to pass between the two length 3 reduced paths from the bottom to the top of this hexagon. That is, a braid diagram for passing between $\sigma\tau\sigma$ and $\tau\sigma\tau$. These paths are depicted in Figure 10.2.

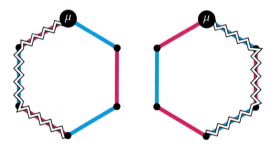

Fig. 10.2: Two reduced paths to $\mu = (1, 3)$ in the weak Bruhat graph of $\mathfrak{S}_3$.

We lift this to a Soergel diagram: we draw the colour sequence for each path as the top and bottom of a rectangle, then draw the naive diagram connecting the top and bottom colour sequences to obtain a new generator as follows:

$$\text{braid}_{\tau\sigma\tau}^{\sigma\tau\sigma} = \vcenter{\hbox{\includegraphics{braid}}} . \qquad (10.1.1)$$

10.1 From paths to diagrammatic algebras (again)

We treat this generator like any other, allowing horizontal and vertical concatenates of the new braid generator with the other diagrams. For example, the braid between the paths S, T in Figure 10.3 is as follows

$$\mathsf{braid}_\mathsf{S}^\mathsf{T} = $$

and it is obtained by horizontal concatenation of the new braid generator together with the idempotent generators in (9.1.1).

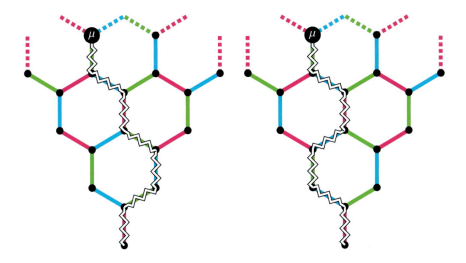

Fig. 10.3: Two reduced paths S and T which differ by a braid.

We generalise our notion of $\mathsf{braid}_\mathsf{T}^\mathsf{S}$ for reduced paths $\mathsf{S}, \mathsf{T} \in \mathrm{Path}(\lambda, -)$ to include our new hexagonal generators. With this in place, the definition of up and down operators (as in Definition 9.1.4) does through unchanged. However, whilst the braids in Chapter 9 were all *invertible*, we will soon see that our new braid-generators are not. Our saving grace is that these generators are invertible *modulo lower terms in the Bruhat ordering*. Before going into this in detail, we provide an illustrate example of how one constructs these braid diagrams.

Example 10.1.2. Let $W = \mathfrak{S}_8$. An example of a quite complicated light leaves basis diagram, d_S, corresponding to the path

$$\mathsf{S} = \mathsf{U}_1^1 \mathsf{U}_2^1 \mathsf{U}_1^1 \mathsf{U}_3^1 \mathsf{U}_2^1 \mathsf{U}_1^1 \mathsf{U}_4^0 \mathsf{D}_3^1 \mathsf{D}_2^1 \mathsf{U}_4^1 \mathsf{U}_3^1 \mathsf{U}_2^0 \mathsf{U}_6^0 \mathsf{U}_5^0 \mathsf{D}_4^1 \mathsf{D}_3^0$$

is given in Figure 10.4. Whilst the braids do take up a lot of superficial space, they are not so important when we come to multiplying diagrams together. Indeed, we shall see that they make almost no impact when we calculate the intersection forms.

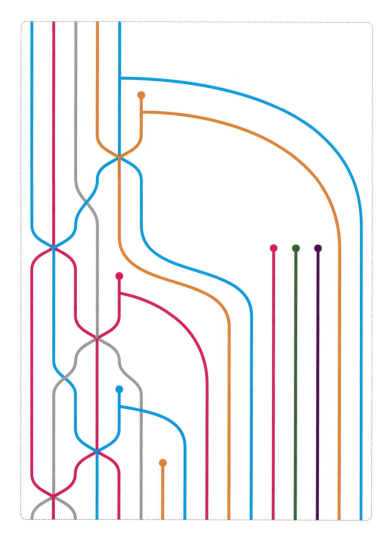

Fig. 10.4: A light leaves basis element $d_S \in \mathcal{H}_W$ for $W = \mathfrak{S}_8$ and S as in Example 10.1.2 and the colouring of reflections as in Figure 9.9 (ignoring the parabolic).

Example 10.1.3. We now construct two basis elements labelled by the two paths $S, T \in \text{Path}(\sigma\tau\rho, T_\mu)$ for T_μ as depicted in Figure 10.6 and T as depicted in Figure 10.5. (We leave picturing S as an exercise for the reader.) The degree 2 basis element can be constructed in exactly the same fashion as in Section 9.1, using only

10.1 From paths to diagrammatic algebras (again)

the fork and spot generators as follows

$$d_S = U^1_\sigma U^0_\tau U^1_\rho U^1_\tau U^0_\sigma D^0_\tau U^0_\rho =$$

and is of degree 2.

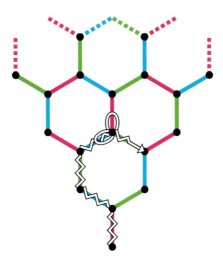

Fig. 10.5: The degree zero folded-path T corresponding to the basis element d_T in Example 10.1.3. This path has weight μ and shape $\lambda = \sigma\tau\rho$ and $\mu = \sigma\tau\rho\sigma\tau\rho$.

The degree 0 element $T \in \text{Path}(\sigma\tau\rho, T_\mu)$ can be constructed with an apparent degree of freedom. The easiest way to construct this path-morphism is using *the minimal number of braid generators required*. In this manner, we produce the diagram d_T corresponding to T in Figure 10.5 as follows

$$d_T = U^1_\sigma U^1_\tau U^1_\rho U^1_\tau U^0_\sigma D^0_\tau D^1_\rho = \qquad (10.1.4)$$

where we apply the $s_1 s_2$-braid after the sixth step of the process, because otherwise we *cannot* apply the green spotted-fork in the seventh step of the process (as the green strand has to be to the right of the blue strand, if we are to connect it to the green spotted-fork).

Notice we could have applied an $s_1 s_2$-braid after the fourth step and *again* after the fifth step. This would have produced a d'_T which *looks* different (because of the "triple braid" in the middle of the diagram) but is, in fact, the same. We discuss this further (after defining the relations) in Section 10.3.

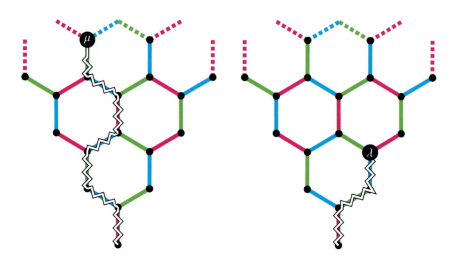

Fig. 10.6: Two paths T_μ and T_λ in the Bruhat graph of $(W, P) = (\widehat{\mathfrak{S}}_3, \mathfrak{S}_3)$. Here $\lambda = \sigma\tau\rho$ and $\mu = \sigma\tau\rho\tau\sigma\tau\rho$. (See also Figure 10.5.)

10.2 Multi-colour relations

We now introduce all the multi-colour relations for the diagrammatic algebra $\mathcal{H}_{(W,P)}$. For $m_{\sigma,\tau} = 3$, the old 2-colour barbell relation goes through unchanged

$$\left(\middle| \mathbf{\dumbbell} - \mathbf{\dumbbell}\middle|\right) = \left(\mathbf{\dumbbell}\middle|\right) - \left(\vdots\right) \tag{H1}$$

however for $m_{\sigma,\varpi} = \infty$ we have a new relation:

$$\left(\middle| \mathbf{\dumbbell} - \mathbf{\dumbbell}\middle|\right) = 2\left(\left(\mathbf{\dumbbell}\middle|\right) - \left(\vdots\right)\right). \tag{H2}$$

10.2 Multi-colour relations

(We remark, for those in the know, that the scalars 1 and 2 on the righthand-side of (H1) and (H2) can be thought of as coming from the off-diagonal entries of the Cartan matrix.)

We now introduce the relations involving the new braid generator. For the remainder of this section, we set $m_{\sigma,\tau} = m_{\sigma,\rho} = 3$ and $m_{\rho,\tau} = m_{\rho,\pi} = m_{\sigma,\pi} = 2$. The first of these new relations is a replacement of S2 from Section 9.3. We have the double-braid relations

(H3)

and we observe that if we set the $\sigma\tau$-braid equal to zero, then the above relation specialises to S2 (and indeed, this was the reason the last chapter was simpler).

Remark 22. The $\sigma\tau$-braid relation involves *all* the degree zero diagrams with northern and southern colour sequence equal to $\sigma\tau\sigma$. We picture the non-identity diagrams in terms of paths in Figures 10.7 and 10.8.

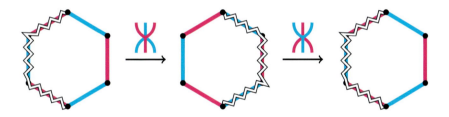

Fig. 10.7: The double braid on the lefthand-side of H3 as a composition of maps between paths.

Fig. 10.8: The non-identity diagram on the righthand-side of H3 as a map between paths.

We now consider the genuinely new relations, which have no counterpart in Section 9.3 (namely, those in which every term has a $\sigma\tau$-braid). We have the fork-braid relation, which essentially says that there is a unique degree -1 way of passing between $\tau\sigma\tau\tau$ and the reduced path $\sigma\tau\sigma$. We picture the fork-braid relation as follows:

$$= \tag{H4}$$

In addition to the existing commutativity relations, we obtain new tri-coloured commutativity relations:

$$\tag{H5}$$

for $m_{\tau\rho} = m_{\tau\pi} = m_{\pi\rho} = m_{\sigma\pi} = 2$ and $m_{\sigma\tau} = 3$. Finally, we require the permutohedron relation (which we have already alluded to in the introduction to this chapter). This relation is as follows,

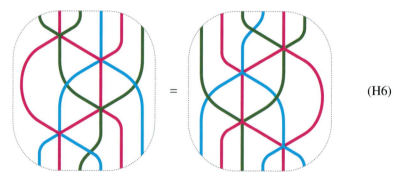

$$\tag{H6}$$

and one should compare the commuting and non-commuting braids with the square and hexagonal faces of the permutohedron respectively, see Figure 10.1. Hence, we see that the diagrams on the left and right of (H6) correspond to "two halves" of the permutohedron.

10.3 Fun with braids

Remark 23. We label these relations by (H1) to (H6) as these are the relations in the case that the Bruhat graphs containing squares and *hexagons*.

For the purposes of completeness, we also restate the non-local "parabolic annihilation" relation

$$\bigg| \otimes D = 0 \qquad \text{(P)}$$

for $\tau \in S_P$, and D any diagram.

Definition 10.2.1. Let (W, P) be such that $m_{\sigma,\tau} \in \{2, 3, \infty\}$ for all $\sigma, \tau \in S_W$ and let $\Bbbk$ be a field. We define $\mathscr{H}_{(W,P)}$ to be the $\Bbbk$-algebra generated by all d_S and their duals with multiplication given by $\circ$ modulo isotopy and the monochrome relations (M1) and (M2), the multi-colour relations (H1) to (H6), together with the non-local relations (C) and (P).

10.3 Fun with braids

We now play around with some examples of diagrams which involve the new braid generator. We will prove a couple of relations which follow from the above ones (such as the Jones–Wenzl and triple-braid relations) and become more familiar with how one can manipulate these diagrams.

Proposition 10.3.1. *Let* $m(\sigma, \tau) = 3$. *We have the following* Jones–Wenzl *relation or* spot-braid *relation:*

Proof. We will consider the two sides of (H3) in turn. We first consider the double-braid side, to which we apply a single spot to obtain the following

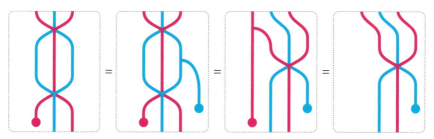

where the first equality follows from the τ-fork-spot relation (we think of this as "growing an unnecessary new arm"); the second equality follows from the fork-braid relation (H4) (making use of the new τ-arm we just grew); and the third follows from using the σ-fork-spot relation to shrink the now-unnecessary σ-arm.

We now reconsider the spotted-double-braid diagram on the lefthand-side above, but this time we apply the double-braid relation (H3) in order to rewrite this diagram as follows:

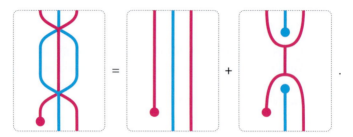

Putting our two equalities together (with a touch of isotopy for aesthetic purposes) we deduce the Jones–Wenzl relation. □

For the most part, we consider the isotopy relation to be the most intuitive of the relations of the diagrammatic algebra $\mathcal{H}_{(W,P)}$. However, it is worth explicitly emphasising the following isotopy, which allows us to rotate braids:

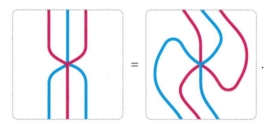

Armed with this rotational isotopy, it is not difficult to deduce the following immediate corollary of Proposition 10.3.1.

Corollary 10.3.2. *Let $m(\sigma, \tau) = 3$. We have the following relation, which can be regarded as a rotation of the previous Jones–Wenzl relation:*

In the previous section we alluded to the idea that we have already considered all possible degree zero maps between paths in $\mathfrak{S}_3$. We now make this more precise, by considering triple products of braid generators, as follows.

10.3 Fun with braids

Proposition 10.3.3. *Let $m(\sigma, \tau) = 3$. We have the following triple-braid relation*

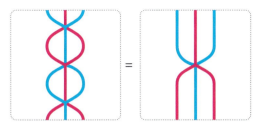

Proof. We begin by applying the double-braid relation to the lefthand-side in order to deduce that

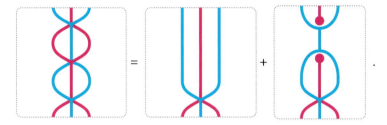

Now, we claim that the second term on the righthand-side is zero. To see this, we apply Corollary 10.3.2 to rewrite the bottom half of the diagram as follows:

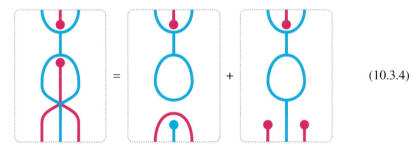

(10.3.4)

and we note that both the terms on the righthand-side are zero by the fork-annihilation relation. Thus the triple-broad relation follows. □

Exercise 10.3.5. Verify that the $\sigma\tau$-double-braid in (H3) follows from the other relations, together with the $\sigma\tau$-Jones–Wenzl relation. In other words, we are free to choose to include either the $\sigma\tau$-double-braid or the $\sigma\tau$-Jones–Wenzl relation in our presentation. (In fact, it is more usual in the literature to include the $\sigma\tau$-Jones–Wenzl relation.)

Exercise 10.3.6. Calculate the scalar $\alpha \in \mathbb{k}$ such that

$$\alpha(1_\sigma \otimes \operatorname{spot}_\emptyset^\tau \otimes 1_\sigma)\operatorname{fork}_\sigma^{\sigma\sigma}\operatorname{fork}_{\sigma\sigma}^\sigma(1_\sigma \otimes \operatorname{spot}_\tau^\emptyset \otimes 1_\sigma)$$

is an idempotent.

Example 10.3.7. We now revisit Example 10.1.3. In this example we discussed the possibility of applying braids unnecessarily and the fact that this was immaterial. To see this, we first apply the double-braid relation to the lefthand-side of Figure 10.9 in order to obtain a sum of two terms (the former of which is d_T of Example 10.1.3. The latter term on the righthand-side of the equality in Figure 10.9 is zero as we already saw in (10.3.4).

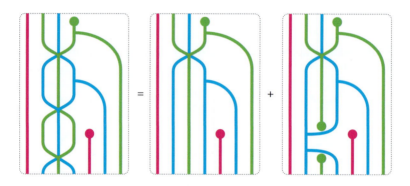

Fig. 10.9: On the left is the over-zealously braided diagram discussed in Example 10.1.3. Applying relation (H3) we obtain the righthand-side. The first term on the right is the diagram from (10.1.4) and we claim that the second term is zero.

10.4 Light leaves and the Kazhdan–Lusztig positivity conjecture

We have already discussed the manner in which one can construct a diagram d_T associated to any path $\text{Path}(\lambda, T_\mu)$ using the *exact same* $\mathsf{U}^0_\sigma, \mathsf{U}^1_\sigma, \mathsf{D}^0_\sigma, \mathsf{D}^1_\sigma$ operators as in Definition 9.1.4, but with our extra "hexagonal" braid generator inserted as required. We now discuss how these diagrams (or "light leaves") play the exact same role as they did in the previous chapter.

Theorem 10.4.1 ([Lib08, EW16, LW22]). *Let $\Bbbk$ be an arbitrary field and fix a preferred choice of reduced path T_λ for each $\lambda \in {}^P W$. The algebra $\mathcal{H}_{(W,P)}$ is a weighted graded weighted cellular algebra with idempotents 1_{T_λ} for $\lambda \in {}^P W$ and basis*
$$\{d_S^* 1_{T_\lambda} d_T \mid S \in \text{Path}(\lambda, T_\mu), T \in \text{Path}(\lambda, T_\nu), \lambda, \mu, \nu \in {}^P W\}$$
and the anti-involution given by flipping a diagram through its horizontal axis and with respect to any total refinement of the strong Bruhat order.

The proof of this theorem is far beyond what we can hope to cover in this book. But we will now discuss the importance of this theorem in the context of the Kazhdan–Lusztig positivity conjecture.

10.4 Light leaves and the Kazhdan–Lusztig positivity conjecture

Corollary 10.4.2. *Let $\Bbbk$ be an arbitrary field. A complete set of non-isomorphic simple $\mathcal{H}_{(W,P)}$-modules, up to isomorphism and grading shift, is provided by*

$$\{L^{\Bbbk}(\lambda) = \Delta^{\Bbbk}(\lambda)/\mathrm{rad}\langle -, -\rangle^{\Bbbk}_{\lambda} \mid \text{for } \lambda \in {}^P W\}$$

where the intersection form $\langle -, -\rangle^{\Bbbk}_{\lambda} : \Delta^{\Bbbk}(\lambda) \times \Delta^{\Bbbk}(\lambda) \to \Bbbk$ is defined in the context of arbitrary cellular algebras in Definition 6.2.11.

By Theorem 10.4.1 the $\mathcal{H}_{(W,P)}$-cell-module, $\Delta^{\Bbbk}(\lambda)$, has basis enumerated by $\cup_{\mu}\mathrm{Path}(\lambda, \mathsf{T}_{\mu})$ the set of paths (of any possible weight μ) in the weak Bruhat graph for (W, P) which terminate at λ. In particular, the matrix

$$\Delta^{(W,P)} := (\Delta_{\lambda,\mu}(q))_{\lambda,\mu \in {}^P W} \qquad \Delta_{\lambda,\mu}(q) = \sum_{S \in \mathrm{Path}(\lambda,\mathsf{T}_{\mu})} q^{\deg(S)}$$

records the graded formal characters of $\mathcal{H}_{(W,P)}$-cell modules. In [EW14, LW22], Elias–Williamson and Libedinsky–Williamson proved that for $\lambda \neq \nu$ the graded $\mathcal{H}^{\mathbb{C}}_{(W,P)}$ composition factor multiplicities occur in strictly positive degree

$$\sum_{k \in \mathbb{Z}} [\Delta^{\mathbb{C}}(\lambda) : L^{\mathbb{C}}(\nu)\langle k\rangle] q^k \in q\mathbb{Z}[q]. \tag{10.4.3}$$

This proof of this result is immensely difficult. On the other hand, graded cellularity immediately tells us that the graded formal characters of the simple $\mathcal{H}^{\mathbb{C}}_{(W,P)}$-module are invariant under swapping q and q^{-1}, that is

$$\dim_q(L^{\mathbb{C}}(\nu)\mathbf{1}_{\mu}) \in \mathbb{Z}[q + q^{-1}] \tag{10.4.4}$$

for all $\mu, \nu \in {}^P W$. Thus, putting (10.4.3) and (10.4.4) together, we can factorise the matrix $\Delta^{(W,P)}$ as a product of (*i*) a uni-triangular matrix with off-diagonal entries in positive degree

$$\sum_{k \in \mathbb{Z}} [\Delta^{\mathbb{C}}(\lambda) : L^{\mathbb{C}}(\nu)\langle k\rangle] q^k \in q\mathbb{Z}[q]$$

for $\lambda \neq \nu$ and (*ii*) a uni-triangular matrix with off-diagonal entries

$$\dim_q(L^{\Bbbk}(\nu)\mathbf{1}_{\mu}) \in \mathbb{Z}[q + q^{-1}].$$

But we have *already defined* the Kazhdan–Lusztig polynomials as the unique uni-triangular matrix (whose off-diagonal entries belonging to $q\mathbb{Z}[q]$) that appears in such a factorisation of $\Delta^{(W,P)}$. Therefore this immediately proves that the graded decomposition multiplicities of simple $\mathcal{H}^{\mathbb{C}}_{(W,P)}$-modules in $\mathcal{H}^{\mathbb{C}}_{(W,P)}$-cell modules are equal to the Kazhdan–Lusztig polynomials $n_{\lambda,\nu}(q)$ and that the graded simple characters are the $b_{\nu,\mu}(q)$ from Definition 8.2.9.

Moreover, this provides an *innately positive* interpretation of the coefficients of the polynomial $n_{\lambda,\nu}(q)$ simply because *no (cell) module can have a simple constituent appearing a negative number of times!* Thus the categorification theorem proves the famous Kazhdan–Lusztig positivity conjecture [EW14] and its parabolic counterpart [LW22] which posited that these polynomials $n_{\lambda,\nu}(q)$ *should* belong to $q\mathbb{Z}_{\geq 0}[q]$ for $\lambda \neq \mu$. That is, we have the following:

Theorem 10.4.5. *Let (W, P) be a parabolic Coxeter system. The associated Kazhdan–Lusztig polynomials $n_{\lambda,\mu}(q)$ admit a structural interpretation as the graded composition factor multiplicities:*

$$n_{\lambda,\mu}(q) = \sum_{k \in \mathbb{Z}} [\Delta^{\mathbb{C}}(\lambda) : L^{\mathbb{C}}(\mu)\langle k \rangle] q^k.$$

For the remainder of this chapter, we will principally be interested in the representation theory of $\mathcal{H}_{(W,P)}$ over finite fields. In light of Theorem 10.4.5, we are now etymologically justified in making the following definition:

Definition 10.4.6. Let (W, P) be a parabolic Coxeter system. and let $\Bbbk$ be a field of characteristic p for any prime $p \geq 0$. We define the associated p-**Kazhdan–Lusztig polynomials** $^p n_{\lambda,\mu}(q)$ as follows

$$^p n_{\lambda,\mu}(q) = \sum_{k \in \mathbb{Z}} [\Delta^{\Bbbk}(\lambda) : L^{\Bbbk}(\mu)\langle k \rangle] q^k.$$

Understanding the p-Kazhdan–Lusztig polynomials is one of the most important problems in Lie theory and modular representation theory. In the remainder of this chapter, we will attempt to calculate these p-Kazhdan–Lusztig polynomials... and see that this is impossibly difficult in full generality.

Remark 24. It is worth noting that, despite the name, the p-Kazhdan–Lusztig polynomials are not necessarily polynomials at all! Unlike over the complex field, it is perfectly possibly that we obtain $^p n_{\lambda,\mu}(q) \in \mathbb{Z}_{\geq 0}[q, q^{-1}]$ with terms in strictly negative degree, however it is a bit of a mouthful to say "p-Kazhdan–Lusztig Laurent polynomials" and so we will perpetuate this misnomer.

10.5 Calculating intersection forms

In Chapter 9 we calculated (p-)Kazhdan–Lusztig polynomials of $(W, P) = (\mathfrak{S}_{m+n}, \mathfrak{S}_m \times \mathfrak{S}_n)$ using (characteristic-free!) counting and grading arguments. Such elementary arguments and combinatorics are not possible in general (indeed, this would imply that all p-Kazhdan–Lusztig polynomials were characteristic-free, which would beg the question as to why we put a p in the notation in the first place!). In this section, we delve deeper into the cellular structure and discuss the intersections forms for several small, but illustrative, families of parabolic Coxeter systems.

Example 10.5.1. We now consider $\mathcal{H}_{\mathfrak{S}_3}$. We have one choice for each reduced path of length 0, 1, or 2 but we have precisely two choices for which reduced path of length 3 (see Figure 10.2); we choose the leftmost of the two paths depicted in Figure 10.2.

10.5 Calculating intersection forms

The algebra $\mathcal{H}_{\mathfrak{S}_3}$ is
$$99 = 7^2 + 5^2 + 4^2 + 2^2 + 2^2 + 1^2$$
dimensional and we shall now illustrate this by constructing each of the six distinct cell modules. The module $\Delta^{\Bbbk}(\emptyset)$ is 7-dimensional with basis

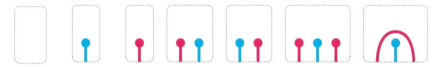

and the degrees of these basis elements are $0, 1, 1, 2, 2, 3, 1$ respectively. The module $\Delta^{\Bbbk}(s_1)$ is 5-dimensional with basis

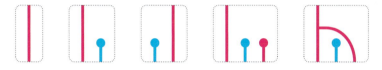

and the degrees of these basis elements are $0, 1, 1, 2, 1$ respectively. The module $\Delta^{\Bbbk}(s_2)$ is 4-dimensional with basis

and the degrees of these basis elements are $0, 1, 1, 2$ respectively. The modules $\Delta^{\Bbbk}(s_1 s_2)$ and $\Delta^{\Bbbk}(s_2 s_1)$ are both $(1+q)$-dimensional with bases

respectively. The module $\Delta^{\Bbbk}(s_1 s_2 s_1)$ is 1-dimensional with basis given by the idempotent $1_\sigma \otimes 1_\tau \otimes 1_\sigma$. Compare the above with the rightmost matrix in Figure 8.8.

When calculating p-Kazhdan–Lusztig polynomials, our first observation is the following, rather trivial, point:

$$\dim(L^{\Bbbk}(\lambda)1_\mu) + \dim(\operatorname{rad}(\langle -, -\rangle_\lambda^{\Bbbk})1_\mu) = \dim(\Delta^{\Bbbk}(\lambda)1_\mu) = \Delta_{\lambda\mu} \qquad (10.5.2)$$

which will actually be the key to disproving Lusztig's conjecture. The postulate of Lusztig's conjecture was that certain $[\Delta^{\Bbbk}(\lambda) : L^{\Bbbk}(\mu)]$ are independent of the characteristic p of the field $\Bbbk$, under certain mild restrictions on p. Of course, the dimension of the cell modules does not depend on the field, and so Lusztig's conjecture is equivalent to the p-independence of the dimensions of the (weight spaces of) simple modules $\dim(L^{\Bbbk}(\lambda)1_\mu)$. Thus by (10.5.2), we are interested in calculating the dimensions of the radicals, $\dim(\text{rad}(\langle -, -\rangle_\lambda^{\Bbbk} 1_\mu)$, of bilinear forms. We illustrate how to compute these intersection forms with a couple of examples.

Example 10.5.3. We wish to show that the structure of $\mathscr{H}_{\mathfrak{S}_3}$ is entirely independent of the underlying field $\Bbbk = \mathbb{F}_p$. That is, we wish to show that

$$\dim(\text{rad}(\Delta^{\Bbbk}(\lambda))1_\mu)$$

is p-independent for all $\lambda, \mu \in \mathfrak{S}_3$. We first consider $\Delta^{\Bbbk}(\lambda)$ for

$$\lambda \in \{s_1 s_2 s_1, s_1 s_2, s_2 s_1, s_2, \emptyset\}.$$

In each of these cases, we note that $\dim(\Delta^{\Bbbk}(\lambda)1_\lambda) = 1$ and $\dim(\Delta^{\Bbbk}(\lambda)1_\mu) \in q\mathbb{Z}_{\geq 0}(q)$, for all $\mu \neq \lambda$. Thus using the grading trick of Section 6.7, we deduce that

$$d_S^* d_T = 0 \times 1_{T_\lambda} + \mathscr{H}_{\mathfrak{S}_3}^{<\lambda}$$

for $S, T \in \text{Path}(\lambda, T_\mu)$ with $\lambda \neq \mu$. (Simply because the lefthand-side is an element of strictly positive degree, whereas the righthand-side is a scalar multiple of an element of degree zero plus lower terms in the Bruhat order.) We now turn to $\lambda = s_1$ and $\mu = s_1 s_2 s_1$ where we encounter our first interesting product,

which is clearly p-independent as the scalar on the righthand-side is -1. Thus $\langle d_S, d_S \rangle = -1$ for $S = \bigcup_\sigma^1 \bigcup_\tau^0 D_\sigma^0$. In particular, the simple module $L^{\Bbbk}(s_1)$ is 2-dimensional with characteristic-free basis given by the pair of diagrams,

We emphasise that the top two basis vectors correspond to the two non-zero entries in the first column of the rightmost matrix in Figure 8.8.

All the examples we have seen so far have been entirely independent of the characteristic of the field. We now provide an example in which the behaviour in characteristics $p = 2, 3$ is genuinely different from characteristic $p \geq 5$.

Example 10.5.4. We set $(W, P) = (\widehat{\mathfrak{S}}_2, \mathfrak{S}_2)$. The Coxeter presentation of $\widehat{\mathfrak{S}}_2$ is $\langle \sigma, \tau \mid \tau^2 = \sigma^2 = 1 \rangle$ and we let $P = \langle \tau \rangle \leq W$ denote the finite parabolic such that the Bruhat graph is as in Figure 8.21. We remark that $m_{\sigma,\tau} = \infty$ and therefore there

10.5 Calculating intersection forms

is a unique reduced path for each $\mu \in {}^P W$. As we do not wish to consider an infinite problem, we first truncate to the finite-dimensional algebra

$$\mathcal{H}_\Pi = 1_\Pi \mathcal{H}_{(\widehat{\mathfrak{S}}_2, \mathfrak{S}_2)} 1_\Pi = 1_{\leqslant \sigma \tau \sigma \tau} \mathcal{H}_{(\widehat{\mathfrak{S}}_2, \mathfrak{S}_2)} 1_{\leqslant \sigma \tau \sigma \tau}$$

obtained by truncating by the saturated set $\Pi = \{\lambda \mid \lambda \leqslant \sigma \tau \sigma \tau\}$. We remark that

$$1_\Pi = 1_{\leqslant \sigma \tau \sigma \tau} = 1_\emptyset + 1_\sigma + 1_{\sigma \tau} + 1_{\sigma \tau \sigma} + 1_{\sigma \tau \sigma \tau}.$$

(See also Section 6.12.) The truncated algebra $\mathcal{H}_\Pi$ is $52 = 3^2 + 5^2 + 4^2 + 2^2 + 1^2$ dimensional. The module $\Delta^k(\emptyset) 1_{\leqslant \sigma \tau \sigma \tau}$ is 3-dimensional with basis

The module $\Delta^k(\sigma) 1_{\leqslant \sigma \tau \sigma \tau}$ is 5-dimensional with basis

The module $\Delta^k(\sigma \tau) 1_{\leqslant \sigma \tau \sigma \tau}$ is 4-dimensional with basis

The module $\Delta^k(\sigma \tau \sigma) 1_{\leqslant \sigma \tau \sigma \tau}$ is 2-dimensional with basis

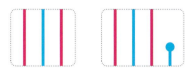

Finally the module $\Delta^k(\sigma \tau \sigma \tau) 1_{\leqslant \sigma \tau \sigma \tau}$ is 1-dimensional with basis

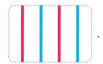.

The degrees of the paths corresponding to each cell module are recorded in the rows of the Δ-matrix as follows:

$\Delta_{\lambda\mu}$	$\sigma\tau\sigma\tau$	$\sigma\tau\sigma$	$\sigma\tau$	σ	$\emptyset$
$\sigma\tau\sigma\tau$	1	$\cdot$	$\cdot$	$\cdot$	$\cdot$
$\sigma\tau\sigma$	q	1	$\cdot$	$\cdot$	$\cdot$
$\sigma\tau$	2	q	1	$\cdot$	$\cdot$
σ	$2q$	1	q	1	$\cdot$
$\emptyset$	$\cdot$	q	$\cdot$	q	1

(10.5.5)

We know that each idempotent belongs to the simple head of its corresponding cell-module. By the grading trick, there are only three other possible elements which might not belong in the radicals of their cell modules:

that is, the three non-idempotent diagrams of degree 0. We claim that the Gram matrices for the intersection forms of $\Delta^k(\sigma)1_{\sigma\tau\sigma}$ and $\Delta^k(\sigma\tau)1_{\sigma\tau\sigma\tau}$ are as follows:

	S
S	-2

	T	U
T	-2	1
U	1	-2

(10.5.6)

We first check the diagonal entry $\langle d_U, d_U \rangle = -2$ (the other diagonal entries can be checked in an identical fashion). We have that

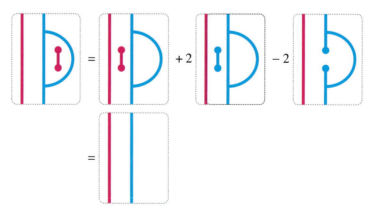

where the first equality follows from (H2) and the second equality follows by the τ-fork-annihilation relation. Thus $\langle d_U, d_U \rangle = -2$ as claimed.

10.5 Calculating intersection forms

Thus it only remains to check the off-diagonal entries, $\langle d_\mathsf{T}, d_\mathsf{U}\rangle$, in the latter matrix in (10.5.6). We have that

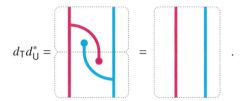

by the fork-spot contraction relation. Therefore $\langle d_\mathsf{T}, d_\mathsf{U}\rangle = 1$, as claimed.

The $p \geqslant 5$ case. The first matrix in (10.5.6) has rank 1 unless $p = 2$ (when it is the zero matrix). The second matrix in (10.5.6) has rank 2 unless $p = 3$ (since $-2 = 1$). Thus for $\Bbbk$ a field of characteristic $p \neq 2, 3$ we can factorise the matrix $\Delta^\Bbbk$ in (10.5.5) as product of the matrices

$N^\Bbbk$	$\sigma\tau\sigma\tau$	$\sigma\tau\sigma$	$\sigma\tau$	σ	$\emptyset$
$\sigma\tau\sigma\tau$	1	·	·	·	·
$\sigma\tau\sigma$	q	1	·	·	·
$\sigma\tau$	·	q	1	·	·
σ	·	·	q	1	·
$\emptyset$	·	·	·	q	1

$B^\Bbbk$	$\sigma\tau\sigma\tau$	$\sigma\tau\sigma$	$\sigma\tau$	σ	$\emptyset$
$\sigma\tau\sigma\tau$	1	·	·	·	·
$\sigma\tau\sigma$	·	1	·	·	·
$\sigma\tau$	2	·	1	·	·
σ	·	1	·	1	·
$\emptyset$	·	·	·	·	1

and we note that the off-diagonal entries of the $N^\Bbbk$-matrix all belong to $q\mathbb{Z}_{\geqslant 0}(q)$.

The $p = 2$ case. Now let $\Bbbk = \mathbb{F}_2$. Over $\mathbb{F}_2$, the leftmost Gram matrix in (10.5.6) has rank 0. Therefore $\mathrm{rad}(\Delta^\Bbbk(\sigma)1_{\sigma\tau\sigma})$ is a 1-dimensional module spanned by d_S. This 1-dimensional submodule must be isomorphic to some $L^\Bbbk(\nu)$ that is (i) a composition factor of $\mathrm{rad}^\Bbbk(\Delta^\Bbbk(\sigma)1_{\leqslant\sigma\tau\sigma\tau})$ and (ii) not annihilated by $1_{\sigma\tau\sigma}$; this implies that $\sigma < \nu \leqslant \sigma\tau\sigma$. We have that $L^\Bbbk(\sigma\tau)1_{\sigma\tau\sigma} = 0$ by degree considerations and hence $\nu = \sigma\tau\sigma$. Therefore $[\Delta^\Bbbk(\sigma) : L^\Bbbk(\sigma\tau\sigma)] = 1$ for $\Bbbk = \mathbb{F}_2$.

Over $\mathbb{F}_2$, the rightmost Gram matrix in (10.5.6) has full rank and so d_T and d_U are both vectors in the simple module $L^\Bbbk(\sigma\tau)$. Putting this altogether we can now factorise the matrix $\Delta^\Bbbk$ in (10.5.5) as the product of the following matrices:

$N^\Bbbk$	$\sigma\tau\sigma\tau$	$\sigma\tau\sigma$	$\sigma\tau$	σ	$\emptyset$
$\sigma\tau\sigma\tau$	1	·	·	·	·
$\sigma\tau\sigma$	q	1	·	·	·
$\sigma\tau$	·	q	1	·	·
σ	·	1	q	1	·
$\emptyset$	·	·	·	q	1

$B^\Bbbk$	$\sigma\tau\sigma\tau$	$\sigma\tau\sigma$	$\sigma\tau$	σ	$\emptyset$
$\sigma\tau\sigma\tau$	1	·	·	·	·
$\sigma\tau\sigma$	·	1	·	·	·
$\sigma\tau$	2	·	1	·	·
σ	·	·	·	1	·
$\emptyset$	·	·	·	·	1

The $p = 3$ case. Over $\mathbb{F}_3$, the rightmost Gram matrix in (10.5.6) has rank 1. Therefore $\mathrm{rad}(\Delta^{\Bbbk}(\sigma\tau)1_{\sigma\tau\sigma\tau})$ has a 1-dimensional submodule spanned by some $v = \alpha_T d_T + \alpha_U d_U$ (one can check that $v = d_T - d_U$ spans this submodule). This 1-dimensional submodule must be isomorphic to some $L^{\Bbbk}(\nu)$ that is (i) a composition factor of $\mathrm{rad}^{\Bbbk}(\Delta^{\Bbbk}(\sigma\tau)1_{\leqslant \sigma\tau\sigma\tau})$ and (ii) not annihilated by $1_{\sigma\tau\sigma\tau}$; this implies that $\sigma\tau < \nu \leqslant \sigma\tau\sigma\tau$. We have that $L^{\Bbbk}(\sigma\tau\sigma)1_{\sigma\tau\sigma\tau} = 0$ by degree considerations and hence $\nu = \sigma\tau\sigma\tau$. Therefore $[\Delta^{\Bbbk}(\sigma\tau) : L^{\Bbbk}(\sigma\tau\sigma\tau)] = 1$ for $\Bbbk = \mathbb{F}_3$. However (unlike the $p = 2$ case) this has a knock-on effect in our table... Regardless of the field $\Bbbk$, we have that

$$\mathrm{rad}^{\Bbbk}(\Delta^{\Bbbk}(\sigma)1_{\leqslant \sigma\tau\sigma\tau}) = \Bbbk \left\{ \begin{array}{c} \text{diagram 1} \end{array}, \begin{array}{c} \text{diagram 2} \end{array} \right\} \subseteq \Delta^{\Bbbk}(\sigma)1_{\leqslant \sigma\tau\sigma\tau}.$$

For $p \neq 3$ this 2-dimensional submodule is isomorphic to $L^{\Bbbk}(\sigma\tau\sigma\tau)\langle 1 \rangle$. However, for $p = 3$ this 2-dimensional submodule is no longer simple; it has a 1-dimensional submodule

$$L^{\Bbbk}(\sigma\tau\sigma\tau)1_{\leqslant \sigma\tau\sigma\tau} \cong \Bbbk \left\{ \begin{array}{c} \text{diagram 1} \end{array} - \begin{array}{c} \text{diagram 2} \end{array} \right\} \subseteq \Delta^{\Bbbk}(\sigma)1_{\sigma\tau\sigma\tau}$$

Thus we factorise the matrix $\Delta^{\Bbbk}$ in (10.5.5) as follows

$N^{\Bbbk}$	$\sigma\tau\sigma\tau$	$\sigma\tau\sigma$	$\sigma\tau$	σ	$\emptyset$
$\sigma\tau\sigma\tau$	1	$\cdot$	$\cdot$	$\cdot$	$\cdot$
$\sigma\tau\sigma$	q	1	$\cdot$	$\cdot$	$\cdot$
$\sigma\tau$	1	q	1	$\cdot$	$\cdot$
σ	q	$\cdot$	q	1	$\cdot$
$\emptyset$	$\cdot$	$\cdot$	$\cdot$	q	1

$B^{\Bbbk}$	$\sigma\tau\sigma\tau$	$\sigma\tau\sigma$	$\sigma\tau$	σ	$\emptyset$
$\sigma\tau\sigma\tau$	1	$\cdot$	$\cdot$	$\cdot$	$\cdot$
$\sigma\tau\sigma$	$\cdot$	1	$\cdot$	$\cdot$	$\cdot$
$\sigma\tau$	1	$\cdot$	1	$\cdot$	$\cdot$
σ	$\cdot$	1	$\cdot$	1	$\cdot$
$\emptyset$	$\cdot$	$\cdot$	$\cdot$	$\cdot$	1

We now extend the above to provide an infinite family of examples which depend heavily on the characteristic of the field.

Proposition 10.5.7. *Let $\Bbbk = \mathbb{F}_p$. There exist infinitely many p-Kazhdan–Lusztig polynomials of $\mathfrak{S}_2 \leqslant \widehat{\mathfrak{S}}_2$ which do depend on the prime $p > 0$.*

Proof. We let $\langle \sigma, \tau \mid \tau^2 = \sigma^2 = 1 \rangle$ and we let $P = \langle \tau \rangle \leqslant W$. For $n \in 2\mathbb{N}$, we set

$$\lambda = \underbrace{\sigma\tau\sigma\tau\sigma\ldots\tau}_{np-1} \qquad \nu = \underbrace{\sigma\tau\sigma\tau\sigma\ldots\tau\sigma}_{np} \qquad \mu = \underbrace{\sigma\tau\sigma\tau\sigma\ldots\tau\sigma\tau}_{np+1} \qquad (10.5.8)$$

The module $\Delta(\lambda)1_\mu$ is $(np-1)$-dimensional with basis $\{f_k \mid 1 \leqslant k < np\}$ given by the following diagrams:

10.5 Calculating intersection forms

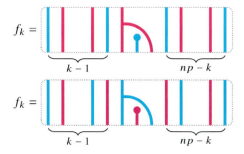

for k odd and k even, respectively. The Gram-matrix of the cell-form of $\Delta(\lambda)1_\mu$ is the $((np-1) \times (np-1))$-matrix

$$\begin{bmatrix} -2 & 1 & 0 & 0 & \cdots \\ 1 & -2 & 1 & 0 & \cdots \\ 0 & 1 & -2 & 1 & \ddots \\ 0 & 0 & 1 & -2 & \ddots \\ \vdots & \vdots & \ddots & \ddots & \ddots \end{bmatrix}$$

via an identical calculation to that in Example 10.5.4. Over $\Bbbk = \mathbb{F}_p$, the rank of this matrix is $np - 2$, whereas over $\mathbb{C}$ this matrix has rank $np - 1$. □

Exercise 10.5.9. Calculate the p-Kazhdan–Lusztig polynomials of $\mathfrak{S}_4$ for any $p \geqslant 0$.

Exercise 10.5.10. We let W_4 denote the Coxeter group

$$W_4 = \langle \sigma, \tau, \rho, \pi \mid (\sigma t)^3 = 1, (st)^2 = 1, \sigma^2 = 1, s^2 = 1, \text{ for } s, t \in \{\tau, \rho, \pi\}\rangle.$$

We claim that there exists p-torsion in this algebra (for some $p \geqslant 2$). Consider the 8-dimensional module $\Delta^k(\tau\rho\pi)1_{\tau\rho\pi\sigma\tau\rho\pi}$ which has light leaves basis consisting of the elements:

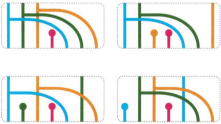

in degrees -2 and $0, 0, 0$ respectively together with the elements:

260   10 Lusztig's conjecture in the diagrammatic algebra $\mathcal{H}_{(W,P)}$

of degrees 2, 2, 2 and 4 respectively. By the grading trick, the Gram matrix is of the intersection form of $\Delta^{\Bbbk}(\tau\rho\pi)1_{\tau\rho\pi\sigma\tau\rho\pi}$ is of the following form

0	0	0	0	?	?	?	0
0	?	?	?	0	0	0	0
0	?	?	?	0	0	0	0
0	?	?	?	0	0	0	0
?	0	0	0	0	0	0	0
?	0	0	0	0	0	0	0
?	0	0	0	0	0	0	0
0	0	0	0	0	0	0	0

where the rows and columns are arranged in terms of increasing degree from left-to-right (from −2 to 0, 0, 0 to 2, 2, 2 to 4). Calculate the ? entries and hence determine the rank of this matrix over any field $\mathbb{F}_p$. For which p does the rank change? (That is, for which p do you encounter p-torsion?)

Exercise 10.5.11. Let $(W, P) = (\widehat{\mathfrak{S}}_3, \mathfrak{S}_3)$ where

$$\mathfrak{S}_3 = \langle \rho, \tau \mid (\tau\rho)^3 = 1, \tau^2 = 1 = \rho^2 \rangle$$
$$\widehat{\mathfrak{S}}_3 = \langle \sigma, \rho, \tau \mid (\sigma\rho)^3 = (\sigma\tau)^3 = (\rho\tau)^3 = 1, \sigma^2 = \tau^2 = \rho^2 = 1 \rangle.$$

Calculate the bases and the intersection forms for the modules $\Delta^{\Bbbk}(\lambda)1_\mu$ for $\lambda, \mu \in \Pi = \{\lambda \mid \lambda \leq \sigma\rho\tau\sigma\rho\tau\}$. Hence calculate the p-Kazhdan–Lusztig polynomials ${}^p n_{\lambda,\mu}(q)$ for $\lambda, \mu \in \Pi$ and all primes $p \geq 0$.

Remark 25. Notice that in Example 10.5.4 we are working by induction on the Bruhat order, just as in the definition of the classical Kazhdan–Lusztig polynomials (see (8.2.10)). In the definition of the Kazhdan–Lusztig polynomials, we calculate

$$\sum_{S \in \text{Path}(\lambda, \mathsf{T}_\mu)} q^{\deg(S)} - \sum_{\lambda < \nu < \mu} n_{\lambda, \nu}(q) b_{\nu, \mu}(q)$$

and then write this uniquely as a sum of a "self-dual" polynomial in $\mathbb{Z}_{\geq 0}[q + q^{-1}]$ and a positive degree polynomial in $q\mathbb{Z}_{\geq 0}[q]$. Through the Elias–Williamson categorification theorem this can be though of as a shadow of the fact that "all composition factors occur with non-negative degree" over $\mathbb{C}$.

When calculating p-Kazhdan–Lusztig polynomials, we must resort to explicitly calculating the intersection forms themselves (as in Example 10.5.4). In a nut-shell, this is the main added difficulty of modular representation theory.

10.6 Counterexamples to Soergel's conjecture in positive characteristic

In this section, we will provide a self-contained and elementary proof that the p-Kazhdan–Lusztig polynomials of $\mathfrak{S}_n$ are not characteristic-free unless p is "large" with respect to the rank n. This was the main result of [Wil17] and we follow the ideas of the proof therein. However, our treatment takes things slowly and makes everything much more diagrammatic and explicit. In Section 10.7 we will "affinize" the examples of this section and hence provide the desired counterexamples to the expected bounds in Lusztig's conjecture.

10.6.1 Picking pairs of permutations

For a given $n \in \mathbb{N}$, we wish to consider $\lambda, \mu \in \mathfrak{S}_n$ such that

$$\dim_q(\Delta^{\Bbbk}(\lambda)1_\mu)|_{q=0} = 1. \tag{10.6.1}$$

In other words, the pairs $\lambda, \mu \in \mathfrak{S}_n$ for which the degree zero component of $\Delta^{\Bbbk}(\lambda)1_\mu$ is 1-dimensional. Why do we wish to do this? Because by the idempotent and grading tricks of Section 6.7, this means we need only consider a (1×1)-Gram matrix $\langle d_S, d_S \rangle$ (for the unique path $S \in \text{Path}(\lambda, T_\mu)$ of degree 0) which is an incredibly easy thing to do. Indeed, it is enough to do one calculation: we must show that p divides $\langle d_S, d_S \rangle$ for certain primes (which contradict Lusztig's proposed bound). We will construct an infinite sequence of such $(\lambda_k, \mu_k)_{k \geq 0}$ and inductively calculate the corresponding Gram matrices.

For the remainder of this chapter, we fix the colouring of the Coxeter graph to be as depicted in Figure 10.10 (we think of this as colouring as dividing the generators into 'blue-ish', 'red-ish', and 'green-ish' reflections, separated into three families by a single black reflection and a single grey reflection).

Fig. 10.10: The colouring of the reflections in the symmetric group.

We now construct a pair of elements

$$\mu_k \in \mathfrak{S}_{3k+5} \qquad \lambda_k \in \mathfrak{S}_{k+1} \times \mathfrak{S}_4 \times \mathfrak{S}_{2k} \leq \mathfrak{S}_{3k+5}$$

for every $k \geq 1$ satisfying the condition of (10.6.1). We roughly divide the reflections into 'blue-ish', 'red-ish', and 'green-ish' reflections and colour the parabolics appropriately. We choose to label our strands in a slightly weird way, so that we can easily embed examples by induction: we set

$$s_1 = (1,2) \quad s_2 = (2,3) \quad s_3 = (3,4)$$

and then embed this copy of $\mathfrak{S}_4$ into larger and larger symmetric groups by adding strand on the left (labelled by non-positive integers) and strands on the right (labelled by positive integers greater than 4).

10.6.1.1 The $k = 0$ case

For $k = 0$ we pick our permutations λ_0 and μ_0 as follows:

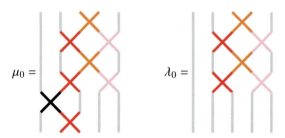

Notice that λ_0 is the longest permutation in $\mathfrak{S}_4$. We choose our reduced path

$$\mathsf{T}_{\mu_0} = (2,3)(1,2)(3,4)(2,3)(1,2)(3,4)(0,1)(1,2)$$

in this case it is completely clear that there is exactly one element (of degree 0)

$$\mathsf{S}_0 = \mathsf{U}_2^1 \mathsf{U}_1^1 \mathsf{U}_3^1 \mathsf{U}_2^1 \mathsf{U}_1^1 \mathsf{U}_3^1 \mathsf{U}_0^0 \mathsf{D}_1^0 \in \mathrm{Path}(\lambda_0, \mathsf{T}_{\mu_0})$$

The corresponding unique degree zero element of $\Delta(\lambda_0)1_{\mu_0}$ is as follows

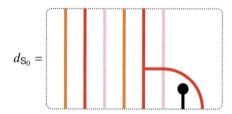

Exercise 10.6.2. Check that $\langle d_{\mathsf{S}_0}, d_{\mathsf{S}_0} \rangle = -1$ (which can be though of as $-f_1 = -1$, the negative of the first Fibonacci number).

10.6.1.2 The $k = 1$ case

Now, for $k = 1$ we set λ_1 to be the longest word in $\mathfrak{S}_2 \times \mathfrak{S}_4 \times \mathfrak{S}_2$. We are going to split μ_0 into two parts and "stick something in the middle", as follows:

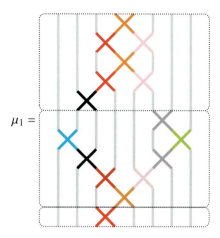

in fact, the grey shaded region in the middle can be thought of as *"inserting the smallest permutation with $(1, 2)(3, 4)(2, 3)$ as a subword, such that the overall product is still reduced"* (in other words, the black, grey, blue and green reflections are merely the necessary "fluff" so that μ_1 is a reduced word). To summarise, we have set

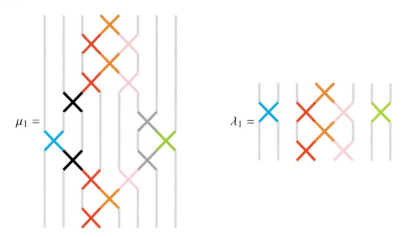

With the permutations in place, we now choose our reduced path

$$T_{\mu_1} = U_2^1 U_1^1 U_3^1 U_2^1 U_1^1 U_3^1 U_0^1 U_{-1}^1 \, U_0^1 U_1^1 U_4^1 U_5^1 U_4^1 U_3^1 U_2^1 U_1^1$$

where the heuristic of this path (should one wish for one) is that we *"first do the longest redish subword, then do the left side, before doing the right side"*.

We now construct the unique degree 0 path $S_1 \in \mathrm{Path}(\lambda_1, T_{\mu_1})$. One should first observe that both λ_1 and μ_1 both have a single blue and green reflection and so both of these blue and green reflection must correspond to U^1-steps (otherwise our path cannot terminate at λ_1). Furthermore, the fact that $\lambda_1 \in \mathfrak{S}_2 \times \mathfrak{S}_4 \times \mathfrak{S}_2$ implies that *all of the black and grey reflections must correspond to U^0-steps* (otherwise our path cannot terminate at λ_1). At this point we notice that the sum over the degrees of the steps prescribed above is equal to 4; therefore we must do (at least) 4 D^0-steps (in the remaining colours yet to be considered: mypink, orange and red) in order to balance the degree (recall that we are constructing the paths of degree $0 = 4-4$). This means that all of the final four (red-ish) steps are D^0-steps. In summary, we have a unique path

$$S_1 = \mathsf{U}^1_2\mathsf{U}^1_1\mathsf{U}^1_3\mathsf{U}^1_2\mathsf{U}^1_1\mathsf{U}^1_3\mathsf{U}^0_0\mathsf{U}^1_{-1}\mathsf{U}^0_0\mathsf{D}^0_1\mathsf{U}^0_4\mathsf{U}^1_5\mathsf{U}^0_4\mathsf{D}^0_3\mathsf{D}^0_2\mathsf{D}^0_1 \in \mathrm{Path}(\lambda_1, T_{\mu_1})$$

of degree 0. The basis element d_{S_1} is of the form depicted in Figure 10.11 (and in an abbreviated "black box form" in Figure 10.12).

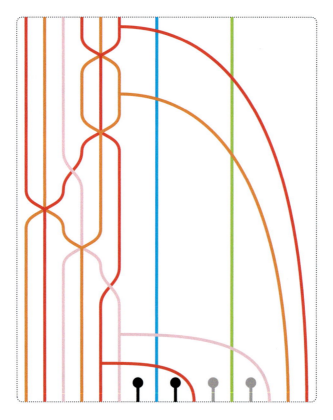

Fig. 10.11: The unique degree zero light leaf basis element for $S_1 \in \mathrm{Path}(\lambda_1, \mu_1)$

10.6 Counterexamples to Soergel's conjecture in positive characteristic

Remark 26. The $k = 1$ case is depicted in full detail in Figure 10.11, but this is already almost too large to draw. However, we will see that the (very large!) braid element on the left is irrelevant to the computation $\langle d_{S_1}, d_{S_1} \rangle \in \Bbbk$. For a preview as to why this is the case: recall that we calculate intersection forms by multiplying a diagram with its flip through the horizontal axis... and for a braid this product is equal to the original idempotent modulo lower order terms in the Bruhat order (which are precisely the terms which are zero in the cell module). This will justify our condensed in which we put the braid in a "black box" as pictured in Figure 10.12.

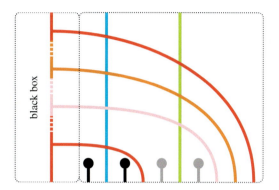

Fig. 10.12: The unique degree zero light leaf basis element for $S_1 \in \text{Path}(\lambda_1, \mu_1 s_1)$ drawn in our "black box" condensed format. We claim that the braid inside the blackbox does not play any role in the proof.

10.6.1.3 The $k = 2$ case

We now return to the business of constructing the pairs (λ_k, μ_k) for $k > 1$. For $k = 2$ we set λ_2 to be the longest word in $\mathfrak{S}_3 \times \mathfrak{S}_4 \times \mathfrak{S}_4$ pictured as follows

Similarly to the previous case, we are going to split μ_1 into two parts and "stick something in the middle". As before the "something in the middle" can be thought of as "inserting the smallest permutation with $(1, 2)(3, 4)(2, 3)$ as a subword, such that the overall product is still reduced" (in other words, the black, grey, blue and green reflections are merely the necessary "fluff" so that μ_1 is a reduced word). Thus μ_2 is as follows:

266   10 Lusztig's conjecture in the diagrammatic algebra $\mathcal{H}_{(W,P)}$

$\mu_2 =$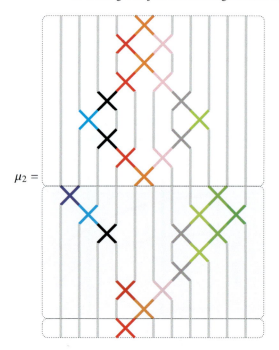

We choose T_{μ_2} so that the unique degree zero path $\mathsf{S}_2 \in \mathrm{Path}(\lambda_2, \mathsf{T}_{\mu_2})$ is as follows

$$\mathsf{U}_2^1\mathsf{U}_1^1\mathsf{U}_3^1\mathsf{U}_2^1\mathsf{U}_1^1\mathsf{U}_3^1\mathsf{U}_0^0\mathsf{U}_{-1}^1\mathsf{U}_0^0\mathsf{D}_1^0\mathsf{U}_4^0\mathsf{U}_5^1\mathsf{U}_4^0\mathsf{D}_3^0\mathsf{D}_2^0\mathsf{U}_{-2}^1\mathsf{U}_{-1}^1\mathsf{U}_0^0\mathsf{D}_1^0\mathsf{U}_6^1\mathsf{U}_5^1\mathsf{U}_4^0\mathsf{U}_7^1\mathsf{U}_6^1\mathsf{U}_5^1\mathsf{U}_4^0\mathsf{D}_3^0\mathsf{D}_2^0\mathsf{D}_1^0$$

and we picture the basis element d_{S_2} in Figure 10.13.

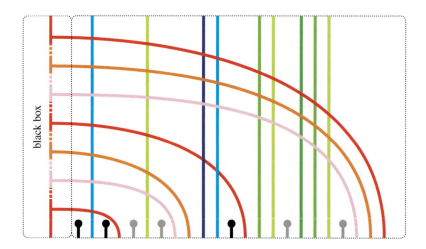

Fig. 10.13: The unique degree zero light leaf basis element for $\mathsf{S}_2 \in \mathrm{Path}(\lambda_2, \mathsf{T}_{\mu_2})$.

10.6 Counterexamples to Soergel's conjecture in positive characteristic

Remark 27. Notice the similarities between the diagram pictured in Figure 10.13 and in Figure 10.12. All blue and green lines propagate, all black and grey strands are spots, and all red, orange, and mypinks strands are forks.

10.6.1.4 The $k > 2$ case

We are now ready to give the general definition of the pairs (λ_k, μ_k) for $k \geq 1$. For $i, j \in \mathbb{Z}$ with $i < j$, we set $w_{[i,j]} = s_i s_{i+1} \ldots s_{j-1}$ the shortest possible permutation which takes the ith northern node to the jth southern node and $w_{[j,i]} = (w_{[i,j]})^*$. We set $w_{[i,i]} = 1$ for $i \in \mathbb{Z}$. In this notation, we have that μ_1 is equal to

$$s_1 s_3 s_2 s_1 s_3 s_0 (s_{[-1,0]} s_0 s_1)(s_{[5,5]} s_4 s_{[6,5]} s_4 s_3 s_2) s_1$$

and μ_2 is equal to

$$s_1 s_3 s_2 s_1 s_3 s_0 (s_{[-1,0]} s_0 s_1)(s_{[5,5]} s_4 s_{[6,5]} s_4 s_3 s_2)(s_{[-2,0]} s_0 s_1)(s_{[7,5]} s_4 s_{[8,5]} s_4 s_3 s_2) s_1$$

More generally, we make the following definition:

Definition 10.6.3. For $k \geq 1$ we set $\nu_k = (w_{[-k,0]} s_0 s_1)(s_{[3+2k,5]} s_4 w_{[4+2k,5]} s_4 s_3 s_2)$ for $k \geq 1$ and we define

$$\mu_k = s_1 s_3 s_2 s_1 s_3 s_0 (\nu_1 \nu_2 \ldots \nu_k) s_1.$$

We set $\lambda_k \in \mathfrak{S}_{k+1} \times \mathfrak{S}_4 \times \mathfrak{S}_{2k} \leq \mathfrak{S}_{3k+5}$ to be the longest element in said subgroup.

Proposition 10.6.4. *For $k \geq 1$, there is a unique degree zero element $\mathsf{S}_k \in \text{Path}(\lambda_k, \mathsf{T}_{\mu_k})$.*

Proof. We fix a reduced path T_{μ_k} to μ_k. We will construct a path $\mathsf{S}_k \in \text{Path}(\lambda_k, \mathsf{T}_{\mu_k})$ of degree zero by considering each colour of reflection in turn (first blue-ish/green-ish, then grey/black, then red-ish). For each colour, we argue as to what type of steps U^0, U^1, D^0, or D^1 of that colour can occur in our path S_k. We shall see that there is no freedom in these choices and so we deduce that S_k is unique.

We first consider reflections from the blue-green subgroup $\mathfrak{S}_{\{-k,\ldots,-1,0\}} \times \mathfrak{S}_{\{-k,\ldots,-1,0\}} \leq \mathfrak{S}_{3k+5}$. Such a reflection appears exactly the same number of times in any choice of reduced path to λ_k as it does to any reduced path to μ_k. Therefore if $\mathsf{S}_k \in \text{Path}(\lambda_k, \mu_k)$, we must have that any blue or green step in S_k must be of the form U^1 or U^1.

We now consider the occurrences of black and grey reflections, s_0 and s_4, within $\mathsf{S}_k \in \text{Path}(\lambda_k, \mu_k)$. Each such step must be a $\mathsf{U}^0/\mathsf{U}^0$-step in S_k as otherwise the path would leave the parabolic $\mathfrak{S}_{\{-k,\ldots,-1,0\}} \times \mathfrak{S}_{\{1,2,3,4\}} \times \mathfrak{S}_{\{-k,\ldots,-1,0\}} \leq \mathfrak{S}_{3k+5}$ and then our path would be unable to terminate at λ_k. Thus there are a total of $3k + 1$ steps which are of the form U^0_0 or U^0_4 (each of which is of degree +1) in our path S_k.

Now we consider the reflections from the red-ish subgroup $\mathfrak{S}_{\{1,2,3,4\}}$. At this point, we emphasise that we are only interested in paths of degree zero in

Path(λ_k, μ_k). In order to balance the positive degree contribution from black/grey reflections above, we must have that at least $3k+1$ of the red-ish steps are of degree -1 (in other words, $3k+1$ D^0-steps). There are a total of $6+(3k-1)$ red-ish steps in S_k and we note that 6 of these must be U^1-steps in order that our path S_k terminates at λ_k; therefore we must have precisely $3k+1$ of the D^0-steps in S_k. In order to perform a D^0-step, we must have first performed a U^1-step (we can't remove what isn't already there!). In particular, the first 6 red-ish steps in S_k must all be U^1-steps and the next $3k+1$ red-ish steps must all be D^1-steps. We have uniquely determined every step in the path S_k and the result follows. □

The proof of the Proposition 10.6.4 involved constructing the unique path explicitly. We hence deduce the following corollary.

Corollary 10.6.5. *The unique degree zero element $\mathsf{S}_k \in \mathrm{Path}(\lambda_k, \mu_k)$ for $k \geq 1$ is constructed inductively as follows. We set*

$$\mathsf{T}_k = (\mathsf{U}^1_{-k} \cdots \mathsf{U}^1_{-1} \mathsf{U}^0_0 \mathsf{D}^0_1)(\mathsf{U}^1_{2+2k} \cdots \mathsf{U}^1_5 \mathsf{U}^0_4)(\mathsf{U}^1_{3+2k} \cdots \mathsf{U}^1_5 \mathsf{U}^0_4)(\mathsf{D}^0_3 \mathsf{D}^0_2)$$

and we set

$$\mathsf{S}_k = \mathsf{U}^1_2 \mathsf{U}^1_1 \mathsf{U}^1 \mathsf{U}^1_2 \mathsf{U}^1_1 \mathsf{U}^1 \mathsf{U}^0_0 (\mathsf{T}_1 \mathsf{T}_2 \ldots \mathsf{T}_k) \mathsf{D}^0_1.$$

Examples of the corresponding light leaf basis element are depicted in Figures 10.11 and 10.13.

10.6.2 Fibonacci numbers as values of intersection forms

We enumerate the Fibonacci numbers by $\mathbb{N}$ as follows

$$f_1 = 1 \quad f_2 = 1 \quad f_3 = 2 \quad f_4 = 3 \quad f_5 = 5 \quad f_6 = 8 \quad f_7 = 13 \ldots$$

We are now ready to recall the main result of [Wil17] in terms of the notation already fixed in this section. The remainder of this section will be dedicated to the proof.

Theorem 10.6.6. *We have that $\langle d_{\mathsf{S}_k}, d_{\mathsf{S}_k} \rangle = (-1)^{k+1} f_{k+1}$ for $k \geq 0$.*

In particular, a necessary condition for the p-Kazhdan–Lusztig polynomials of $\mathfrak{S}_{5+3k}$ to be p-independent is that p does not divide any of the first k Fibonacci numbers. By Carmichael's theorem, the first $n \gg 0$ Fibonacci numbers have at least n distinct prime factors [Car14]. By the prime number theorem, the number of prime divisors of n grows at least as fast as $n \log n$. Therefore we have the following:

Corollary 10.6.7. *For $n \in \mathbb{N}$, there is no linear bound $b(n)$ for which the p-Kazhdan–Lusztig polynomials ${}^p n_{\lambda, \mu}$ for $\lambda, \nu \in \mathfrak{S}_n$ are independent of the prime p for all $p > b(n)$.*

10.6 Counterexamples to Soergel's conjecture in positive characteristic

We thus obtain counterexamples to the following "toy version" of Lusztig's conjecture (we remark that Soergel did not explicitly conjecture the following statement).

Conjecture 10.6.8 ("Soergel's conjecture in positive characteristic"). *The p-Kazhdan–Lusztig polynomials $^p n_{\lambda,\mu}$ for $\lambda, \mu \in \mathfrak{S}_n$ are independent of the prime p for $p \geq n$.*

In the next section we will discuss how the counterexamples to Soergel's conjecture in positive characteristic serve to also disprove Lusztig's conjecture. It is worth emphasising that Soergel never formally stated the above conjecture, but rather he emphasised the "toy version" of Lusztig's conjecture above as an important stepping stone to tackling the Lusztig conjecture in full generality. (Remember, it was almost universally assumed that Lusztig's conjecture *was true*... almost everyone was trying to *prove* this conjecture, not disprove it.)

We now introduce some notation and prove a few lemmas about moving linear combinations of barbells through a diagram.

Lemma 10.6.9. *Given $\sigma, \rho \in S_W$ such that $m_{\sigma,\rho} = 3$ we have that*

[diagram: barbell equation = − strand]

Proof. We begin by using the (σ, ρ)-barbell-strand relation to move the barbell rightwards as follows:

[diagram showing equality with multiple terms]

and then the second equality follows by the fork-annihilation and fork-spot contraction relations. □

Lemma 10.6.10. *Given $\rho \in S_W$, we have that*

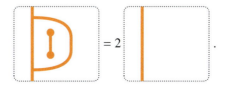

Proof. We begin by using relation (M2') to move the barbell rightwards as follows:

where the second equality follows by (M1). □

We now introduce a small piece of diagrammatic notation. We often wish to consider a linear combination of almost identical diagrams, which differ by some barbell diagrams in a localised region. In order to do this, we place a dotted frame around the polynomial in the barbell relations. For example

Lemma 10.6.11. *Given $\sigma, \tau, \rho \in S_W$ such that $m_{\sigma,\rho} = 3 = m_{\tau,\rho}$ we have that*

Proof. We begin by using the (τ, ρ)-barbell-strand relation to move the barbell rightwards as follows:

the second equality follows from Lemma 10.6.9 and the third follows from (M1). □

10.6.3 The base case of the proof

We first consider the $k = 1$ case. Technically speaking, this need not be done separately (as will become clear in Subsection 10.6.5). However, we believe that this will serve as a nice warm-up to the ideas and will help the reader understand the structure of the proof — particularly for those readers who have never seen such a diagrammatic proof before. We make an aside that $\ell(\lambda_1) = 8$

Proposition 10.6.12. *The base case for induction holds: that is $\langle d_{S_1}, d_{S_1} \rangle = 1$. In other words, we have that*

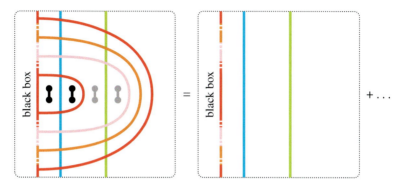

where the ... are terms from lower in the Bruhat order on the cellular algebra. (In particular, these terms are linear combinations of diagrams with at most 7 propagating strands.)

Proof of the base case. We observe that we can pull the green strand all the way to the right at the expense of an error term as follows:

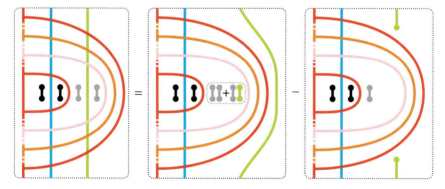

and we notice that the second diagram on the righthand-side belongs to a lower order cell-ideal (in particular it can be written as a linear combination of diagrams with at most 7 propagating strands). Therefore we can ignore the second term from now on as it makes no contribution to the calculation $\langle d_{S_1}, d_{S_1} \rangle$. We reminder the reader that the first diagram depicted on the righthand-side of the equality is actually a sum of two elements (the diagram with two grey barbells and the diagram with a grey-green pair of barbells).

We now repeat the above argument to move the blue strand all the way to the right (at the expenses of an ignorable error term) and hence obtain the following

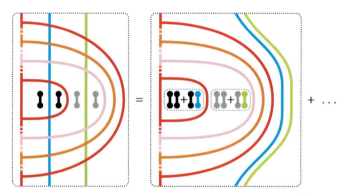

where the ... are terms belonging to lower order cell ideals (namely, those in which we remove a green or blue strand). We emphasise that the righthand-side consists of $4 = 2 \times 2$ diagrams (as well as the non-pictured terms which are denoted by ...)

Now, we recall that the "black box" on the lefthand edge of each diagram consists solely of red orange and pink strands. The blue and green barbells commute with the red orange and pink strands (see Figure 10.10 for the Coxeter graph) and so we can pull these barbells leftward (through the blackbox) until they are annihilated by the cyclotomic relation. We hence obtain the following

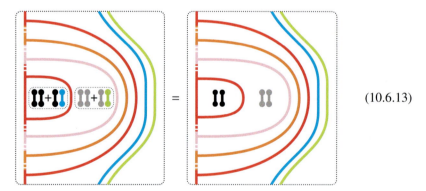

(10.6.13)

We notice that the diagram on the right of the equality is *precisely* the diagram which we started with, just with the blue and green strands moved all the way to the right. In other words, the calculation of $\langle d_{\mathsf{S}_1}, d_{\mathsf{S}_1} \rangle$ does not "see" these blue and green strands at all.

We now apply Lemma 10.6.11 to the black barbells within the red fork to the diagram on the righthand-side of (10.6.13) to obtain a linear combination of black and red barbells. We repeat this argument to each resulting diagram: we apply

10.6 Counterexamples to Soergel's conjecture in positive characteristic

Lemma 10.6.11 to the grey barbells within the pink fork and we hence obtain a linear combination of pink and grey barbells as follows:

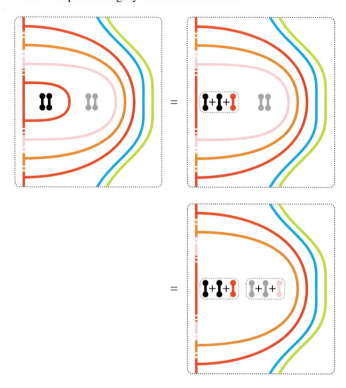

We now expand out the $9 = 3 \times 3$ terms in the rightmost diagram. We gather them together and hence obtain the following:

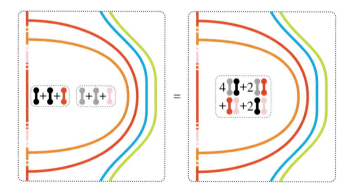

We now pull each of these four pairs of coloured barbells through the orange circle. For the red-grey barbell term and the orange circle we apply the commuting relations and Lemma 10.6.9. Similarly, for mypink-black barbell term and the orange circle, we apply the commuting relations and Lemma 10.6.9. The pair of black/grey barbells commute with the orange strand and so we can apply for orange fork-annihilation relation. We hence obtain:

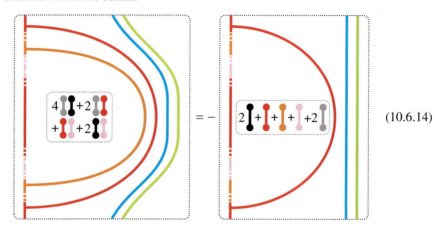 (10.6.14)

Finally, we can apply the Lemmas 10.6.9 and 10.6.10 and the commutativity relations to move the barbells through the last red strand and hence obtain

$$-(2\times -1 + 1\times 2 + 1\times -1 + 1\times 0 + 2\times 0)\, 1_{21_321_3} \otimes 1_{-1} \otimes 1_5 + \cdots = 1_{\lambda_1} + \ldots$$

where the important point is the coefficient $+1 = (-1)^2$ of the idempotent on the righthand-side, as claimed. Here we have used the bold colours to highlight the scalars obtained from applying Lemmas 10.6.9 and 10.6.10, the non-bold scalars are simply the coefficients of the barbells in the diagram on the righthand-side of (10.6.14). □

10.6.4 An observation and inductive reformulation

The purpose of this section is to illustrate that we have, in some sense, already done most of the work! We will reformulate the $k > 1$ case in an inductive manner and we illustrate this inductive approach by explicitly calculating the $k = 2$ case (using the $k = 1$ case calculated in the previous section).

We first require some notation. We let $\mathbb{U}^1_\sigma$ and $\mathbb{U}^0_\sigma$ denote the diagrams obtained by performing U^1_σ and U^0_σ symmetrically, in the following manner:

10.6 Counterexamples to Soergel's conjecture in positive characteristic

We let $\mathbb{D}^0_\sigma$ and $\mathbb{D}^1_\sigma$ denote the diagrams obtained by performing D^0_σ and D^1_σ symmetrically, in the following manner:

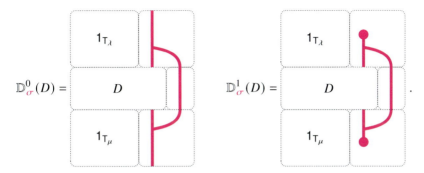

Observe that the blue and green vertical strands played absolutely no role in the proof of the base case. More generally, pulling green vertical strands to the right only produces error terms which factor through idempotents labelled by words *lower in the Bruhat ordering* than λ_k for $k \geq 1$. This is simply because these error terms involve "breaking a blue or green strand" as in the base case (and thus is equal to zero *in the cell module*) or involve blue or green barbells (and thus is equal to zero *in the algebra* by the commuting and cyclotomic relations). In particular, all these "error" diagrams are zero in our cell module and so can be ignored for the purposes of computing the intersection forms on these modules. (Compare Figure 10.13 versus Figure 10.14.)

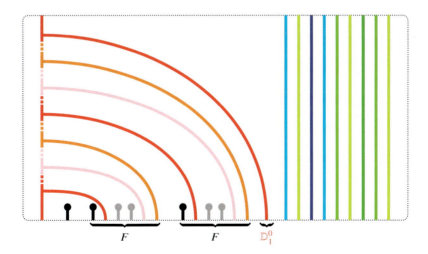

Fig. 10.14: Redrawing of the diagram of Figure 10.13 in order to highlight the inductive approach of (10.6.15) (we have not draw the braid in (10.6.15)). Notice that the same six operators in F repeat.

This leads us to define the following sequences, which are obtained from those of Corollary 10.6.5 by deleting all occurrences of blue and green operators. We set

$$F = \mathbb{U}_0^0 \mathbb{U}_1^0 \mathbb{U}_4^1 \mathbb{D}_4^0 \mathbb{D}_3^0 \mathbb{D}_2^0$$

and we notice that this is precisely the sequence applied to S_k to get to S_{k+1} *modulo* blue and green operators. Since it doesn't matter which order we apply the blue and green operators, we have the following

$$(((\mathbb{U}_2^1 \mathbb{U}_1^1 \mathbb{U}_3^1 \mathbb{U}_2^1 \mathbb{U}_1^1 \mathbb{U}_3^1 \mathbb{U}_0^0)F^k))\mathbb{D}_1^0 \otimes 1_{\mathsf{U}_k} = \langle d_{S_k}, d_{S_k}\rangle 1_{\lambda_k} \qquad (10.6.15)$$

where U_k is our fixed choice of reduced word for the longest element of $\mathfrak{S}_{\{-k,\ldots,-1\}} \times \mathfrak{S}_{\{5,6,\ldots 2k+4\}}$. For $k = 2$ this process is highlighted in Figure 10.14 (where the F-operators are depicted) and Figure 10.15 (where we draw the "symmetric version").

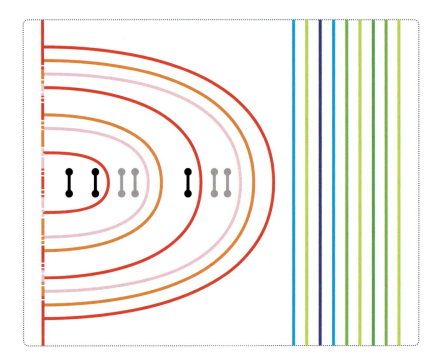

Fig. 10.15: The element on the righthand-side of (10.6.15) for $k = 2$. This is obtained by multiplying the diagram in Figure 10.14 by its dual.

Thus we now have an inductive approach to our problem, building on our case $k = 1$ calculation from the previous section. The $\mathbb{U}_0^0$ step in (10.6.15) produces a black barbell. In (10.6.14) we already calculated the effect on the black barbell as follows:

10.6 Counterexamples to Soergel's conjecture in positive characteristic 277

$$\left(1_{21\cdot21\cdot} \otimes \begin{array}{c}\bullet\\|\\\bullet\end{array}\right) F = -1_{21\cdot21\cdot} \otimes \left(2\begin{array}{c}\bullet\\|\\\bullet\end{array} + \begin{array}{c}\bullet\\|\\\bullet\end{array} + \begin{array}{c}\bullet\\|\\\bullet\end{array} + \begin{array}{c}\bullet\\|\\\bullet\end{array} + 2\begin{array}{c}\bullet\\|\\\bullet\end{array}\right) \quad (10.6.16)$$

and so in order to calculate F^2 is suffices to consider the effect of F on each of these 5 coloured barbells. In fact, we will see in the next section that F^k will always take a linear combination of these 5 barbells to another linear combination of these 5 barbells! Finally, we must then use the two-colour barbell relations in order to pull the linear combination of barbells (produced by the F^k step) through the final red strand (the $\mathbb{D}_1^0$ step in (10.6.15)) and we are done.

Before we explicitly calculate the effect of F on an arbitrary barbell, we now illustrate how our inductive approach is used to calculate the $k = 2$ case. We already calculated the effect of applying the operator F to the black barbell (see (10.6.14)). We then recalled this statement explicitly in (10.6.15) in terms of our new language. So, we can apply the F operator to the diagram in Figure 10.15 using (10.6.15) (if you cannot see how this applies, looks at the F-subscripts on the braced terms of Figure 10.14). We depict this simplified diagram in Figure 10.16.

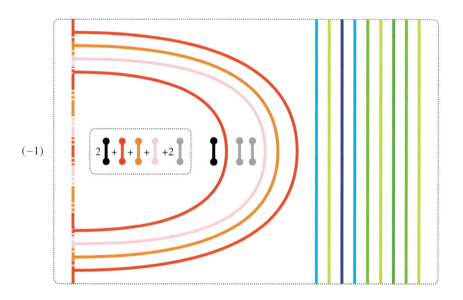

Fig. 10.16: The diagram obtained from that of Figure 10.15 by simplifying using (10.6.16). (In other words, applying the F operator once to the leftmost black barbell in the diagram in Figure 10.15.

Now, we wish to simplify the diagram in Figure 10.16. We can do this by calculating the effect of the operator F to each of the barbells (black, red, orange, pink and grey) in turn. We have not yet seen how to do this, but we will do shortly in Proposition 10.6.17. In the mean time, we simply ask the reader to think of this as magic trick (or to do the calculation themselves). Either way, applying F to these barbells we obtain the diagram depicted in Figure 10.17.

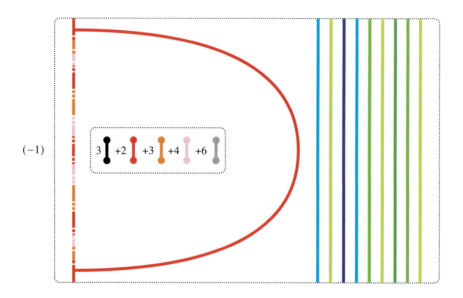

Fig. 10.17: The diagram obtained from that of Figure 10.16 by applying (10.6.16). (In other words, applying the F operator once to the linear combination of barbells in the diagram in Figure 10.16 (using Proposition 10.6.17 to deduce the coefficients of barbells in the dotted rectangle, or by hand).

Finally, we can apply the Lemmas 10.6.9 and 10.6.10 and the commutativity relations to move the barbells in the diagram in Figure 10.17 through the last red strand in and hence obtain

$$- (3 \times \mathbf{-1} + 2 \times \mathbf{2} + 3 \times \mathbf{-1} + 4 \times \mathbf{0} + 6 \times \mathbf{0}) \, 1_{21321} \otimes 1_{-1} \otimes 1_5 \otimes 1_{-2-1} \otimes 1_{65765} = -21_{\lambda_1}$$

where here we have used the bold colours to highlight the scalars obtained from the barbell relations, the non-bold scalars are simply the coefficients of these barbells from (10.6.14) (and T_2 is a longest word for the longest element of $\mathfrak{S}_{\{-3,\dots,-1\}} \times \mathfrak{S}_{\{5,6,7,8\}}$). The scalar on the righthand-side is the third Fibonacci number, $f_3 = 2$, as claimed.

Remark 28. When writing down $(1_{21321} \otimes \text{bar}(\sigma))F$, we will sometimes not mention the idempotent explicitly. We do this for ease of readability (compare (10.6.16) versus (10.6.18)). We remark there is no ambiguity here, as the F operator cannot be applied to the barbell without the idempotent.

10.6.5 Counterexamples to "Soergel's conjecture"

Finally, we now calculate the effect of the operator F on any given barbell. This is the final diagrammatic calculation we require, reducing the calculation of $\langle d_{S_k}, d_{S_k} \rangle_{\lambda_k}^{\underline{k}}$ to an elementary linear algebraic problem.

Proposition 10.6.17. *We have the following rules for applying the Fibonacci operator to a barbell.*

$$\left(\begin{array}{c}\vdots\end{array}\right) F = -2\;\vdots\; - \;\vdots\; - \;\vdots\; - \;\vdots\; - 2\;\vdots \tag{10.6.18}$$

$$\left(\begin{array}{c}\vdots\end{array}\right) F = 2\;\vdots\; + \;\vdots\; + \;\vdots\; + \;\vdots\; + 2\;\vdots \tag{10.6.19}$$

$$\left(\begin{array}{c}\vdots\end{array}\right) F = -\;\vdots\; - \;\vdots\; - \;\vdots\; - \;\vdots \tag{10.6.20}$$

$$\left(\begin{array}{c}\vdots\end{array}\right) F = \;\vdots\; +2\;\vdots\; +2\;\vdots \tag{10.6.21}$$

$$\left(\begin{array}{c}\vdots\end{array}\right) F = -\;\vdots\; -2\;\vdots\; -3\;\vdots \tag{10.6.22}$$

(using the conventions of Remark 28). We summarise this in the following matrix

$$F = \begin{bmatrix} -2 & -1 & -1 & -1 & -2 \\ 2 & 1 & 1 & 1 & 2 \\ -1 & -1 & -1 & -1 & 0 \\ 0 & 0 & 1 & 2 & 2 \\ 0 & 0 & -1 & -2 & -3 \end{bmatrix} \tag{10.6.23}$$

Proof. We have already proven (10.6.18) in great detail in Subsection 10.6.3. We refer to (10.6.14) for the explicit statement (as our emphasis was slightly different in that section). The other barbells can be handled in a similar fashion. □

Exercise 10.6.24. Prove each of (10.6.19) to (10.6.22) using the techniques and ideas of this chapter.

We recall, from the introduction, Binet's formula for the kth Fibonacci number:

$$f_k = \frac{\left(\frac{1+\sqrt{5}}{2}\right)^k - \left(\frac{1-\sqrt{5}}{2}\right)^k}{\sqrt{5}}$$

which will be the final ingredient in the disproof of "Soergel's conjecture" over $\mathbb{F}_p$.

Corollary 10.6.25. Let $(\lambda_k, \mu_k)_{k \geq 1}$ be as in Definition 10.6.3 and $S_k \in \text{Path}(\lambda_k, \mu_k)$ as in Corollary 10.6.5. The value of the corresponding intersection form $\langle d_{S_k}, d_{S_k} \rangle_{\lambda_k}^{k}$ is as follows:

$$\begin{bmatrix} 1 & 0 & 0 & 0 & 0 \end{bmatrix} \begin{bmatrix} -2 & -1 & -1 & -1 & -2 \\ 2 & 1 & 1 & 1 & 2 \\ -1 & -1 & -1 & -1 & 0 \\ 0 & 0 & 1 & 2 & 2 \\ 0 & 0 & -1 & -2 & -3 \end{bmatrix}^k \begin{bmatrix} -1 \\ 2 \\ -1 \\ 0 \\ 0 \end{bmatrix} = (-1)^{k+1} f_{k+1} \quad (10.6.26)$$

where f_{k+1} is the $(k+1)$th Fibonacci number.

Proof. We have that $\begin{bmatrix} 1 & 0 & 0 & 0 & 0 \end{bmatrix} F^k$ calculates the coefficients of the the barbells in the linear combination of diagrams produced by

$$\left(1_{213213} \otimes \;\;\substack{\bullet \\ \bullet}\;\; \right) F^k.$$

Now post-multiplying by the column matrix calculates the effect of pushing the barbells (produced in the previous step) through the final red forking strands (note the entries are $0, -1, 2$ for reflections which commute with/don't commute with/or are equal to the red reflection respectively). Thus the value of $\langle d_{S_k}, d_{S_k} \rangle$ given by the diagrammatic calculation

$$\left(\left(1_{213213} \otimes \;\;\substack{\bullet \\ \bullet}\;\; \right) F^k \right) \mathbb{D}_1^0 \otimes 1_{U_k} = \langle d_{S_k}, d_{S_k} \rangle 1_{\lambda_k}$$

is indeed the integer obtained by performing the matrix-product on the lefthand-side of (10.6.26).

So, now it's time to calculate the aforementioned product of matrices. The matrix F from (10.6.23) is diagonalisable and has eigenvalues

$$-1, 0, -\tfrac{1+\sqrt{5}}{2}, -\tfrac{1-\sqrt{5}}{2}.$$

Thus we can write $F = PDP^{-1}$ where

$$P = \begin{bmatrix} 0 & -1 & 0 & \tfrac{1+\sqrt{5}}{2} & \tfrac{1-\sqrt{5}}{2} \\ 0 & 1 & 1 & -\tfrac{1+\sqrt{5}}{2} & -\tfrac{1-\sqrt{5}}{2} \\ -2 & 0 & -2 & \tfrac{1-\sqrt{5}}{2} & \tfrac{1+\sqrt{5}}{2} \\ 0 & 0 & 1 & -\tfrac{3-\sqrt{5}}{2} & -\tfrac{3+\sqrt{5}}{2} \\ 1 & 0 & 0 & 1 & 1 \end{bmatrix},$$

10.7 Affinization: from Soergel's to Lusztig's conjecture

$$D = \begin{bmatrix} -1 & 0 & 0 & 0 & 0 \\ 0 & -1 & 0 & 0 & 0 \\ 0 & 0 & 0 & 0 & 0 \\ 0 & 0 & 0 & -\frac{1+\sqrt{5}}{2} & 0 \\ 0 & 0 & 0 & 0 & -\frac{1-\sqrt{5}}{2} \end{bmatrix},$$

$$P^{-1} = \begin{bmatrix} -1 & -1 & -1 & -1 & -1 \\ 0 & 1 & 2 & 3 & 4 \\ 1 & 1 & 0 & 0 & 0 \\ \frac{5+\sqrt{5}}{10} & \frac{5+\sqrt{5}}{10} & \frac{5+3\sqrt{5}}{10} & \frac{1+\sqrt{5}}{2} & 1 + \frac{3}{\sqrt{5}} \\ \frac{5-\sqrt{5}}{10} & \frac{5-\sqrt{5}}{10} & \frac{5-3\sqrt{5}}{10} & \frac{1-\sqrt{5}}{2} & 1 - \frac{3}{\sqrt{5}} \end{bmatrix}.$$

Now the value of the intersection form is given by $-(F^k)_{11} + 2(F^k)_{12} - (F^k)_{13}$. Using $F^k = PD^k P^{-1}$, we have

$$(F^k)_{11} = (-1)^k \left(\frac{5+\sqrt{5}}{10}\right) \left(\frac{1+\sqrt{5}}{2}\right)^{k+1} + (-1)^k \left(\frac{5-\sqrt{5}}{10}\right) \left(\frac{1-\sqrt{5}}{2}\right)^{k+1}$$

$$(F^k)_{12} = (-1)^{k+1} + (-1)^k \left(\frac{5+\sqrt{5}}{10}\right) \left(\frac{1+\sqrt{5}}{2}\right)^{k+1} + (-1)^k \left(\frac{5-\sqrt{5}}{10}\right) \left(\frac{1-\sqrt{5}}{2}\right)^{k+1}$$

$$(F^k)_{13} = 2(-1)^{k+1} + (-1)^k \left(\frac{5+3\sqrt{5}}{10}\right) \left(\frac{1+\sqrt{5}}{2}\right)^{k+1} + (-1)^k \left(\frac{5-3\sqrt{5}}{10}\right) \left(\frac{1-\sqrt{5}}{2}\right)^{k+1}$$

Pre- and post-multiplying F^k by the row and column matrices in (10.6.26) we obtain the following

$$-(F^k)_{11} + 2(F^k)_{12} - (F^k)_{13} = (-1)^{k+1} \left(\frac{2\sqrt{5}}{10} \left(\frac{1+\sqrt{5}}{2}\right)^{k+1} - \frac{2\sqrt{5}}{10} \left(\frac{1-\sqrt{5}}{2}\right)^{k+1}\right)$$

$$= (-1)^{k+1} \frac{1}{\sqrt{5}} \left(\frac{1+\sqrt{5}}{2}\right)^{k+1} - \left(\frac{1-\sqrt{5}}{2}\right)^{k+1}$$

$$= (-1)^{k+1} f_{k+1}$$

where the final equality follows by Binet's formula, as required. □

10.7 Affinization: from Soergel's to Lusztig's conjecture

So far in this chapter we have focussed on "Soergel's conjecture" and the calculation of p-Kazhdan–Lusztig polynomials for finite symmetric groups. We now wish to affinize and hence pass from the domain of Soergel's to that of Lusztig's conjecture. Whilst we have already shown that the Kazhdan–Lusztig polynomials are compatible with this affinization (when we are "sufficiently far away from the parabolic"), we do *not yet* know whether or not the *p-Kazhdan–Lusztig polynomials* are compatible with this procedure. We will now show that the p-Kazhdan–Lusztig polynomials are compatible with this process; thus lifting a bijection between graded cellular bases of our diagrammatic algebras to the level of an algebra isomorphism.

Theorem 10.7.1. *Set* $(W, P) = (\widehat{\mathfrak{S}}_n, \mathfrak{S}_n)$. *Suppose that* $\pi \in {}^P W$ *is below a* σ-*Steinberg. We let* $\mathrm{St}(\pi) = \{\pi w \mid w \in \langle \tau \mid \sigma \neq \tau \in S_W \rangle\}$. *We have a bijective map,* $\mathrm{st} : \mathfrak{S}_n \to \mathrm{St}(\pi)$ *which respects the Bruhat ordering. Under this map, the finite and affine p-Kazhdan–Lusztig polynomials coincide, that is*

$$^P n_{\lambda\mu} = {}^P n_{\mathrm{st}(\lambda),\mathrm{st}(\mu)}$$

for $\lambda, \mu \in \mathfrak{S}_n = \langle \tau \mid \sigma \neq \tau \in S_W \rangle \leqslant \widehat{\mathfrak{S}}_n$.

Proof. We have already proven most of this in Theorem 8.3.10 and it is stated here merely for notation purposes. The main point from the proof of Theorem 8.3.10 is that we can choose our reduced paths $\mathsf{T}_{\pi w}$ for $\pi w \in \mathrm{St}(\pi)$ so that they are all of the form

$$\mathsf{T}_{\pi w} = \mathsf{T}_\pi \otimes \mathsf{T}_w$$

for any choice of reduced path T_π terminating at π and any $w \in \mathfrak{S}_n$. Therefore we have a closed subquotient algebra $\mathcal{H}_{\mathrm{St}(\pi)}$ of $\mathcal{H}_{(W,P)}$ with basis

$$\{d_{\mathsf{ST}} \mid \mathsf{S} \in \mathrm{Path}(\pi z, \mathsf{T}_{\pi x}), \mathsf{T} \in \mathrm{Path}(\pi z, \mathsf{T}_{\pi y}), x, y, z \in \mathfrak{S}_n\}.$$

The algebra $\mathcal{H}_{\mathrm{St}(\pi)}$ is the subquotient of $\mathcal{H}_{(W,P)}$ generated by the diagrams

$$1_{\mathsf{T}_\pi} \otimes D \qquad \text{modulo } \mathcal{H}^{<\pi}_{(W,P)}$$

for D a Soergel diagram with no edges coloured by $\sigma \in S_W$. By construction, the map $\mathrm{st} : \mathcal{H}_{\mathfrak{S}_n} \to \mathcal{H}_{\mathrm{St}(\pi)}$ given by $\mathrm{st}(D) = 1_{\mathsf{T}_\pi} \otimes D$ is a graded bijection (because we have matched up the cellular bases). Thus it only remains to check that the map is a $\Bbbk$-algebra homomorphism.

All of the local relations go through immediately by definition and so it only remains to check that the image of (C) holds in the subquotient $\mathcal{H}_{\mathrm{St}(\pi)}$. Let $\tau \in S_W$, by the barbell and cyclotomic relations we have that

$$1_{\mathsf{T}_\pi} \otimes \;\vcenter{\hbox{$\big\vert$}}\; = \sum_{\substack{\alpha < \pi \\ \mathsf{U},\mathsf{V} \in \mathrm{Path}(\alpha,\pi)}} a_{\mathsf{UV}} d_{\mathsf{UV}} \in \mathcal{H}^{<\pi}_{(W,P)}$$

for scalars $a_{\mathsf{UV}} \in \Bbbk$ and $(W, P) = (\widehat{\mathfrak{S}}_n, \mathfrak{S}_n)$. Thus we obtain the required isomorphism. We conclude that the map $\mathrm{st} : \mathfrak{S}_n \to \mathrm{St}(\pi)$ is a $\mathbb{Z}$-graded $\Bbbk$-algebra isomorphism which respects the cellular structures; the result follows. $\square$

Thus calculating the p-Kazhdan–Lusztig polynomials for $(W, P) = (\widehat{\mathfrak{S}}_n, \mathfrak{S}_n)$ subsumes the problem of calculating the p-Kazhdan–Lusztig polynomials of $\mathfrak{S}_n$. In particular, Corollary 10.6.7 provides counterexamples to the following conjecture:

10.7 Affinization: from Soergel's to Lusztig's conjecture

Conjecture 10.7.2 (The toy version of Andersen's and Lusztig's conjectures). *Let $(W, P) = (\widehat{\mathfrak{S}}_n, \mathfrak{S}_n)$. For $\lambda, \mu \in \mathfrak{S}_n$ the p-Kazhdan–Lusztig polynomials ${}^p n_{\mathrm{St}(\lambda), \mathrm{St}(\mu)}$ are independent of the prime p for $p > 2n - 4$.*

We say that this is the "toy version" of Andersen's and Lusztig's conjectures, this will be made more precise in Part 5 of this book. In a nutshell: Lusztig and Andersen posed two conjectures (in "Ringel dual" frameworks) for calculating certain decomposition numbers (the former was for general linear groups and the latter for symmetric groups) in terms of (p-)Kazhdan–Lusztig polynomials. These "dual conjectures" coincide within a Steinberg region and so specialise to the "toy version" stated above (although the statements differ from each other *away* from the Steinberg region).

Conjecture 10.7.2 was referred to by experts as the "toy version" of these conjectures as it was presumed easier to *prove to be true*. Whilst even the "toy" conjectures turned out to be false, this insight of focussing in the Steinberg region was key to the resolution of these conjectures.

Remark 29. We note that, for a fixed pair $\lambda, \mu \in {}^P W$, the sequence of polynomials

$$\lim_{p \to \infty} {}^p n_{\lambda,\mu}(q) \tag{10.7.3}$$

does eventually stabilise for large primes $p \gg 0$. To see this, first set $\Pi = \{\nu \mid \lambda \leqslant \nu \leqslant \mu\}$. Consider the Gram matrices of the intersection forms on the modules $\Delta^{\mathbb{Q}}(\nu) 1_\mu$ for $\nu \in \Pi$. These are a finite collection of matrices with integer-valued entries. Thus, we can set b to be the the largest value obtainable as a determinant of one of these Gram matrices. Trivially, we note that stability is achieved for all $p > b$. That the stable limits in (10.7.3) are equal to the Kazhdan–Lusztig polynomials was proven in [EW14, LW22].

Remark 30. Lusztig originally posed his conjecture under the restriction that $p > 2n - 4$, see [Jan03, Section 8.22] for a discussion. Lusztig's conjecture is known to be true for $p \gg n$ due to work of Andersen–Jantzen–Soergel [AJS94] with an explicit bound given in [Fie12].

Remark 31. Let $(W, P) = (\widehat{\mathfrak{S}}_n, \mathfrak{S}_n)$. Andersen's conjecture concerns the stable limits in (10.7.3) for certain pairs λ, μ (those in the "first p^2-alcove, this will be discussed in the next section). Andersen stated that the the stable limits should be achieved providing $p \geqslant n$, [And98]. Andersen's conjecture is false by Conjecture 10.7.2.

It is not known whether Andersen's conjecture is true or false for $p \gg n$ and nowadays people often speak of "Andersen's conjecture" in this limiting sense. Notice this asymmetry between Lusztig's and Andersen's conjectures for large primes, this comes from the existence of a "Steinberg endomorphism" on only one side of the aforementioned Ringel duality.

10.8 What might the future hold?

In this short section, we discuss where the future might lead, highlighting the conjectures which arose in the aftermath of Williamson's examples of explosive p-torsion, and surveying the state-of-the-art. We begin by presenting a replacement analogue of "Soergel's conjecture", that is, a bound for p-torsion for the finite Weyl groups. As we have just seen, any such bound needs to grow at least exponentially with the rank. A tantalisingly simple and natural suggestion for such a bound has been provided by He–Williamson:

He–Williamson's Hope 10.8.1. *Let W be a finite Weyl group and let $\Bbbk$ be a field of characteristic $p > |W|$. The graded decomposition numbers of $\mathcal{H}_W$ are independent of the prime p and are equal to the classical Kazhdan–Lusztig polynomials, that is*

$$\sum_{k \in \mathbb{Z}} [\Delta(\lambda) : L(\mu)\langle k \rangle] = n_{\lambda,\mu}(q)$$

for all $\lambda, \mu \in W$.

We now turn to the affine Weyl groups, focussing on the affine symmetric group. We will organise the problem of understanding p-Kazhdan–Lusztig polynomials for $(W, P) = (\widehat{\mathfrak{S}}_n, \mathfrak{S}_n)$ along two axes: the rank $n \in \mathbb{Z}_{>0}$ of the group; and the "p-translation distance" into $^P W$ (an idea which we will soon make more precise!).

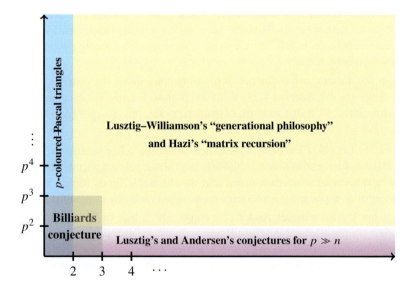

Fig. 10.18: The two axes of complexity within modular Lie theory and what is known/conjectured. The y-axis represents the level of p-dilation and the x-axis represents the rank $n \in \mathbb{Z}_{>0}$ of the group $\widehat{\mathfrak{S}}_n$.

10.8 What might the future hold?

Recall that the affine symmetric group $\widehat{\mathfrak{S}}_n$ can be visualised in terms of permutations on a cylinder as in Figure 8.16. For each $1 \leq k \leq n$ we can picture a "translation permutation" t_k on the cylinder as in Figure 10.19. Each translation permutation t_k for $1 \leq k \leq n$ sends each $j \neq k$ to itself (trivially), but takes the number k *clockwise all the way around the cylinder* before returning back to its starting position k. Each t_k has an inverse diagram, t_k^{-1}, which looks similar, but where the kth strand goes *anti-clockwise* around the cylinder.

The translation permutations (and their inverses) *cannot be written* as a product of the Coxeter generators $s_0, s_1, \ldots s_{n-1}$ of the affine symmetric group (that is $t_k \notin \widehat{\mathfrak{S}}_n$). If one wishes, one can define the **extended affine symmetric group** to be the group generated by the Coxeter generators together with these translations. However, we will not need this here. Instead we will make use of the following theorem.

Theorem 10.8.2. *Let W be the affine symmetric group $W = \widehat{\mathfrak{S}}_n$ with Coxeter presentation*

$$W = \langle s_0, s_1, \ldots, s_{n-1} \mid s_i^2 = 1, (s_i s_{i+1})^3 = 1, (s_i s_j)^2 = 1 \text{ for } i - j \neq 0, \pm 1 \rangle$$

with subscripts read modulo n. The group W contains all translations of the form $t_j t_k^{-1} \in W$ for $1 \leq j, k \leq n$. Moreover, W is generated by the "finite" reflections $s_1, \ldots, s_{n-1}$ together with any translation permutation $t_j t_k^{-1} \in W$ for $1 \leq j \neq k \leq n$.

Proof. The proof is left as an exercise for the reader! □

Fig. 10.19: The translation permutations $t_1, t_2, t_3 \notin \widehat{\mathfrak{S}}_6$ pictured from the bottom of the cylinder on which they are drawn (with some artistic license).

Definition 10.8.3. Let W be the affine symmetric group $W = \widehat{\mathfrak{S}}_n$ with Coxeter graph as depicted in Figure 8.17. We fix a prime number, p, and we set

$$\tau := t_1^p t_{n-1}^{-p}$$

and we define the pth **fractal subgroup**, W_p, as follows $W_p = \langle s_1, \ldots, s_{n-1}, \tau \rangle \leq W$.

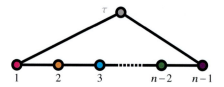

Fig. 10.20: The Coxeter graph of W_p.

Theorem 10.8.4. *The group W_p is a Coxeter group with Coxeter graph as depicted in Figure 10.20.*

Proof. The proof is again left as an exercise for the reader! □

The word "fractal" in Definition 10.8.3 might seem unnecessary (borderline meaningless!) until one considers the definition in light of Theorem 10.8.4. The takeaway from Theorem 10.8.4 is that we can iterate the above construction, defining a subgroup $W_{p^2} \leqslant W$ and so on until we do obtain a "fractal" chain of subgroups

$$\ldots \leqslant W_{p^4} \leqslant W_{p^3} \leqslant W_{p^2} \leqslant W_p \leqslant W \tag{10.8.5}$$

all of which exert influence on the modular representation theory of $\mathcal{H}_{(W,P)}$ over a field $\mathbb{F}_p$. We seek a language for codifying this influence…

In order to discuss this fractal-like behaviour, we must first write our elements of $W = \widehat{\mathfrak{S}}_n$ in a way that is compatible with this chain of subgroups. We can write any element of W in the form

$$w = \ldots w_{p^4} w_{p^3} w_{p^2} w_p w_0 \tag{10.8.6}$$

where $w_0 \in W_p \backslash W$ and $w_p \in W_{p^2} \backslash W_p$ and $w_{p^2} \in W_{p^3} \backslash W_{p^2}$ et cetera are minimal length coset representatives (although $w \in W$ will not be of minimal length when written in this form). If the expansion of w as in (10.8.6) is given by

$$w = w_{p^k} w_{p^{k-1}} \ldots w_{p^3} w_{p^2} w_p w_0$$

then we say that w belongs to the first p^{k+1}-alcove.

Remark 32. The conjectures of Andersen and Lusztig concern formal characters of modules from the first p^2-alcove. In the notation of (10.8.6), Andersen's conjecture can be stated as follows: The decomposition numbers of $\mathcal{H}_{(W,P)}$ from the first p^2-alcove can be calculated in terms of Kazhdan–Lusztig polynomials as follows:

$$\sum_k [\Delta(x_0) : L(y_0)\langle k\rangle] q^k = n_{x_0, y_0}(q)$$

for $x_0, y_0 \in {}^P W$ (using the notation (10.8.6)). We recall that Andersen's and Lusztig's conjecture are both false for small primes $p > 2n-4$. Lusztig's conjecture is known to be true for $p \gg n$ by [AJS94] (with explicit bounds known [Fie12]) and Andersen's

10.8 What might the future hold?

conjecture remains open for $p \gg n$. In summary: this pair of conjectures (with Lusztig's known, and Andersen's unknown) allow us to understand p-Kazhdan–Lusztig polynomials of arbitrary rank within the first p^2-alcove providing the prime $p \gg n$ is enormous (perhaps $p > n!$ will suffice?). This region of the general problem is shaded in pink in Figure 10.18.

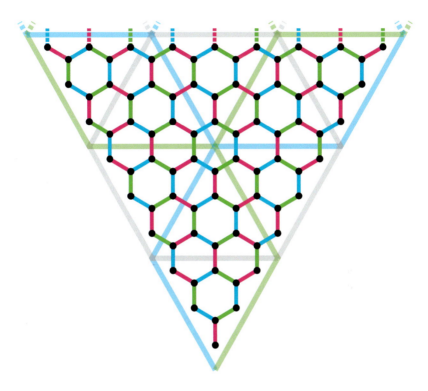

Fig. 10.21: For $p = 3$ we depict the first p^3-alcove for $(W, P) = (\widehat{\mathfrak{S}}_3, \mathfrak{S}_3)$. We highlight the action of W_p on the Bruhat graph of $^P W$. Notice that W_p is generated by the finite reflections s_1 and s_2 together with $\tau = s_0 s_1 s_2 s_0 s_1 s_2 s_0$.

Remark 33. For $W = \widehat{\mathfrak{S}}_2$ all the (parabolic) p-Kazhdan–Lusztig polynomials can be calculated explicitly in terms of p-modular colourings on Pascal's triangle! We encountered this flavour of combinatorics earlier in the book, see for example Figure 6.5. For (dual) maximal parabolic Kazhdan–Lusztig polynomials this was done by Carter–Cline, Deriziotis, and James [CC76, Der81, Jam76]. For non-parabolic Kazhdan–Lusztig polynomials this was done by Cox–Graham–Martin in [CGM03]. In summary: we understand p-Kazhdan–Lusztig polynomials of rank $n = 2$ for all p^k-alcoves $k \in \mathbb{Z}_{\geqslant 0}$. This region of the general problem is shaded in blue in Figure 10.18. Thus our understanding clings to the axes in Figure 10.18.

Let us choose to be hopeful that Andersen's conjecture is true for $p \gg n$. In which case, we could further hope that for enormous primes $p \to \infty$ the p-Kazhdan–Lusztig polynomials are *all capable of being understood combinatorially!* In light of Remarks 32 and 33 there is a logical first step to take: we must try and understand the p-Kazhdan–Lusztig polynomials for rank $n = 3$ lying within the first p^3-alcove (this region is highlighted in grey in Figure 10.18). Surely this cannot be too hard?!

Conjecture 10.8.7 (Jensen–Lusztig–Williamson [LW18a, Jen21]). *The p-Kazhdan–Lusztig polynomials*

$$^p n_{x,y}(q) := \sum_k [\Delta(x) : L(y)\langle k \rangle] q^k$$

for $x = x_p x_0, y = y_p y_0$ in the first p^3-alcove can be calculated by an explicit combinatorial algorithm in terms of bouncing billiards.

The above conjecture was presented at Williamson's 2018 ICM address [Wil18] which is available on YouTube, complete with an animation of these "bouncing billiards". We do not discuss this combinatorics explicitly here, except to say that it's not easy and that perhaps we were a bit overly-optimistic a moment ago.

Undeterred, Lusztig and Williamson proposed a "generational philosophy" whereby the influence exerted by the chain of subgroups in (10.8.5) gives rise to a series of formal characters, which one hopes can be lifted to the level of some concrete representation theoretic objects for $p \gg n$. The "first generation" should be given by classical Kazhdan–Lusztig polynomials and the "second generation" should be calculable using the billiards algorithm for $n = 3$.

Thus far, we have only considered questions and conjectures. We have not discussed any possible mode of tackling these conjectures or even a language for discussing the general problem. To the best of our knowledge, there is only one suggested framework for tackling this problem: Amit Hazi's theory of *matrix recursion* [Haz]. Hazi's matrix recursion gives a structural relationship between 'large' objects and 'smaller' ones in the diagrammatic Hecke category. More precisely, matrix recursion is a categorification of a certain non-trivial faithful representation

$$\xi : \mathbb{Z}W \longrightarrow \text{Mat}_{p^r}(\mathbb{Z}W)$$

of the group ring $\mathbb{Z}W$, in terms of $p^r \times p^r$ matrices over itself. The matrix recursion functor maps each object in the (ungraded) diagrammatic Hecke category to a matrix of 'smaller' objects in a similar (ungraded) diagrammatic Hecke category, all over characteristic p.

Hazi's matrix recursion is the unique possible explanation we possess for understanding the p-fractal behaviour that characterises modular representation theory. This, to our minds, is a pinnacle of the categorification programme to-date.

Further Reading. Obviously we invite the reader to enjoy for themselves the original paper in which Williamson discovered the counterexamples to the expected bounds in Lusztig's conjecture [Wil17]. Williamson also discusses in that paper the earlier discoveries of torsion which inspired his search for counterexamples. In the appendix

10.8 What might the future hold?

of that paper, Kontorovich–McNamara–Williamson go further and show that no polynomial bound will suffice — this should be of independent interest.

The definition of $\mathcal{H}_{(W,P)}$ for an arbitrary parabolic Coxeter system is due to Elias–Williamson [EW14], this built on earlier work of Elias–Khovanov in type A [EK10] and of Elias for dihedral groups [Eli16]. The only result in this section for which we did not provide a proof was the light leaves cellular basis theorem. This was originally proven by Libedinsky [Lib08] (for (W, P) with P the trivial subgroup) using localisation methods, and is later reproven in the diagrammatic context by Elias–Williamson in [EW16]. Libedinsky–Williamson then later extended this result to arbitrary parabolic subgroups in [LW22].

For more examples of p-Kazhdan–Lusztig polynomials calculated using intersection forms, we refer to [JW17] and for the implementation of these intersection forms in MAGMA, we refer to [GJW23].

Part V
Group theory versus diagrammatic algebra

> "Mathematics is the art of giving the
> same name to different things."
>
> Henri Poincaré

In the previous chapter, we claimed that mathematicians had long-expected the Hecke category to be (mostly) torsion-free, a belief we referred to as "Lusztig's conjecture". In actuality we were taking some artistic license: the Hecke category is a more modern construction which acts on classical algebraic objects, such as categories of representations of general linear groups. It was in this latter, group theoretic, context that Lusztig's conjecture was originally posed. In Part 5 of this book we show that these two, apparently different, things are the same and that they can hence both share the name "Lusztig's conjecture".

In more detail: our intention with this book was to bring the reader to a place where they could understand the counterexamples to Lusztig's conjecture. Thus far, we have chosen to do this entirely within the setting of the diagram algebra $\mathscr{H}_{(W,P)}$. This was in order to get the reader up-to-speed with the diagram algebras $\mathscr{H}_{(W,P)}$ as quickly as possible without having to get to grips with the historical development of the research area. Lusztig's and Andersen's conjectures were not originally phrased in our diagrammatic setting, but rather as conjectures concerning the decomposition matrices of symmetric groups and their Schur algebras. Thus the expert reader might feel like they want their money back unless we explain *exactly how* the diagram algebra $\mathscr{H}_{(W,P)}$ comes into this group theoretic picture: this is what the final part of the book is dedicated toward.

As we venture beyond the Hecke category, we embark on a journey through the various actions of Hecke categories on more "classical" algebraic objects. At the time when Geordie Williamson first proved his torsion explosion counter examples, much of the theory had yet to be fully developed (but he was able to focus on the "toy model" using results from [Soe00]). It was only after Geordie's counterexamples that the Hecke categories were shown to control the structure of (principal blocks of) general linear and symmetric groups and their Schur algebras through various equivalences. The main result of Part 5 of this book is this famous equivalence of Riche–Williamson [RW18] later reproven and generalised by Elias–Losev, Bowman–Cox–Hazi, Achar–Makisumi–Riche–Williamson [EL, AMRW19, BCHM22, BCH23].

Part V Group theory versus diagrammatic algebra

The proof that we choose to focus on in these chapters is that of Anton Cox, Amit Hazi, and myself. Naturally, opportunities to emphasise one's own contributions should never be passed up! However our *main* reason for doing this is because this proof is entirely elementary: it proceeds via a (graded) $\Bbbk$-algebra isomorphism between the relevant cellular subquotients of $\Bbbk \mathfrak{S}_r$ and $\mathscr{H}_{(W,P)}$. It also affords us the opportunity to revisit the beautiful work of Lascoux–Leclerc–Thibon and to take a whirlwind tour through the quiver Hecke algebras.

The overarching idea of Part 5 of this book is that the group algebras of finite symmetric groups always had the diagrammatic algebras $\mathscr{H}_{(W,P)}$ "living inside them all along" — our task is merely to chip away the excess to reveal the beautiful core. This story has its origins in the mid 1990s when Lascoux–Leclerc–Thibon provided a conjectural "LLT algorithm" for computing the decomposition matrix of a "quantum analogue" of the symmetric group over the complex field [LLT96], proven by Ariki in [Ari96]. Later Khovanov–Lauda–Rouquier, Hu–Mathas, and Kleshchev–Nash established that the group algebra of the symmetric group admits a tableaux-theoretic graded cellular basis and they hence recast the LLT algorithm as a natural calculation within this basis [Rou, KL09, HM10, KN10]. Thus by 2010, we already have a graded structure that allows us to formulate the conjectures of Lusztig, Andersen, and James directly on the cellular basis of the group algebra $\Bbbk \mathfrak{S}_r$, it remains to start chipping away to the core...

The connection between Hecke categories and the KLR diagrammatics was probed over the next decade by David Plaza and his collaborators [Pla13, PRH14, EP19, LPRH21] and culminated in a conjecture posed by Libedinsky–Plaza in [LP20]. This conjecture (although phrased in a slightly different manner, due to the authors interest in algebraic statistical mechanics) essentially chipped down to the beautiful core of the the quiver Hecke algebra and posited that it should be isomorphic to the corresponding $\mathscr{H}_{(W,P)}$. This conjecture was proven by the author, Anton Cox, and Amit Hazi in [BCH23].

Chapter 11
Reformulating Lusztig's and Andersen's conjectures

In this chapter, we phrase Lusztig and Andersen's conjectures in their original group theoretic context. All of the material we consider here is "classical" in the sense that it is over 25 years old and makes no mention of categorification whatsoever, it is all entirely group theoretic. We recall the most salient points regarding the representation of the symmetric groups and their Schur algebras, both in the language of tableaux and the language of alcove geometries, hence arriving at the conjectures of Andersen and Lusztig within the latter context.

11.1 The symmetric group and its representation theory

We let $\mathbb{T}_{n,r}$ denote the set of all sequences of length r consisting solely of the numbers $1, \ldots, n$, for example, $\mathbb{T}_{3,2} = \{11, 22, 33, 12, 13, 23, 21, 31, 32\}$. We define the weight of a sequence, $\pi \in \mathbb{T}_{n,r}$, to be $\mu = (\mu_1, \mu_2, \ldots, \mu_n) \in \mathbb{Z}_{\geq 0}^n$ where μ_k is the total number of ks in the sequence π. We let $\Gamma_{n,r}$ denote the set of all possible weights of sequences from $\mathbb{T}_{n,r}$. We let $\Gamma_{n,r}^+ \subseteq \Gamma_{n,r}$ denote the subset of weights $(\lambda_1, \lambda_2, \ldots, \lambda_n)$ such that $\lambda_1 \geq \lambda_2 \geq \ldots \geq \lambda_n$ and we refer to such weights as the n-row partitions of r. We place a partial ordering on the set $\Gamma_{n,r}^+$ by saying that λ dominates μ and writing $\lambda \trianglerighteq \mu$ if

$$\sum_{i \leq j} \lambda_i \geq \sum_{i \leq j} \mu_i$$

for all $1 \leq j \leq n$. We set $\Gamma_r^+ := \Gamma_{r,r}^+$.

Example 11.1.1. There are 5 weights in $\Gamma_{4,4}^+$ given by

$$(4, 0, 0, 0), (3, 1, 0, 0), (2, 2, 0, 0), (2, 1, 1, 0), (1, 1, 1, 1) \tag{11.1.2}$$

and there are many more weights in $\Gamma_{4,4}$. For example $1121, 2322, 3331, 4244$ have weights equal to

$(3, 1, 0, 0), (0, 3, 1, 0), (1, 0, 3, 0), (0, 1, 0, 3)$

respectively. The set of sequences of weight $(1, 0, 2, 1)$ is given by

$\{1334, 4331, 3134, 4131, 1343, 4313, 3143, 3413, 3314, 3341, 1433, 4133\}.$

The partitions of 4 are totally ordered from left to right as in (11.1.2).

Exercise 11.1.3. Find a pair $\lambda, \mu \in \Gamma^+_{6,3}$ which are not ordered under $\triangleright$.

The symmetric group, $\mathfrak{S}_r$, is generated by the elements $s_i = (i, i+1)$ for $1 \leq i < r$ modulo the relations $s_i s_{i+1} s_i = s_{i+1} s_i s_{i+1}$ and $s_i^2 = 1$. This group acts on the set $\mathbb{T}_{n,r}$ by letting s_i swap the ith and $(i+1)$th terms in the sequence $\pi \in \mathbb{T}_{n,r}$.

Given $\lambda \in \Gamma^+_{n,r}$, we define a λ-tableau to be a filling of the boxes of λ with the numbers $1, \ldots, r$. We say that a λ-tableau is **standard** if the rows (respectively columns) are strictly increasing from left to right (respectively top to bottom). Let $\text{Std}(\lambda)$ be the set of standard λ-tableaux and we write $\text{Shape}(t) = \lambda$ for $t \in \text{Std}(\lambda)$.

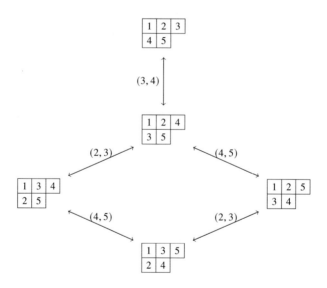

Fig. 11.1: The dominance order on $\text{Std}(3, 2)$. The reading words (reading from top-to-bottom, left-to-right) are $(1, 1, 1, 2, 2)$, $(1, 1, 2, 1, 2)$, $(1, 2, 1, 1, 2)$, $(1, 1, 2, 2, 1)$, and $(1, 2, 1, 2, 1)$.

For $1 \leq k \leq n$, we let $t\!\downarrow_{\{1,\ldots,k\}}$ denote the restriction of t to this subset of boxes containing thee entries $1, \ldots, k$. We extend the dominance ordering to $\text{Std}(\lambda)$ by writing $s \triangleright t$ if

$$\text{Shape}(t\!\downarrow_{\{1,\ldots,k\}}) \trianglerighteq \text{Shape}(t\!\downarrow_{\{1,\ldots,k\}})$$

for all $1 \leq k \leq n$.

11.1 The symmetric group and its representation theory

Example 11.1.4. The Hasse diagram of the partial order on Std(3, 2) is recorded in Figure 11.1. We decorate each edge between a pair of tableaux with the Coxeter generator s_i which swaps the corresponding entries $i, i + 1$ in these tableaux.

Definition 11.1.5. For a given partition $\lambda \in \Gamma_{n,r}$, we let $t_\lambda \in \text{Std}(\lambda)$ denote the minimal element under the dominance ordering. Explicitly, t is the λ-tableau in which the integers $1, \ldots, r$ are entered in increasing order top-to-bottom along the successive columns of $[\lambda]$. We let y_λ denote the signed sum over all permutations which preserve the columns of t_λ.

Definition 11.1.6. Let $\lambda \in \Gamma_{n,r}^+$. Given $s \in \text{Std}(\lambda)$ with $s[r_k, c_k] = k \in \lambda$ we define the reading word of s to be given by the ordered sequence $\text{read}(s) = \{(r_k)_{k \geqslant 0}\}$.

Example 11.1.7. The reading words of the tableaux from Figure 11.1 are recorded in the caption to Figure 11.1. For example the tableau $t_{(3,2)}$ has reading word $\text{read}(t_{(3,2)}) = (1, 2, 1, 2, 1)$.

Remark 34. A reading word uniquely determines its corresponding standard tableau (since specifying the row in which to add a box uniquely determines the column in which the box is added).

Definition 11.1.8. Let $\lambda \in \Gamma_{n,r}^+$. Given $s, t \in \text{Std}(\lambda)$ we define a permutation $w_t^s \in \mathfrak{S}_n$ as follows. First, we write $\text{read}(s)$ and $\text{read}(t)$ along the northern and southern edges of a frame and we then set w_t^s to be the unique label-preserving permutation with the minimal number of crossings.

Example 11.1.9. Continuing with Example 11.1.7 we consider the tableaux

$$t_{(3,2)} = \begin{array}{|c|c|c|} \hline 1 & 3 & 5 \\ \hline 2 & 4 \\ \cline{1-2} \end{array} \qquad s = \begin{array}{|c|c|c|} \hline 1 & 3 & 4 \\ \hline 2 & 5 \\ \cline{1-2} \end{array} \qquad t = \begin{array}{|c|c|c|} \hline 1 & 2 & 3 \\ \hline 4 & 5 \\ \cline{1-2} \end{array}$$

from Std(3, 2) (depicted in Figure 11.1). These tableaux have reading words

$$(1, 2, 1, 2, 1) \qquad (1, 2, 1, 1, 2) \qquad (1, 1, 1, 2, 2)$$

respectively. The permutations $w_t^{t_{(3,2)}}$ and w_t^s are as follows:

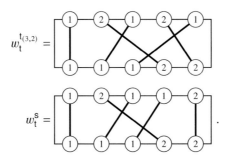

Notice that if two strands have the same step-label, then they do not cross (by the minimality condition on the number of crossings).

Example 11.1.10. We now consider a more complicated example. We choose s, t ∈ Std((12, 4)) as follows:

$$\mathsf{s} = \begin{array}{|c|c|c|c|c|c|c|c|c|c|c|c|} \hline 1 & 3 & 5 & 7 & 9 & 10 & 11 & 12 & 13 & 14 & 15 & 16 \\ \hline 2 & 4 & 6 & 8 \\ \cline{1-4} \end{array} \qquad (11.1.11)$$

$$\mathsf{t} = \begin{array}{|c|c|c|c|c|c|c|c|c|c|c|c|} \hline 1 & 2 & 3 & 4 & 5 & 6 & 7 & 8 & 13 & 14 & 15 & 16 \\ \hline 9 & 10 & 11 & 12 \\ \cline{1-4} \end{array} \qquad (11.1.12)$$

The corresponding (unique) step-preserving permutation with minimal number of crossings is given by

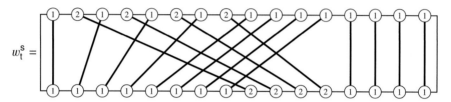

Again, notice that if two strands have the same step-label, then they do not cross (by the minimality condition on the number of crossings). We have that

$$y_{(12,4)} = (\mathrm{id} - (1,2))(\mathrm{id} - (3,4))(\mathrm{id} - (5,6))(\mathrm{id} - (7,8)).$$

Definition 11.1.13. For $\mathsf{t} \in \mathrm{Std}(\lambda)$, we write $d_\mathsf{t} := w_\mathsf{t}^{\mathsf{t}^\lambda}$.

The following theorem generalises our example of $\Bbbk \mathfrak{S}_3$ from Section 6.4 to all symmetric groups. This result is due to Murphy [Mur92, Mur95].

Theorem 11.1.14. *The group algebra of the symmetric group has cellular basis*

$$\{y_{\mathsf{st}} := d_\mathsf{s}^* y_\lambda d_\mathsf{t} : \mathsf{s}, \mathsf{t} \in \mathrm{Std}(\lambda) \text{ for } \lambda \in \Gamma_r^+\}$$

with respect to the anti-involution $* : g \to g^{-1}$ *and any total refinement of the dominance ordering on* Γ_r^+.

Example 11.1.15. We let $\mathsf{s}, \mathsf{t} \in \mathrm{Std}(3, 2)$ be as in Example 11.1.9. Thus $d_\mathsf{t} = (2, 3)(4, 5)(3, 4)$ and $d_\mathsf{s} = d_\mathsf{s}^* = (4, 5)$. We have that

$$y_{\mathsf{st}} = (4, 5)\bigl(\mathrm{id} - (1, 2) - (3, 4) + (1, 2)(3, 4)\bigr)(2, 3)(4, 5)(3, 4).$$

Definition 11.1.16. Let $\lambda \in \Gamma_r^+$. For any $\mathsf{s} \in \mathrm{Std}(\lambda)$ we set $y_\mathsf{s} := y_\lambda d_\mathsf{s} + \Bbbk\mathfrak{S}_r^{<\lambda}$. We define the right Specht module of $\Bbbk\mathfrak{S}_r$ to be

$$\Delta_{\Bbbk\mathfrak{S}_r}(\lambda) = y_\lambda \Bbbk\mathfrak{S}_r / (\Bbbk\mathfrak{S}_r^{<\lambda} \cap y_\lambda \Bbbk\mathfrak{S}_r)$$

with basis given by $\{y_\mathsf{s} \mid \mathsf{s} \in \mathrm{Std}(\lambda)\}$.

11.1 The symmetric group and its representation theory

We now investigate these Specht modules in a bit more detail. From a modern point-of-view, the key to understanding these modules is the action of the so-called Jucys–Murphy elements. First, we define the Jucys–Murphy elements to be the

$$X_i = (1, i) + (2, i) + \ldots + (i-1, i) \in k\mathfrak{S}_r$$

for $2 \leq i \leq n$, we set $X_1 = 0$. These elements satisfy the relations

$$s_i(X_i X_{i+1}) = (X_i X_{i+1})s_i \qquad s_i(X_i + X_{i+1}) = (X_i + X_{i+1})s_i \qquad X_i X_j = X_j X_i$$

$$s_i X_{i+1} = 1 + X_i s_i \qquad s_i X_k = X_k s_i \text{ for } i \neq k-1, k.$$

In particular, notice that any symmetric polynomial in the Jucys–Murphy elements commutes with the Coxeter generators $s_i \in k\mathfrak{S}_r$ and therefore with all of $k\mathfrak{S}_r$. In order to understand the importance of Jucys–Murphy elements, we first need some extra "content" combinatorics. For $\lambda \in \Gamma_r$ we define the content of a box $[x, y] \in \lambda$ to be equal to $\text{ct}[x, y] = y - x$. Given $s \in \text{Std}(\lambda)$ we define its associated content vector, $\text{ct}(s)$, to be the ordered r-tuple

$$\text{ct}(s) = (\text{ct}(s^{-1}(1)), \text{ct}(s^{-1}(2)), \ldots, \text{ct}(s^{-1}(r))).$$

We observe that if $s, t \in \text{Std}(\lambda)$ then $\text{ct}(s), \text{ct}(t)$ differ only by permuting their entries. In fact, we have the following:

Lemma 11.1.17. *Let* $s \in \text{Std}(\lambda)$, $t \in \text{Std}(\mu)$ *for* $\lambda, \mu \in \Gamma_r^+$. *If* $\text{ct}(s) = \text{ct}(t)$ *then* $s = t$ *and* $\lambda = \mu$.

Proof. Two nodes have the same content if and only if they lie on the same "diagonal" of nodes $\{(i+k, j+k) \mid k \in \mathbb{Z}\}$. Therefore by adding the boxes $t^{-1}(k)$ for $k = 1, \ldots, n$ in turn, we uniquely determine the tableau and the result follows. □

Theorem 11.1.18 ([Mur92, Section 4]). *Let* $\lambda \in \Gamma_{n,r}^+$ *and* $s \in \text{Std}(\lambda)$ *and* $1 \leq k \leq r$. *The action of the Jucys–Murphy elements of* $\mathbb{Q}\mathfrak{S}_r$ *is given by*

$$y_s X_k = \text{ct}_s(k) y_s + \sum_{\substack{t \in \text{Std}(\lambda) \\ t \triangleleft s}} a_t y_t$$

for some scalars $a_t \in \mathbb{Z}$.

Example 11.1.19. Let $r = 3$ and $\lambda = (2, 1)$. We have two standard λ-tableaux,

$$s = \begin{array}{|c|c|} \hline 1 & 3 \\ \hline 2 \\ \cline{1-1} \end{array} \qquad t = \begin{array}{|c|c|} \hline 1 & 2 \\ \hline 3 \\ \cline{1-1} \end{array}$$

with corresponding content vectors $\text{ct}(s) = (0, -1, 1)$, and $\text{ct}(t) = (0, 1, -1)$. The basis of $\Delta_{k\mathfrak{S}_3}((2, 1))$ is given by

$$y_s = (1 - (1, 2)), \qquad y_t = (1 - (1, 2))(2, 3)$$

and the action of the Jucys–Murphy elements is given by

$$y_s X_2 = (1 - (1,2))(1,2) = -(1 - (1,2)) = -y_s$$
$$y_s X_3 = (1 - (1,2))((2,3) + (1,3))$$
$$= (2,3) + (1,3) - (1,2,3) - (1,3,2)$$
$$= (1 - (1,2)) - (1 - (1,2) - (2,3) - (1,3) + (1,2,3) + (1,3,2))$$
$$\in y_s + \Bbbk \mathfrak{S}_3^{\triangleleft (2,1)}$$

Exercise 11.1.20. Construct the bases of the modules $\Delta_{\Bbbk \mathfrak{S}_4}((3,1))$ and $\Delta_{\Bbbk \mathfrak{S}_4}((2,2))$ and calculate the action of the Jucys–Murphy elements on these modules.

Theorem 11.1.18 has the following immediate corollary.

Corollary 11.1.21. *The algebra $\mathbb{Q}\mathfrak{S}_r$ has trivial decomposition matrix; the Specht modules $\{\Delta_{\mathbb{Q}\mathfrak{S}_r}(\lambda) \mid \lambda \in \Gamma_r^+\}$ provide a complete set of non-isomorphic simple $\mathbb{Q}\mathfrak{S}_r$-modules.*

Proof. We will use (11.1.18) to show that there are no non-trivial $\mathbb{Q}\mathfrak{S}_r$-homomorphisms between these Specht modules (and hence deduce the result). Suppose, for a contradiction, that there exists a non-trivial $\mathbb{Q}\mathfrak{S}_r$-homomorphism

$$\varphi : \Delta_{\mathbb{Q}\mathfrak{S}_r}(\lambda) \to \Delta_{\mathbb{Q}\mathfrak{S}_r}(\mu)$$

which is determined on the generator $y_{t_\lambda} \in \Delta_{\mathbb{Q}\mathfrak{S}_r}(\lambda)$ by $\varphi(y_{t_\lambda}) = (\sum_{s \in \text{Std}(\mu)} a_s y_s)$ for some scalars $a_s \in \mathbb{Q}$. Now, since t_λ is minimal in the dominance ordering on $\text{Std}(\lambda)$, we have that

$$\varphi(\text{ct}(t_\lambda^{-1}(k))y_{t_\lambda}) = \varphi(y_{t_\lambda} X_k)$$
$$= \varphi(y_{t_\lambda}) X_k$$
$$= \sum_{s \in \text{Std}(\mu)} a_s y_s X_k$$
$$= \sum_{s \in \text{Std}(\mu)} a_s \text{ct}(s^{-1}(k))y_s$$

for each $1 \leq k \leq r$. Therefore the result follows from Lemma 11.1.17. □

We momentarily fix $\Bbbk = \mathbb{Q}$. We wish to straighten the Murphy basis so that the Jucys–Murphy elements act as scalars, i.e. to obtain a version of Theorem 11.1.18 in which all the scalars $a_t = 0$ for $t \neq s$ (that is, there are no "error terms"). We define

$$F_s := \prod_{\{c \mid -n < c < n\}} \prod_{\{i \mid \text{ct}_s(i) \neq c\}} \frac{X_i - c}{\text{ct}_s(i) - c} \in \mathbb{Q}\mathfrak{S}_r. \tag{11.1.22}$$

and we set $f_s = y_s F_s$ and $f_{st} = F_s y_{st} F_t$.

Example 11.1.23. Continuing with Example 11.1.19 the corresponding Murphy operators are as below

$$F_\mathsf{s} = \left(\frac{X_2+1}{-2}\right)\left(\frac{X_3-1}{2}\right) \qquad F_\mathsf{t} = \left(\frac{X_2-1}{2}\right)\left(\frac{X_3+1}{-2}\right).$$

Theorem 11.1.24 ([Mur92, Section 5]). *The algebra $\mathbb{Q}\mathfrak{S}_r$ is a weighted cellular algebra with respect to the idempotents* id $= \sum_{\mathsf{t}\in\mathrm{Std}(\lambda)} F_\mathsf{t}$ *and the cellular basis*

$$\{f_{\mathsf{st}} \mid \mathsf{s},\mathsf{t} \in \mathrm{Std}(\lambda) \text{ for some } \lambda \in \Gamma_r^+\}.$$

Given $\mathsf{t} \in \mathrm{Std}(\lambda)$ *and* $1 \leq i < j \leq r$, *we define the axial distance,* $a(i,j) = \mathrm{ct}_\mathsf{t}(i) - \mathrm{ct}_\mathsf{t}(j)$. *With respect to this basis the Coxeter generators of $\mathbb{Q}\mathfrak{S}_r$ act as follows*

$$f_\mathsf{t} s_i = \begin{cases} \frac{1}{a(i,i+1)} f_\mathsf{t} + \left(1 + \frac{1}{a(i,i+1)}\right) f_{s_i(\mathsf{t})} & \text{if } \mathsf{t}_{i\leftrightarrow i+1} \in \mathrm{Std}(\lambda) \\ \frac{1}{a(i,i+1)} f_\mathsf{t} & \text{otherwise} \end{cases}$$

where $\mathsf{t}_{i\leftrightarrow i+1}$ *is the tableau obtained from* t *by swapping the ith and* $(i+1)$*th entries.*

For our purposes, the more important aspect of the above theorem is the construction of the idempotents. This construction involves many fractions and so cannot be immediately lifted to fields $\mathbb{F}_p$. Later in this chapter we will discuss a clever summation of these elements which *cancels out these fractions* so that we obtain nice integral elements which can be realised in $\mathbb{F}_p\mathfrak{S}_r$. This will be a key step in providing a grading structure on $\mathbb{F}_p\mathfrak{S}_r$.

Definition 11.1.25. Let $\Bbbk = \mathbb{F}_p$. A partition $\lambda \in \Gamma_r^+$ is *p-regular* if $\lambda_i > \lambda_{i+p-1}$ or $\lambda_i = 0$ for all $1 \leq i \leq r$. We let $\Gamma_r^\Bbbk$ denote the set of p-regular partitions of r.

Example 11.1.26. The 2-regular partitions of 6 are

$$\Gamma_r^{\mathbb{F}_2} = \{(6),(5,1),(4,2),(3,2,1)\}.$$

Theorem 11.1.27. *Let* $\Bbbk = \mathbb{F}_p$. *The algebra* $\Bbbk\mathfrak{S}_r$ *has a complete set of pairwise non-isomorphic simple modules given by*

$$\{L_{\Bbbk\mathfrak{S}_r}(\lambda) = \Delta_{\Bbbk\mathfrak{S}_r}(\lambda)/\mathrm{rad}\langle -,-\rangle_\lambda \mid \lambda \in \Gamma_r^\Bbbk\}.$$

11.2 Generalised Temperley–Lieb algebras

We wish to consider the subset of the $(\mathbb{T}_{n,r} \times \mathbb{T}_{n,r})$-matrices which are generated by the permutation operators, s_i, which act by swapping the ith and $(i+1)$th terms in a given sequence $\pi \in \mathbb{T}_{n,r}$. It is easy to see that the symmetric group does not act faithfully on $\mathbb{T}_r$ and so our aim is to understand the kernel and image of this action. We illustrate this problem with an example.

Example 11.2.1. Consider $\mathbb{T}_{2,3} = \{111, 222, 112, 121, 211, 122, 212, 221\}$. The generators of $\mathfrak{S}_3$ act as follows:

$$(1,2) \mapsto \begin{bmatrix} 1 & \cdot & \cdot & \cdot & \cdot & \cdot & \cdot & \cdot \\ \cdot & 1 & \cdot & \cdot & \cdot & \cdot & \cdot & \cdot \\ \cdot & \cdot & 1 & \cdot & \cdot & \cdot & \cdot & \cdot \\ \cdot & \cdot & \cdot & 1 & \cdot & \cdot & \cdot & \cdot \\ \cdot & \cdot & \cdot & 1 & \cdot & \cdot & \cdot & \cdot \\ \cdot & \cdot & \cdot & \cdot & \cdot & 1 & \cdot & \cdot \\ \cdot & \cdot & \cdot & \cdot & \cdot & 1 & \cdot & \cdot \\ \cdot & \cdot & \cdot & \cdot & \cdot & \cdot & \cdot & 1 \end{bmatrix} \quad (2,3) \mapsto \begin{bmatrix} 1 & \cdot & \cdot & \cdot & \cdot & \cdot & \cdot & \cdot \\ \cdot & 1 & \cdot & \cdot & \cdot & \cdot & \cdot & \cdot \\ \cdot & \cdot & \cdot & 1 & \cdot & \cdot & \cdot & \cdot \\ \cdot & \cdot & 1 & \cdot & \cdot & \cdot & \cdot & \cdot \\ \cdot & \cdot & \cdot & \cdot & \cdot & \cdot & 1 & \cdot \\ \cdot & \cdot & \cdot & \cdot & \cdot & \cdot & \cdot & 1 \\ \cdot & \cdot & \cdot & \cdot & 1 & \cdot & \cdot & \cdot \\ \cdot & \cdot & \cdot & \cdot & \cdot & 1 & \cdot & \cdot \end{bmatrix}$$

and it is easy to see that the element

$$\text{id} - (1,2) - (2,3) + (1,2,3) + (1,3,2) - (1,3)$$

acts as the zero matrix (in other words, it lies in the kernel of the map $\Psi : \Bbbk \mathfrak{S}_r \to \text{Mat}(\mathbb{T}_{n,r})$). In fact, this is the only element of $\Bbbk \mathfrak{S}_3$ which acts as zero. Thus

$$\Psi_{2,3}(\Bbbk \mathfrak{S}_3) \cong \Bbbk \mathfrak{S}_3 / \Bbbk \{\text{id} - (1,2) - (2,3) + (1,2,3) + (1,3,2) - (1,3)\}$$

is 5-dimensional. We cometimes

Definition 11.2.2. Given $n, r \in \mathbb{Z}_{\geq 0}$, we define the **generalised Temperley–Lieb algebra** to be the $\Bbbk$-algebra $\text{TL}_{n,r} = \Psi_n(\Bbbk \mathfrak{S}_r) \subseteq \text{Mat}_{\mathbb{T}_{n,r}}(\Bbbk)$.

Exercise 11.2.3. Show that $\{\Psi(\text{id}), \Psi(1,2), \Psi(2,3), \Psi(1,2,3), \Psi(1,3,2)\}$ forms a basis of the (generalised) Temperley–Lieb algebra $\text{TL}_{2,3}$.

More generally, we have the following:

Theorem 11.2.4 ([H99]). *The kernel of the map $\Psi_n : \Bbbk \mathfrak{S}_r \to \text{Mat}_{\mathbb{T}_{n,r}}(\Bbbk)$ is the ideal of $\Bbbk \mathfrak{S}_r$ generated by the y_λ for $\lambda \notin \Gamma_{n,r}^+$. In other words, $\text{TL}_{n,r}$ has cellular basis*

$$\{d_{\mathsf{s}}^* y_\lambda d_{\mathsf{t}} : \mathsf{s}, \mathsf{t} \in \text{Std}(\lambda) \text{ for } \lambda \in \Gamma_{n,r}^+\}$$

with respect to the anti-involution $*$ and the poset $\Gamma_{n,r}^+ \subseteq \Gamma_r^+$. For $\lambda \in \Gamma_{n,r}^+$, we let $\Delta_{\text{TL}_{n,r}}(\lambda)$ denote the corresponding right cell-module with basis $\{y_\lambda d_{\mathsf{t}} \mid \mathsf{t} \in \text{Std}(\lambda)\}$.

Exercise 11.2.5. Prove that $\Psi_2(\Bbbk \mathfrak{S}_r)$ is isomorphic to the classical Temperley–Lieb algebra with parameter $\delta = 2 \in \Bbbk$ (this justifies our name "generalised Temperley–Lieb algebra"). Hint: the isomorphism is defined via : $1 - s_i \mapsto e_i$.

Exercise 11.2.6. Let $\Bbbk$ be arbitrary. Calculate the dimension of the $\Bbbk$-algebra $\text{TL}_{3,4} = \Psi_3(\Bbbk \mathfrak{S}_4)$.

Exercise 11.2.7. Let $\Bbbk = \mathbb{F}_3$. Calculate the Gram matrices of the intersection forms of the cell modules $\Delta_{\text{TL}_{2,6}}(6), \Delta_{\text{TL}_{2,6}}(5,1), \Delta_{\text{TL}_{2,6}}(4,2), \Delta_{\text{TL}_{2,6}}(3,3)$ for $\text{TL}_{2,6}$ and hence determine the decomposition matrix of this algebra.

11.3 Schur algebras

We wish to consider all $(\mathbb{T}_{n,r} \times \mathbb{T}_{n,r})$-matrices generated by "raising" and "lowering" operators. For $i < j$, the raising operator $f := f_{i \to j}$ takes a sequence $\pi \in \mathbb{T}_{n,r}$ to the sum over all sequences which can be obtained from π by replacing an i with a j. In fact, we further require the operator $f_{i \to j}^{[k]}$ which takes a sequence π to the sum over all sequences which can be obtained from π by replacing k of the is with js.

For $i < j$, we require the lowering operator $e_{j \to i}^{[k]}$ which takes a sequence π to the sum over all sequences which can be obtained from π by replacing k of the js with is. We also require the idempotent maps 1_λ which send a sequence π of weight λ to itself and annihilate all sequences of weight $\mu \neq \lambda$.

Definition 11.3.1. The $\Bbbk$-algebra $S_{n,r}^{\Bbbk}$ is the associative $\Bbbk$-algebra given by generators 1_λ ($\lambda \in \Gamma_{n,r}$), $f_{i \to i+1}^{[k]}, e_{i+1 \to i}^{[k]}$ ($1 \leq i < r$) subject to idempotent relations

$$1_\lambda 1_\mu = \delta_{\lambda\mu} 1_\lambda, \qquad \sum_{\lambda \in \Gamma_r} 1_\lambda = 1$$

$$f^{[k]} 1_\lambda = 1_{\lambda + k\varepsilon_i - k\varepsilon_{i+1}} f^{[k]} \qquad e^{[k]} 1_\lambda = 1_{\lambda - k\varepsilon_i + k\varepsilon_{i+1}} e^{[k]}$$

(with the convention that any 1_λ for $\lambda \notin \Gamma_{n,r}$ is zero), the summing power relations

$$e_{i+1 \to i}^{[a]} e_{i+1 \to i}^{[b]} = \frac{(a+b)!}{a!b!} e_{i+1 \to i}^{[a+b]} \qquad e_{i+1 \to i}^{[0]} = 1$$

$$f_{i \to i+1}^{[a]} f_{i \to i+1}^{[b]} = \frac{(a+b)!}{a!b!} f_{i \to i+1}^{[a+b]} \qquad f_{i \to i+1}^{[0]} = 1$$

and the relations for passing divided powers through one another as follows:

$$f_{i \to i+1}^{[a]} 1_\lambda e_{i+1 \to i}^{[b]} = \sum_{k \geq 0} \binom{a+b+\lambda_i - \lambda_{i+1}}{k} e_{i+1 \to i}^{[b-k]} 1_{\lambda + (a+b-k)(\varepsilon_i - \varepsilon_{i+1})} f_{i \to i+1}^{[a-k]}$$

$$e_{i+1 \to i}^{[b]} 1_\lambda f_{i \to i+1}^{[a]} = \sum_{k \geq 0} \binom{a+b-\lambda_i + \lambda_{i+1}}{k} f_{i \to i+1}^{[a-k]} 1_{\lambda + (a+b-k)(\varepsilon_{i+1} - \varepsilon_i)} e_{i+1 \to i}^{[b-k]}$$

and $e_{i \to i+1}^{[a]} f_{j+1 \to j}^{[b]} = f_{j+1 \to j}^{[b]} e_{i \to i+1}^{[a]}$ for $i \neq j$.

Example 11.3.2. In $\mathbb{T}_{4,8}$ we have that $(12432232) f_{2 \to 4}^{[2]}$ is equal to the element

1<u>4</u>43<u>4</u>232 + 1<u>4</u>432<u>4</u>32 + 1243<u>4</u><u>4</u>32 + 1<u>4</u>4322<u>3</u>4 + 1243<u>4</u>2<u>3</u>4 + 12432<u>4</u>3<u>4</u>.

Given $\lambda \in \Gamma_{n,r}^+$ and $\mu \in \Gamma_{n,r}$, we define a λ-tableau of weight μ to be a map $\mathsf{T} : [\lambda] \to \{1, \dots, n\}$ such that $\mu_i = |\{x \in [\lambda] : \mathsf{T}(x) = i\}|$ for $i \geq 1$. If T is a λ-tableau of weight μ, we say that T is **semistandard** if the rows are weakly increasing from left to right and the columns are strictly increasing from top to bottom. The set of all semistandard tableaux of shape λ and weight μ is denoted $\mathrm{SStd}_n(\lambda, \mu)$ and we let $\mathrm{SStd}_n(\lambda,-) := \cup_{\mu \in \Gamma_{n,r}} \mathrm{SStd}_n(\lambda,\mu)$. We let T_λ denote the unique element of $\mathrm{SStd}_n(\lambda,\lambda)$. For example, we have the following semistandard λ-tableaux,

1	1	1	1
2	2	2	
3			

1	1	1	3
2	2	3	
3			

1	1	2	2
2	3	3	
3			

1	1	2	4
2	3	4	
3			

for $\lambda = (4,3,1) \in \Gamma^+_{4,8}$ (and we note that the first of these tableau is equal to T_λ).

Definition 11.3.3. Given $1 \leq i, j \leq n$ and $\mathsf{T} \in \mathrm{SStd}_n(\lambda,\mu)$, we let $\mathsf{T}(i,j)$ denote the number of entries equal to j lying in the ith row of T. Since T is semistandard we have that $\mathsf{T}(i,j) = 0$ for $i > j$ and $\sum_{1 \leq i \leq d} \mathsf{T}(i,j) = \mu_j$.

Definition 11.3.4. Given $\mathsf{S}, \mathsf{T} \in \mathrm{SStd}_n(\lambda,\mu)$ we let

$$\xi_{\mathsf{S}\lambda} = \prod_{i=d}^{1}\left(\prod_{j=1}^{d} e_{i,j}^{[\mathsf{S}(i,j)]}\right) \qquad \xi_{\lambda\mathsf{T}} = \prod_{i=1}^{d}\left(\prod_{j=1}^{d} f_{i,j}^{[\mathsf{T}(i,j)]}\right)$$

(notice the ordering on these products) and we define $\xi_{\mathsf{ST}} = \xi_{\mathsf{S}\lambda} 1_\lambda \xi_{\lambda\mathsf{T}}$.

Example 11.3.5. Let $\lambda = (3,3)$, $\mu = (2,2,1,1)$, and $\nu = (2,1,2,1)$. We set

$$\mathsf{S} = \begin{array}{|c|c|c|}\hline 1 & 1 & 3 \\\hline 2 & 2 & 4 \\\hline\end{array} \in \mathrm{SStd}_4(\lambda,\mu) \qquad \mathsf{T} = \begin{array}{|c|c|c|}\hline 1 & 1 & 2 \\\hline 3 & 3 & 4 \\\hline\end{array} \in \mathrm{SStd}_4(\lambda,\nu).$$

We have that $\mathsf{S}(2,4) = 1$, $\mathsf{S}(2,2) = 2$, $\mathsf{S}(1,3) = 1$, $\mathsf{S}(1,1) = 2$, and $\mathsf{S}(i,j) = 0$ otherwise. Similarly, $\mathsf{T}(2,4) = 1$, $\mathsf{T}(2,3) = 2$, $\mathsf{T}(1,2) = 1$, $\mathsf{T}(1,1) = 2$, and $\mathsf{T}(i,j) = 0$ otherwise. Therefore,

$$\xi_{\mathsf{ST}} = e_{1,3}^{[1]} e_{2,4}^{[1]} 1_\lambda f_{2,4}^{[1]} f_{2,3}^{[2]} f_{1,2}^{[1]}.$$

Theorem 11.3.6 ([Gre93]). *The Schur algebra $S^{\Bbbk}_{n,r}$ is a weighted cellular algebra with respect to the idempotents $1 = \sum_{\lambda \in \Gamma_{n,r}} 1_\lambda$ and basis*

$$\{\xi_{\mathsf{ST}} \mid \mathsf{S} \in \mathrm{SStd}_n(\lambda,\mu), \mathsf{T} \in \mathrm{SStd}_n(\lambda,\nu) \text{ for } \lambda \in \Gamma^+_{n,r}, \mu,\nu \in \Gamma_{n,r}\}.$$

and anti-automorphism $* : S^{\Bbbk}_{n,r} \to S^{\Bbbk}_{n,r}$ *determined by*

$$(e^{[k]}_{i+1 \to i})^* = f^{[k]}_{i \to i+1} \qquad (f^{[k]}_{i \to i+1})^* = e^{[k]}_{i+1 \to i} \qquad 1^*_\lambda = 1_\lambda$$

for $\lambda \in \Gamma_{n,r}$ and any total refinement of the opposite of the order $\rhd$ on $\Gamma^+_{n,r}$.

Definition 11.3.7. Let $\lambda \in \Gamma^+_{n,r}$. For any $\mathsf{S} \in \mathrm{Path}(\lambda,-)$ we set $\xi_\mathsf{S} := 1_\lambda \xi_{\lambda \mathsf{S}} + S^{<\lambda}_{n,r}$. We define the right Weyl module of $S^{\Bbbk}_{n,r}$ to be

$$\Delta_{S^{\Bbbk}_{n,r}}(\lambda) = 1_\lambda S^{\Bbbk}_{n,r}/(S^{<\lambda}_{n,r} \cap 1_\lambda S^{\Bbbk}_{n,r})$$

with basis given by $\{\xi_\mathsf{S} \mid \mathsf{S} \in \mathrm{SStd}_n(\lambda,\mu), \mu \in \Gamma_{n,r}\}$.

11.3 Schur algebras

Exercise 11.3.8. The action of the f-generators on the $S^{\Bbbk}_{3,5}$-module $\Delta_{S^{\Bbbk}_{3,5}}(3, 1^2)$ is given in Figure 11.2. Calculate the action of the e-generators on this module using the relations from Definition 11.3.1.

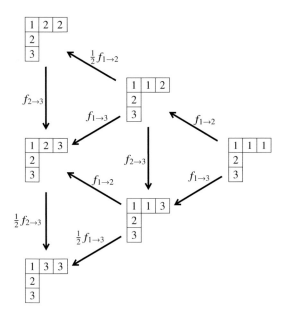

Fig. 11.2: The action of the f-generators on the $S^{\Bbbk}_{3,5}$-module $\Delta_{S^{\Bbbk}_{3,5}}(3, 1^2)$. The action of the e-generators is given by reversing the arrows (and computing the corresponding scalars). The action of the idempotents is easy to deduce. Compare this with Example 6.1.27.

Proposition 11.3.9. *Let* $\Bbbk = \mathbb{F}_p$. *The Schur algebra* $S^{\Bbbk}_{n,r}$ *has a complete set of pairwise non-isomorphic simple modules given by*

$$\{L_{S^{\Bbbk}_{n,r}}(\lambda) = \Delta_{S^{\Bbbk}_{n,r}}(\lambda)/\mathrm{rad}\langle -, -\rangle_\lambda \mid \lambda \in \Gamma^+_{n,r}\}.$$

Exercise 11.3.10. Use your answer to Exercise 11.3.8 in order to prove that $S^{\Bbbk}_{3,5}$-module $\Delta_{S^{\Bbbk}_{3,5}}(3, 1^2)$ has a 3-dimensional submodule for $\Bbbk = \mathbb{F}_2$.

Exercise 11.3.11. Calculate the decomposition matrix of $S^{\Bbbk}_{3,5}$ over $\Bbbk = \mathbb{F}_2$.

Exercise 11.3.12. Calculate the Gram matrices of the cell modules of $S^{\Bbbk}_{p,p}$ for $p = 3, 5$. Hence calculate the decomposition matrices of these algebras.

11.4 Lusztig's and Andersen's conjectures

We wish to rephrase character-theoretic questions for symmetric groups and their Schur algebras in terms of (p)-Kazhdan–Lusztig theory. We begin by identifying the parabolic Coxeter system which we believe should "control" the representation theory of symmetric groups and their Schur algebras (spoiler alert: it is $(W, P) = (\widehat{\mathfrak{S}}_n, \mathfrak{S}_n)$). We assume throughout this section that $p > n$. For each $1 \leq i \leq n$ we let ε_i denote a formal symbol, and define an n-dimensional real vector space

$$\mathbb{E}_n = \bigoplus_{1 \leq i \leq n} \mathbb{R}\varepsilon_i$$

and we set $\overline{\mathbb{E}}_n$ to be the quotient of this space by the one-dimensional subspace spanned by $\sum_{1 \leq i \leq n} \varepsilon_i$ (this is a trivial technicality which is unimportant, it merely allows us to visualise the $n = 3$ case in 2-dimensional space, as in Figure 11.3). We set

$$\widehat{\mathbb{E}}_n^+ = \{(x_1, \ldots, x_n) \mid x_i - x_{i+1} \geq -1 \text{ for } 1 \leq i < n\} \subseteq \mathbb{E}_n.$$

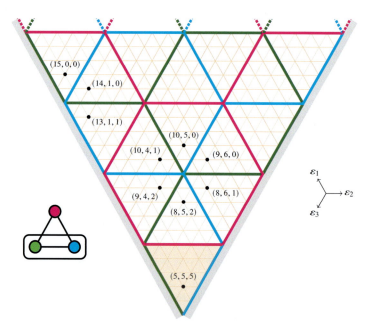

Fig. 11.3: The fundamental region $\mathcal{F}_3^5$ is highlighted in orange. We colour the walls of $\mathcal{F}_3^5$ by $\mathbb{E}_{(1,2)}, \mathbb{E}_{(2,3)}$, and $\mathbb{E}_{(3,1)}$. We have tiled the region $\widehat{\mathbb{E}}_3^+$, which is pictured as one sixth of $\mathbb{R}^2$ and highlighted the principal block $\mathcal{B}_{1,3,5}$. Compare with Figure 11.5.

11.4 Lusztig's and Andersen's conjectures

We define the fundamental region to be the "bottom-most" subset of points

$$\mathcal{F}_n^p = \{(x_1, \ldots, x_n) \mid x_i - x_{i+1} \geq -1 \text{ for } 1 \leq i < n \text{ and } \lambda_1 - \lambda_n \leq p + 1 - n\}.$$

We colour the edges of the fundamental region as follows. We place a distinct colour on each of the n "walls" of the fundamental region:

$$\mathbb{E}_{(i,i+1)} = \{x \in \mathcal{F}_n^p \mid x_i - x_{i+1} = -1\}$$

for each $1 \leq i < n$ and we introduce one final "affine wall",

$$\mathbb{E}_{(n,1)} = \{x \in \mathcal{F}_n^p \mid x_1 - x_n = p + 1 - n\}$$

each of these "walls" is simply a hyperplane intersected with $\mathcal{F}_n^p$. By reflecting through these walls we can tile all of $\widehat{\mathbb{E}}_n^+$ with copies of $\mathcal{F}_n^p$. We colour the walls of these regions in a similar fashion. This is illustrated in Figures 11.3 and 11.4.

We say that a point $x \in \widehat{\mathbb{E}}_n^+$ is **singular** if it lies on one of these coloured walls/hyperplanes and **regular** otherwise. In other words, the singular points are those belonging to

$$\{(x_1, \ldots, x_n) \mid x_i - x_{i+1} \in -1 + p\mathbb{Z} \text{ for some } 1 \leq i < n \text{ or } x_1 - x_n \in 1 - n + p\mathbb{Z}\}.$$

The parabolic Coxeter graph of $(W, P) = (\widehat{\mathfrak{S}}_n, \mathfrak{S}_n)$ is labelled by the "finite reflections" $(1, 2), (2, 3), \ldots$ together with the "affine reflection" $(n, 1)$ — the last of which we visualise as a transposition on a cylinder. We colour these reflections in accordance with the colouring of the walls of $\mathcal{F}_n^p$. The group W acts on $\mathbb{E}_n$ on the right via these reflections. Given $x \in \mathbb{E}_n$, we let xW denote the orbit of the point x under this action of W.

Definition 11.4.1. We identify $\lambda \in \Gamma_{n,r}^+$ with an element of the integer lattice inside $\widehat{\mathbb{E}}_n^+$ via the map

$$\lambda \longmapsto \sum_{1 \leq i \leq n} \lambda_i \varepsilon_i. \tag{11.4.2}$$

We refer to the set of partitions within a given orbit, $\Gamma_{n,r}^+ \cap \lambda W$, as a **block** of $\mathbb{F}_p \mathfrak{S}_r$. We note that if $\pi = (dp, dp, \ldots, dp) \in \Gamma_{n,dnp}^+$ is rectangular partition, then it lies within the fundamental region and *does not lie on any hyperplanes* under our assumption that $p > n$. We refer to $\Gamma_{n,dnp}^+ \cap \pi W$, as the **principal block**, $\mathcal{B}_{d,n,p}$, of $\mathbb{F}\mathfrak{S}_{dnp}$.

Example 11.4.3. For $p = 5$, we depict the principal block $\Gamma_{3,15}^+ \cap (5, 5, 5)W$ as the set of labelled points in Figure 11.3. For $p = 7$, we depict the principal block $\Gamma_{3,21}^+ \cap (7, 7, 7)W$ as the set of labelled points in Figure 11.4.

Theorem 11.4.4. *Let $\lambda, \mu \in \Gamma_{n,r}^+$. A necessary condition for*

$$[\Delta_{S_{n,r}^\Bbbk}(\lambda) : L_{S_{n,r}^\Bbbk}(\mu)]_{S^\Bbbk(n,r)} \neq 0 \quad \text{or} \quad [\Delta_{\Bbbk\mathfrak{S}_r}(\lambda) : L_{\Bbbk\mathfrak{S}_r}(\mu)]_{\Bbbk\mathfrak{S}_r} \neq 0$$

is that $\mu \in \lambda W$. In other words, λ and μ must belong to the same block.

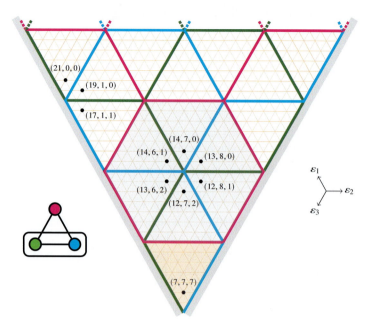

Fig. 11.4: The elements of $\Gamma_{21,3}^+$ from the principal block $\mathcal{B}_{1,3,7}$ are pictured within $\frac{1}{6}$th of the plane $\mathbb{R}^2$. Compare with Figure 11.4. In grey we highlight the "region around the σ-Steinberg".

Notice that if we reflect any regular point $x \in \mathcal{F}_n^P$ through the "affine hyperplane/wall" of $\mathcal{F}_n^P$ we obtain a regular point *belonging to* the region $\widehat{\mathbb{E}}_n^+$. On the other hand, if we reflect any regular point $x \in \mathcal{F}_n^P$ through one of the "finite" hyperplanes/walls of $\mathcal{F}_n^P$ then we obtain a regular point lying *outside* the region $\widehat{\mathbb{E}}_n^+$. This allows us to identify regular points in $\mathbb{E}_n^+$ with cosets $^P W$ for $(W, P) = (\widehat{\mathfrak{S}}_n, \mathfrak{S}_n)$ by first associating any regular point $x \in \mathcal{F}_n^P$ to the identity coset $\varnothing \in {}^P W$ and then labelling every $y \in xW$ by any word $s_{i_1} s_{i_2} \ldots s_{i_k}$ such that $(x) s_{i_1} s_{i_2} \ldots s_{i_k} = y$.

Example 11.4.5. Compare the Coxeter graph in Figure 11.5 with generators $(1, 2)$, $(2, 3)$ and $(3, 1)$ versus the colouring of the walls of $\mathcal{F}_n^P$ in Figure 11.3. Reflecting $(5, 5, 5)$ through the hyperplane $x_2 - x_3 = -1$ we obtain the point $(5, 4, 6) \notin \mathbb{E}_3^+$. Similarly, reflecting $(5, 5, 5)$ through the hyperplane $x_1 - x_2 = -1$ we obtain the point $(4, 5, 6) \notin \mathbb{E}_3^+$. Whereas reflecting $(5, 5, 5)$ through the hyperplane $x_1 - x_3 = 3$ we obtain the point $(8, 5, 2) \in \mathbb{E}_3^+$. We can then subsequently reflect $(8, 5, 2) \in \mathbb{E}_3^+$ through the hyperplane $x_2 - x_3 = 4$ to obtain $(8, 6, 1)$. Thus we write $(5, 5, 5)\sigma\tau = (8, 6, 1)$.

For the principal block, we identify the rectangular partition $(dp, dp, \ldots, dp) \in \mathcal{B}_{d,n,p}$ for $d, n \in \mathbb{Z}_{\geq 0}$ with the identity coset $\varnothing \in {}^P W$. Given $\lambda \in \mathcal{B}_{d,n,p}$ we let $w_\lambda \in {}^P W$ be such that $(dp, dp, \ldots, dp) w_\lambda = \lambda$.

11.4 Lusztig's and Andersen's conjectures

Example 11.4.6. Let $p = 7$ and consider $\Gamma^+_{21,3}$. We have that

$w_{(7,7,7)} = \emptyset$, $w_{(12,7,2)} = \sigma$, $w_{(13,6,2)} = \sigma\rho$, $w_{(12,8,1)} = \sigma\tau$,
$w_{(14,6,1)} = \sigma\rho\tau$, $w_{(13,8,0)} = \sigma\tau\rho$, $w_{(14,7,0)} = \sigma\rho\tau\rho$, $w_{(17,1,1)} = \sigma\rho\tau\sigma$,
$w_{(20,1,0)} = \sigma\rho\tau\sigma\rho$, $w_{(21,0,0)} = \sigma\rho\tau\sigma\rho\tau$.

These reflections are highlighted in Figure 11.5, see also Figure 11.4.

The map : $\lambda \mapsto w_\lambda$ gives us a "characteristic-free" way of understanding the principal blocks $\mathscr{B}_{d,n,p}$ of symmetric groups as we let p vary (this is highlighted in Figures 11.3 and 11.4). Lusztig's and Andersen's conjectures posited that certain composition factor multiplicities for modules of the principal block $\mathscr{B}_{d,n,p}$ should stabilise for p any reasonable prime number.

Conjecture 11.4.7 (Andersen's conjecture [And98]). *Let $\Bbbk = \mathbb{F}_p$ with $p \geq n$. Let $\lambda, \mu \in \Gamma^+_{n,r}$ be such that $\lambda_1 - \lambda_n, \mu_1 - \mu_n < p(p-2)$. The corresponding decomposition number of the symmetric group can be calculated as follows*

$$[\Delta_{\Bbbk\mathfrak{S}_r}(w_\lambda) : L_{\Bbbk\mathfrak{S}_r}(w_\mu)]_{\Bbbk\mathfrak{S}_r} = n_{w_\lambda,w_\mu}(q)|_{q=1}$$

where $n_{w_\lambda,w_\mu}(q)$ is the Kazhdan–Lusztig polynomial associated to $w_\lambda, w_\mu \in {}^PW$ for $(W, P) = (\widehat{\mathfrak{S}}_n, \mathfrak{S}_n)$. In particular, these decomposition numbers are independent of the prime providing $p \geq n$.

Conjecture 11.4.8 (Lusztig's conjecture [Lus80]). *Let $\lambda, \mu \in \Gamma^+_{n,r}$ be such that $\lambda_1 - \lambda_n, \mu_1 - \mu_n < p(p-2)$. The corresponding decomposition number of the Schur algebra*

$$[\Delta_{S^\Bbbk_{n,r}}(w_\mu) : D(w_\lambda)]_{S^\Bbbk_{n,r}}$$

is independent of the prime providing $p > 2n - 4$ and moreover these multiplicities can be explicitly calculated using the Kazhdan–Lusztig polynomials in which the parabolic plays a "dual" role to that considered in this book.

Specialising to the case that the partitions are "around the Steinberg partition" these "dual" statements can be brought together into one uniform statement (as we can forget about the "dual" roles played by the parabolic subgroup). This allows us to discuss both statements in the context of the combinatorics developed earlier in this book.

Definition 11.4.9. We define the Steinberg partition to be $(p-1)\rho_n$ where ρ_n is the triangular partition $\rho_n = (n-1, n-2, \ldots, 1, 0)$ of the triangle number $\binom{n}{2} \in \mathbb{Z}_{\geq 0}$.

Example 11.4.10. For $n = 3$ and $p = 7$ the Steinberg partition is $6 \times (2, 1, 0) = (12, 6, 0)$ and this partition is located at the centre of the grey region in Figure 11.4. We refer to the region in grey as the region "around the Steinberg partition". Notice that the highlighted partitions in this region are obtained from one another by applying reflections from the finite group

$$\mathfrak{S}_3 = \langle (1,2), (2,3) \mid ((1,2)(2,3))^3 = (1,2)^2 = (2,3)^2 = \mathrm{id}\rangle.$$

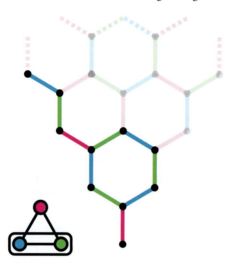

Fig. 11.5: The embedding of the principal block of of $\Gamma^+(21,3)$ for $p = 7$ as cosets of $^P W$. We have highlighted the reflections which connect partitions of size 21 from Figure 11.4. We have greyed out the remaining (infinite) set of minimal coset representatives.

We observe that $(p-1)\rho_n + x\rho_n \in \mathbb{E}_n^+$ is a regular partition for all $x \in \mathfrak{S}_n$. (Compare this with Definition 8.3.7.)

Conjecture 11.4.11 (The "toy version" of Lusztig's and Andersen's conjectures). *Let $p > 2n - 4$. The decomposition numbers of the symmetric group and Schur algebras "around the Steinberg partition" can be calculated as follows*

$$n_{x,y}(q)|_{q=1} = [\Delta_{S_{n,r}^{\Bbbk}}((p-1)\rho_n + y\rho_n) : L_{S_{n,r}^{\Bbbk}}((p-1)\rho_n + x\rho_n)]$$
$$= [\Delta_{\Bbbk\mathfrak{S}_r}((p-1)\rho_n + x\rho_n) : L_{\Bbbk\mathfrak{S}_r}((p-1)\rho_n + y\rho_n)]$$

where $n_{x,y}(q)$ is the Kazhdan–Lusztig polynomial associated to $x, y \in \mathfrak{S}_n$. Thus these polynomials can be calculated with the algorithm from Part 3 of this book.

Further Reading. In this book we have perhaps treated the semisimple representation theory of symmetric groups and their Schur algebras as "old news". It is worth emphasising here that there are a few very old, famous, and difficult open problems in this field. Two of the most famous of these are the Kronecker and plethysm problems. The Kronecker problem asks for a combinatorial understanding of the "tensor product" of two simple modules of a given symmetric group. The plethysm problem ask for an understanding of "induced modules" from wreath product subgroups. These problems have intimate connections with the geometric complexity programme [IP17] (an approach that seeks to settle the P ≠ NP problem) and quantum information theory [AK08, CHM07, BCI11a]. The importance of these coefficients in geometric complexity theory derives from their frequent ap-

11.4 Lusztig's and Andersen's conjectures

pearance in formulas for multiplicities in coordinate rings: the orbit of the product of variables [Lan17], the permanent polynomial [BLMW11], the power sum polynomial and the unit tensor [FI20]. There is an extensive combinatorial literature on this subject (see for example, [Dvi93, BK99, BOR10, BCI11b, Ste12, SS16, Liu17, Bla16, BB17, Bes18, BP20, BBS21, BBP22, BB23] and references therein) and these problems have been tackled from the diagrammatic viewpoint of this book [BDO15, BDE19, BDVE21, BP24, BPW25].

Should the reader wish to understand how the (modular) representation theory of the symmetric groups and Schur algebras were developed in their original context (without passing to the diagrammatic algebras of this book) we would recommend the books of Gordon James, Sandy Green, and Stuart Martin [Jam78, Gre07, Mar93] and also the Lie algebra theoretic material of [FH91]. The presentation of the Schur algebra we use here was developed in a pair of papers by Doty and Doty–Giaquinto [DG02, Dot03].

For purposes of not over-burdening the reader, we have side-stepped the theory of tilting modules for the Schur algebra. This is the context in which Andersen's conjecture was originally formulated in [And98], but the reformulation we provide here in terms of symmetric groups is well-known [Erd97, Section 4.1]. For more details on tilting theory for Schur algebras (of algebraic groups), we refer to Donkin's seminal paper [Don93]. Examples of explicit calculations of the structure, Alperin diagrams, and formal characters of tilting modules of Schur algebras in small rank can be found in [DH05, BDM11, BDM15]. It is difficult to overstate the importance of Donkin's tilting theory for algebraic groups, and it is unfortunate that we are unable to cover it here — we merely remark that tilting theory is ubiquitous in the study of Hecke categories (particularly of affine Weyl groups) and far beyond.

Chapter 12
Hidden gradings on symmetric groups

It is, unfortunately, already time for us to dispense with group theory. In this chapter we uncover the hidden gradings lying behind this group theoretic picture: first on the level of tableaux and then on the level of the group algebra of the symmetric group (via the quiver Hecke algebra). The combinatorial highlight of this chapter is the reconciliation of the tableaux-grading (on $\Bbbk\mathfrak{S}_r$) with the Kazhdan–Lusztig theoretic grading on paths in the Bruhat graph — this is the first hint that we are "on track" toward proving that the decomposition numbers of symmetric groups are equal to p-Kazhdan–Lusztig polynomials. Going further down the rabbit hole, we introduce the graded generators, relations, and bases for $\Bbbk\mathfrak{S}_r$ and help the reader get to grips with these ideas by providing plenty of examples.

12.1 The tableaux-theoretic LLT algorithm

In this section, we construct a grading on standard λ-tableaux for $\lambda \in \Gamma_r^+$. This will allow us to define a uni-triangular matrix of Laurent polynomials, $(\nabla_{\lambda,\mu})_{\lambda,\mu \in \Gamma_r^+}$, with rows and columns indexed by partitions where the (λ, μ)th entry is given by summing over the degrees of tableaux. We will then factorise this matrix in order to define the main objects of interest in this section: the LLT polynomials $m_{\lambda\nu}(q) \in q\mathbb{Z}[q]$ for $\lambda \neq \nu$ of Lascoux–Leclerc–Thibon [LLT96]. This should remind the reader of the manner in which we defined the Δ-matrix in Part 3 (which we then factorised in order to define the the Kazhdan–Lusztig polynomials). In the section after this one, we will resolve this apparent coincidence.

In order to construct our desired grading, we will first need a way to colour partitions and tableaux according to $\mathbb{F}_p$. We depict this colouring on $\mathbb{F}_p$ graphically as in Figure 12.1. For the remainder of this section, unless otherwise stated all colourings will be for the prime $p = 5$ as in Figure 12.1.

Definition 12.1.1. Let $\lambda \in \Gamma_r^+$. Given p a prime number, we colour the boxes $[x, y] \in \lambda$ according to the reduction of $\text{ct}[x, y]$ modulo p, which we refer to as the **residue modulo** p.

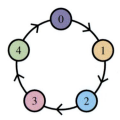

Fig. 12.1: A colouring on the field of 5 elements, $\mathbb{F}_5$.

Given $\lambda \in \Gamma_{n,r}^+$ we say that $[x, y] \notin \lambda$ is **addable** if $\lambda \cup \{[x, y]\}$ is a partition. We say that $[x, y] \in \lambda$ is **removable** if $\lambda - \{[x, y]\}$ is a partition. We refine this by residue: we say that a node $[x, y]$ is i-addable (respectively i-removable) if it is addable (respectively removable) and of residue $y - x = i \in \mathbb{Z}/p\mathbb{Z}$.

Example 12.1.2. Let $p = 5$. The partition $\lambda = (8, 5)$ has two addable 3-nodes and one removable 3-node which we highlight as follows

and no addable or removable nodes of residue $3 \neq i \in \mathbb{Z}/5\mathbb{Z}$.

Let $\lambda \in \Gamma_{n,r}^+$ and $\mathsf{t} \in \text{Std}(\lambda)$. We let $\mathsf{t}^{-1}(k)$ be the node in t containing the integer $k \in \{1, \ldots, r\}$. Given $1 \leq k \leq r$, we let $\mathcal{A}_\mathsf{t}(k)$, (respectively $\mathcal{R}_\mathsf{t}(k)$) denote the set of all addable $\text{res}(\mathsf{t}^{-1}(k))$-nodes (respectively all removable $\text{res}(\mathsf{t}^{-1}(k))$-nodes) of the partition $\text{Shape}(\mathsf{t}\!\downarrow_{\{1,\ldots,k\}})$ which are above $\mathsf{t}^{-1}(k)$, i.e. those in an earlier row. Let $\lambda \in \Gamma_{n,r}^+$ and $\mathsf{t} \in \text{Std}(\lambda)$. We define the **degree** of t as follows:

$$\deg(\mathsf{t}) = \sum_{k=1}^r \left(|\mathcal{A}_\mathsf{t}(k)| - |\mathcal{R}_\mathsf{t}(k)| \right).$$

Given $\mathsf{t} \in \text{Std}(\lambda)$ we define the **residue sequence** of t as follows:

$$\text{res}(\mathsf{t}) = (\text{res}(\mathsf{t}^{-1}(1)), \text{res}(\mathsf{t}^{-1}(2)), \ldots, \text{res}(\mathsf{t}^{-1}(r))) \in (\mathbb{Z}/p\mathbb{Z})^r.$$

Example 12.1.3. We depict two standard tableaux in Figure 12.2. The overall degree of the first tableau is

$$0 + 0 + 0 + 0 + 1 + 0 + 0 - 1 + 0 + 0 + 0 - 1 + 2 + 1 = 2$$

12.1 The tableaux-theoretic LLT algorithm

and the overall degree of the second tableau is

$$0 + 0 + 0 + 0 + 1 + 0 + 0 - 1 + 0 + 0 + 0 - 1 + 1 + 0 = 0$$

where in both cases the sum is given by summing over the degrees of the first 14 boxes, in order. The residue sequence of the former tableau is given by $0, 4, 3, 2, 1, 1, 0, 0, 4, 3, 4, 3, 2$.

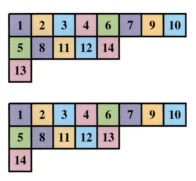

Fig. 12.2: Two coloured tableaux s, t for $p = 5$. The former is of degree 2 and the latter is of degree 0.

Definition 12.1.4. Let $t \in \text{Std}(\mu)$ be a tableau with the property that if $\text{res}(s) = \text{res}(t)$ for some $s \in \text{Std}(\lambda)$ (not necessarily equal to μ) then $t \trianglerighteq s$ (in particular, $\mu \trianglerighteq \lambda$). In which case we say we say that t is a **reduced tableau** for μ and that $\text{res}(t)$ is a **reduced residue sequence** for μ.

Example 12.1.5. Of the two tableaux in Figure 12.2, neither is reduced. The two tableaux in Figure 12.3 are both reduced.

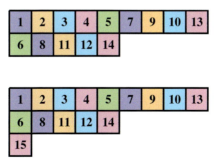

Fig. 12.3: Two reduced (coloured) tableaux u, v for $p = 5$.

Definition 12.1.6. Given t ∈ Std(μ) a reduced tableau, we define Std(λ, t) to be the set of standard tableaux s ∈ Std(λ) such that res(s) = res(t).

Example 12.1.7. We have that s, t ∈ Std((8, 5, 1), u) for s, t, u as in Figures 12.2 and 12.3.

For any reduced tableau t ∈ Std(μ) we have that Std(λ, t) = ∅ if $\lambda \rhd \mu$, by definition. Moreover, every s ∈ Std(λ) belongs to Std(λ, t) for some reduced tableau t of shape $\mu \unrhd \lambda$. There are many ways of constructing reduced tableaux, the most famous of which is James' "ladder tableau" construction.

Definition 12.1.8. Given $[x, y] \in \mu$, we define the ladder number $L_p[x, y] = y - x + p(1 - y)$. For each p-regular μ we define the μ-ladder tableau to be s_μ the maximal tableau in the dominance ordering on Std(μ) such that

$$L_p(\mathsf{s}_\mu^{-1}(1)), L_p(\mathsf{s}_\mu^{-1}(2)), L_p(\mathsf{s}_\mu^{-1}(3)), \ldots, L_p(\mathsf{s}_\mu^{-1}(r))$$

is weakly decreasing. We set CStd(λ, μ) := Std(λ, s_μ) and refer to these as the μ-coloured standard tableaux.

Example 12.1.9. For $p = 5$, the ladder numbers are depicted in Figure 12.4.

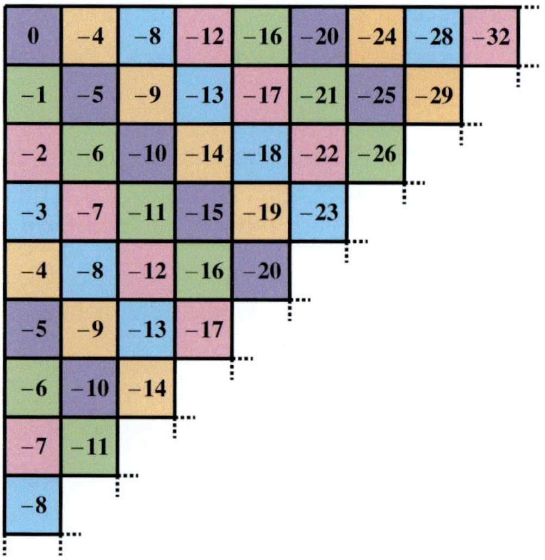

Fig. 12.4: The ladder numbers for $p = 5$ (the colouring of boxes is as in Figure 12.1). Notice that the ladder numbers are all non-positive and so it is easy to distinguish between ladder numbers versus standard tableaux.

Proposition 12.1.10. *For μ a p-regular partition, the μ-ladder tableau is reduced.*

Proof. This was proven in [Jam76] and another excellent account is given in [Mat99, Lemma 3.40]. □

12.1 The tableaux-theoretic LLT algorithm

Thus ladder tableaux provide us with an analogy of reduced paths (from the Kazhdan–Lusztig setting) in our new tableaux-theoretic setting. Moreover the μ-coloured standard tableaux $\mathsf{CStd}(\lambda, \mu)$ will provide us with an analogy of "folded up paths". Motivated by the analogous situation for Kazhdan–Lusztig theory, we first consider the tableaux-counting polynomials

$$\sum_{\mathsf{s} \in \mathsf{CStd}(\lambda,\mu)} q^{\deg(\mathsf{s})}. \tag{12.1.11}$$

We will see that this is not an exact fit for our needs, but nevertheless it serves as a good warm-up to the problem.

Example 12.1.12. Let $p = 5$ and $n = 5$. The ladder tableaux are depicted in Figure 12.5. We record the polynomials $\sum_{\mathsf{s} \in \mathsf{CStd}(\lambda,\mu)} q^{\deg(\mathsf{s})}$ for $\lambda, \mu \in \Gamma_5^+$ in the following matrix

	(5)	(4,1)	(3,2)	(3,1²)	(2²,1)	(2,1³)
(5)	1	0	0	0	0	0
(4,1)	q	1	0	0	0	0
(3,2)	0	0	1	0	0	0
(3,1²)	0	q	0	1	0	0
(2²,1)	0	0	0	0	1	0
(2,1³)	0	0	0	q	0	1
(1⁵)	0	0	0	0	0	q

(12.1.13)

It is incredibly rare that we can calculate the entries of this matrix so easily. In particular, this matrix is uni-triangular with all non-diagonal entries belonging to $q\mathbb{Z}[q]$. We will see that neither of these things happens in general (but this is okay!).

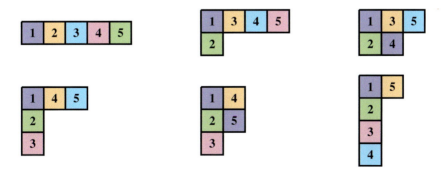

Fig. 12.5: The ladder tableaux for $n = 5$ and $p = 5$. Each of these corresponds to an entry equal to 1 in the matrix in (12.1.13).

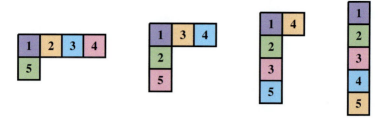

Fig. 12.6: The tableaux belonging to CStd(λ, μ) for $\lambda \neq \mu$ for $n = 5$ and $p = 5$. Each of these corresponds to an entry equal to q in the matrix in (12.1.13).

Example 12.1.14. For $p = 5$, the partition $\lambda = (8, 6, 4, 3^2, 2, 1^3)$ is p-regular and the ladder tableau is depicted in Figure 12.7. The ladder residue sequence for λ is

$$(0, 4, 3, 2, 1, 1, 0, 0, 4, 4, 3, 3, 2, 2, 2, 1, 1, 0, 4, 3, 3, 2, 1, 4, 3, 0, 4, 1, 2).$$

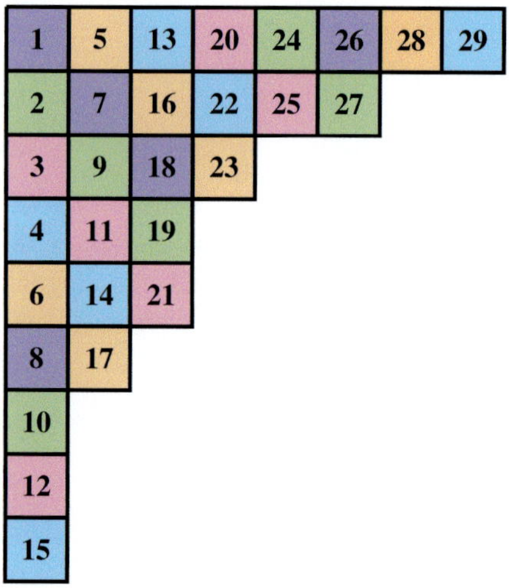

Fig. 12.7: The ladder tableau of shape $\lambda = (8, 6, 4, 3^2, 2, 1^3)$ for $p = 5$.

Recall that in the Kazhdan–Lusztig-setting, the matrix of path-counting polynomials was uni-triangular and that we factorised this matrix as a product of uni-triangular matrices. This is not the case for our matrices of coloured-tableaux-counting polynomials as we will now illustrate in an example.

Example 12.1.15. Let $p = 2$, Notice that the ladder numbers are very easy in this case: $L_2[x, y] = 2 - x - y$. The ladder tableaux for $n = 6$ are depicted in Figure 12.8

12.1 The tableaux-theoretic LLT algorithm

and the corresponding ladder sequences are

$$(0, -1, -2, -3, -4, -5) \quad (0, -1, -1, -2, -3, -4)$$
$$(0, -1, -1, -2, -2, -3) \quad (0, -1, -1, -2, -2, -2)$$

respectively. Notice that the condition that the ladder numbers are weakly decreasing ensures that we fill in the tableau *"one north-easterly to south-westerly diagonal at a time"* and that the maximality condition (in the dominance ordering) ensures that we fill in these boxes starting from the north-east and finishing at the south-west.

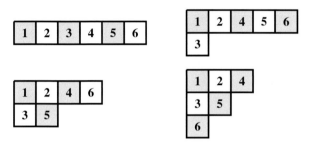

Fig. 12.8: The ladder tableaux for $n = 6$ and $p = 2$.

We notice that there are a total of 4 tableaux in $\mathsf{Std}(4, 2)$ whose residue sequences are the same as that of the ladder tableau $\mathsf{s}_{(4,2)}$, these tableaux are pictured in Figure 12.9 and have degrees $-2, 0, 0, 2$ respectively. Therefore the coloured-tableaux-counting polynomial of shape $(4, 2)$ and weight $(4, 2)$ is equal to

$$\sum_{\mathsf{s} \in \mathsf{CStd}((4,2),(4,2))} q^{\deg(\mathsf{s})} = q^2 + 2 + q^{-2} = (q + q^{-1})^2.$$

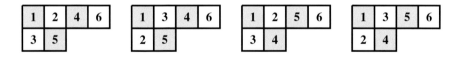

Fig. 12.9: The four tableaux in $\mathsf{CStd}((4, 2), (4, 2))$. Notice that these tableaux are obtained from one another by swapping the second and third entries and the fourth and fifth entries.

We similarly calculate the coloured-tableaux-counting polynomials

$$\sum_{\mathsf{s} \in \mathsf{CStd}((3,2,1),(3,2,1))} q^{\deg(\mathsf{s})} = q^4 + 3q^2 + 4 + 3q^{-2} + q^{-4} = (q + q^{-1})^2(q^2 + 1 + q^{-2}),$$

the corresponding tableaux are depicted in Figure 12.10. Finally, we have that

$$\sum_{\mathsf{s}\in\mathsf{CStd}((6),(6))} q^{\deg(\mathsf{s})} = 1 \qquad \sum_{\mathsf{s}\in\mathsf{CStd}((5,1),(5,1))} q^{\deg(\mathsf{s})} = q + q^{-1}.$$

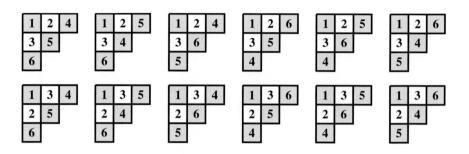

Fig. 12.10: The 12 tableaux in $\mathsf{CStd}((3,2,1),(3,2,1))$ differ by permuting the entries $4,5,6$ amongst themselves and the entries $2,3$ amongst themselves.

Exercise 12.1.16. Let $p = 2$. Calculate the ladder tableau s_μ for $\mu = (9,6,5,4,2,1)$ and the polynomial $\sum_{\mathsf{s}\in\mathsf{CStd}(\mu,\mu)} q^{\deg(\mathsf{s})}$.

From the above examples, we conclude that the matrix of coloured-tableaux-counting polynomials of (12.1.11) is far from uni-triangular (in contrast with the analogous matrix of path-counting polynomials in Definition 8.2.9). We will now provide a closed combinatorial interpretation for the diagonal entries of this matrix.

Definition 12.1.17. For μ a p-regular partition, we let $\mathsf{s}_\mu \in \mathsf{Std}(\mu)$ denote the ladder tableau of shape μ with ladder sequence

$$\underbrace{0}_{\nu_1}, \underbrace{-1,-1,\dots,-1}_{\nu_2}, \underbrace{-2,-2,\dots,-2}_{\nu_3}, \underbrace{-3,-3,\dots,-3}_{\nu_4}, \dots$$

where $\mathrm{Lad}_p(\mu) := (\nu_1, \nu_2, \dots)$ counts the multiplicity of each distinct ladder number in s_μ in order in which they appear.

Example 12.1.18. For the tableaux in Example 12.1.15 with $p = 2$, we have that

$$[\mathrm{Lad}_2(6)] = (1,1,1,1,1,1) \qquad [\mathrm{Lad}_2(5,1)] = (1,2,1,1,1,0)$$
$$[\mathrm{Lad}_2(4,2)] = (1,2,2,1,0) \qquad [\mathrm{Lad}_2(3,2,1)] = (1,2,3,0,0,0).$$

Definition 12.1.19. We define the self-dual quantum integers and factorials by setting $[0]_q = 1$ and for $n \in \mathbb{Z}_{\geqslant 0}$, we set

$$[n]_q = q^{n-1} + q^{n-3} + \dots + q^{3-n} + q^{1-n} \qquad [n]_q! = [n]_q [n-1]_q \dots [2]_q [1]_q.$$

For $\mu = (\mu_1, \mu_2, \dots)$, we set $[\mu]_q! = [\mu_1]_q! [\mu_2]_q! \dots [\mu_r]_q!$.

12.1 The tableaux-theoretic LLT algorithm

Example 12.1.20. For $\mu = (8, 6, 4, 3^2, 2, 1^3)$ and with $p = 5$, the ladder tableau is depicted in Example 12.1.14 and we have that

$$[\text{Lad}_5(\mu)] = (1, 1, 1, 1, 2, 2, 2, 2, 3, 2, 1, 1, 2, 1, 1, 0, 1, 1, 0, 0, 1, 1, 0, 0, 1, 0, 0, 0, 1)$$

and therefore

$$[\text{Lad}_5(\mu)]_q! = [3]_q!([2]_q!)^6 = (q^2 + 1 + q^{-2})(q + q^{-1})^7.$$

Example 12.1.21. Continuing with Example 12.1.18, we have that

$$[\text{Lad}_2(6)]_q! = [1]_q![1]_q![1]_q![1]_q![1]_q![1]_q! = 1$$
$$[\text{Lad}_2(5, 1)]_q! = [1]_q![2]_q![1]_q![1]_q![1]_q! = q + q^{-1}$$
$$[\text{Lad}_2(4, 2)]_q! = [1]_q![2]_q![2]_q![1]_q! = (q + q^{-1})^2$$
$$[\text{Lad}_2(3, 2, 1)]_q! = [1]_q![2]_q![3]_q! = (q + q^{-1})^2(q^2 + 1 + q^{-2}).$$

which the reader should compare against the polynomials in Example 12.1.15.

Exercise 12.1.22. Let $\Bbbk = \mathbb{F}_p$ and $\mu \in \Gamma_r^+$. Prove that

$$[\text{Lad}_p(\mu)]_q! = \sum_{\mathsf{s} \in \text{CStd}(\mu, \mu)} q^{\deg(\mathsf{s})}.$$

We are now ready to provide the main definition of this section, one should compare the following with Definition 8.2.9.

Definition 12.1.23. We define

$$(\nabla_{\lambda,\mu}(q))_{\lambda \in \Gamma_r^+, \mu \in \Gamma_r^k} \qquad \nabla_{\lambda,\mu}(q) = \frac{1}{[\text{Lad}_p(\mu)]_q!} \sum_{\mathsf{s} \in \text{CStd}(\lambda, \mu)} q^{\deg(\mathsf{s})}$$

which is a lower uni-triangular matrix. This matrix can be factorised *uniquely* as a product $\nabla = MA$ of lower uni-triangular matrices

$$M := (m_{\lambda,\nu}(q))_{\lambda \in \Gamma_r^+, \mu \in \Gamma_r^k} \qquad A := (a_{\nu,\mu}(q))_{\nu \in \Gamma_r^+, \nu \in \Gamma_r^k}$$

such that $m_{\lambda,\nu}(q) \in q\mathbb{Z}[q]$ for $\lambda \neq \nu$ and $a_{\nu,\mu}(q) \in \mathbb{Z}[q + q^{-1}]$. A recursive algorithm for this matrix factorisation is given by setting $a_{\lambda,\lambda}(q) = 1 = m_{\lambda,\lambda}(q)$ and defining the polynomials

$$a_{\lambda,\mu}(q) \in \mathbb{Z}[q + q^{-1}] \qquad m_{\lambda,\mu}(q) \in q\mathbb{Z}[q]$$

by induction on the dominance order $\triangleleft$ as follows

$$a_{\lambda,\mu}(q) + m_{\lambda,\mu}(q) = \frac{1}{[\text{Lad}_p(\mu)]_q!} \sum_{\mathsf{s} \in \text{CStd}(\lambda, \mu)} q^{\deg(\mathsf{s})} - \sum_{\lambda \triangleleft \nu \triangleleft \mu} m_{\lambda,\nu}(q) a_{\nu,\mu}(q).$$

The polynomials $m_{\lambda,\mu}(q)$ are called the **LLT polynomials** associated to λ, μ.

Over the next few sections we will discuss the role played by the LLT polynomials in the representation theory of symmetric groups. For now, we simply give a few illustrative examples of this construction.

Example 12.1.24. Notice that the factorisation of the matrix from Example 12.1.12 is easy: we have that $(\nabla_{\lambda,\mu}(q))_{\lambda \in \Gamma_r^+, \mu \in \Gamma_r^k} = (M_{\lambda,\mu}(q))_{\lambda \in \Gamma_r^+, \mu \in \Gamma_r^k}$.

Example 12.1.25. We record the uni-triangular matrix of polynomials

$$\frac{1}{[\text{Lad}_2(\mu)]_q!} \sum_{t \in \text{CStd}(\lambda,\mu)} q^{\deg(t)} \tag{12.1.26}$$

for $\lambda \in \Gamma_6^+$ and $\mu \in \Gamma_6^+$ in Table 12.1.

	(6)	(5, 1)	(4, 2)	(3, 2, 1)
(6)	1	0	0	0
(5, 1)	q	1	0	0
(4, 2)	1	q	1	0
(3, 2, 1)	0	0	0	1
(4, 1²)	$2q$	q^2	q	0
(3²)	q	0	q	0
(2³)	q^2	0	q^2	0
(3, 1³)	$2q^2$	q	q^2	0
(2², 1²)	q^3	q^2	q^3	0
(2, 1⁴)	q^2	q^3	0	0
(1⁶)	q^3	0	0	0

Table 12.1: The polynomials from (12.1.26) recorded in matrix form.

In more detail, let's consider the 2nd column of the matrix in Table 12.1 indexed by the partition $(5, 1)$. Each non-zero entry corresponds to a pair of tableaux of a given shape λ, both of which with residue sequence equal to 0, 1, 1, 0, 1, 0. For example, the intersection of $(4, 2)$th row and $(5, 1)$th column corresponds to the pair

$$\begin{array}{|c|c|c|c|} \hline 1 & 3 & 4 & 6 \\ \hline 2 & 5 \\ \cline{1-2} \end{array} \qquad \begin{array}{|c|c|c|c|} \hline 1 & 2 & 4 & 6 \\ \hline 3 & 5 \\ \cline{1-2} \end{array}$$

which have degree 2 and 0 respectively and we have that

$$\sum_{t \in \text{CStd}((4,2),(5,1))} q^{\deg(t)} = q^2 + q^0 = q(q + q^{-1}) = q \cdot [\text{Lad}_2(5, 1)]_q!$$

and we hence we obtain a q in the $(4, 2)$th row intersected with the $(5, 1)$th column.

12.1 The tableaux-theoretic LLT algorithm

For another example, the intersection of the $(2^2, 1^2)$th row with the $(5, 1)$th column corresponds to the pair

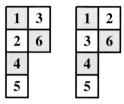

which have degree 3 and 1 respectively and we have that

$$\sum_{t \in \mathrm{CStd}((2^2,1^2),(5,1))} q^{\deg(t)} = q^1 + q^3 = q^2(q + q^{-1}) = q^2 \cdot [\mathrm{Lad}_2(5,1)]_q!$$

and so we obtain a q^2 in the $(2^2, 1^2)$th row intersected with the $(5, 1)$th column.

Exercise 12.1.27. Compute all the tableaux in $\mathrm{CStd}(\lambda, \mu)$ which contribute entries to the second column of the matrix in Table 12.1. Notice that they come in pairs which differ by swapping the entries 2 and 3. Do the same for the third column of the matrix in Table 12.1, but notice that the tableaux now come in quadruplets which differ by swapping the entries 2 and 3 or swapping the entries 4 and 5.

Example 12.1.28. Continuing with Example 12.1.25, the factorisation of the ∇-matrix is as a product of a matrix of self-dual polynomials:

	(6)	(5,1)	(4,2)	(3,2,1)
(6)	1	0	0	0
(5,1)	0	1	0	0
(4,2)	1	0	1	0
(3,2,1)	0	0	0	1

and a matrix whose off-diagonal entries are all of strictly positive degree:

	(6)	(5,1)	(4,2)	(3,2,1)
(6)	1	0	0	0
(5,1)	q	1	0	0
(4,2)	0	q	1	0
(3,2,1)	0	0	0	1
(4,1²)	q	q^2	q	0
(3²)	0	0	q	0
(2³)	0	0	q^2	0
(3,1³)	q^2	q	q^2	0
(2²,1²)	q^3	q^2	q^3	0
(2,1⁴)	q^2	q^3	0	0
(1⁶)	q^3	0	0	0

whose entries are the LLT polynomials for $\mathbb{F}_2\mathfrak{S}_6$, by definition.

Exercise 12.1.29. Let $n = 4$ and $p = 2$. Construct the ladder tableaux and factorise the ∇-matrix in order to show that the LLT polynomials are given by

	(4)	(3,1)	(2,2)
(4)	1	0	0
(3,1)	q	1	0
(2,2)	0	0	1
(2,1²)	q	q^2	0
(1⁴)	q^2	0	0

The following conjecture was first formulated by Gordon James in [Jam90] (in more generality than we discuss here). For now, the reader should merely observe that James' and Andersen's conjectures both propose that certain polynomials control the structure of $k\mathfrak{S}_r$.

Conjecture 12.1.30. *The decomposition numbers of $\mathbb{F}_p\mathfrak{S}_r$ for $p > \sqrt{r}$ are given by the (evaluations at 1 of the) LLT polynomials.*

Exercise 12.1.31. Using you answer to Exercise 12.1.29, verify that the decomposition numbers of $\mathbb{F}_2\mathfrak{S}_4$ are given by the LLT polynomials (with $q \mapsto 1$).

All of the above is compatible with truncating to any closed subset of Γ_r^+ (in much the same way as in the Kazhdan–Lusztig-setting).

Exercise 12.1.32. Calculate the coloured-tableaux corresponding to the entries in the matrix in Table 12.2.

	(15,0,0)	(14,1,0)	(13,1,1)	(10,5,0)	(9,6,0)	(10,4,1)	(9,4,2)	(8,6,1)	(8,5,2)	(5³)
(15,0,0)	1	·	·	·	·	·	·	·	·	·
(14,1,0)	q	1	·	·	·	·	·	·	·	·
(13,1,1)	·	q	1	·	·	·	·	·	·	·
(10,5,0)	1	q	·	1	·	·	·	·	·	·
(9,6,0)	$2q$	q^2	q	q	1	·	·	·	·	·
(10,4,1)	q	·	·	q	·	1	·	·	·	·
(9,4,2)	$2q^2$	q	·	$1+q^2$	q	q	1	·	·	·
(8,6,1)	q^2	·	·	q^2	q	q	·	1	·	·
(8,5,2)	q^3	q^3	q	$q+q^3$	q^2	q^2	q	q	1	·
(5³)	·	·	q^2	·	·	·	·	·	q	1

Table 12.2: The ∇^Π matrix for $\Pi \subset \Gamma_{3,15}$ obtained by intersecting with the p-block containing the partition $(15, 0, 0)$ for $p = 5$.

12.2 Bringing together LLT and Kazhdan–Lusztig polynomials

In the previous section we highlighted the similarities in the definitions of the Kazhdan–Lusztig and LLT polynomials. In this section, we explain this apparent coincidence by proving that the Kazhdan–Lusztig polynomials of type $(W, P) = (\widehat{\mathfrak{S}}_n, \mathfrak{S}_n)$ are equal to the LLT polynomials for $\mathbb{F}_p \mathfrak{S}_r$ labelled by $\Gamma^+_{n,r} \subseteq \Gamma^+_r$ providing that $p > n$. Notice the different symmetric groups involved: the LLT polynomials are defined in terms of the tableaux combinatorics of a finite symmetric group of rank $r \in \mathbb{Z}_{\geqslant 0}$; the Kazhdan–Lusztig polynomials are defined in terms of the path combinatorics of an *affine* symmetric group of rank $n \in \mathbb{Z}_{\geqslant 0}$. We begin with a trivial, but important observation:

Lemma 12.2.1. *Let* $\Bbbk = \mathbb{F}_p$ *with* $p > n$. *The ladder tableau of any* $\mu \in \Gamma^+_{n,r}$ *is equal to the least dominant tableau* $\mathsf{t}_\mu \in \mathrm{Std}(\mu)$. *Moreover,* $[\mu]_q! = 1$ *for any* $\mu \in \Gamma^+_{n,r}$.

The upshot of this lemma is that our combinatorics becomes slightly easier. We celebrate this in the only way we know how: more examples!

Example 12.2.2. Let $p = 5$ and $\mu = (13, 1, 1) \in \Gamma^+(3, 13)$. The four tableaux belonging to $\mathrm{Std}(\lambda, \mathsf{t}_{(13,1,1)})$ for $\lambda \in \Gamma^+(3, 15)$ are depicted in Figure 12.11. The resulting path-polynomials are

$$\sum_{\mathsf{t} \in \mathrm{CStd}(\lambda, (13, 1^2))} q^{\deg(\mathsf{t})} = \begin{cases} 1 & \text{if } \mu = (13, 1, 1) \\ q & \text{if } \mu = (10, 4, 1) \text{ or } (8, 5, 2) \\ q^2 & \text{if } \mu = (5, 5, 5). \end{cases} \quad (12.2.3)$$

We now wish to shift our focus by re-imagining what a tableau is for us. Instead of thinking of $\mathsf{t} \in \mathrm{Std}(\lambda)$ for $\lambda \in \Gamma^+_{n,r}$ as a filling of boxes of λ, we can think of it as the series of partitions

$$\mathrm{Shape}(\mathsf{t}\!\downarrow_{\{1,\ldots,k\}})_{1 \leqslant k \leqslant r} \in \Gamma^+_{n,r}.$$

For example, in this manner the tableau $\mathsf{t}_{(10,5,0)}$ can be though of as the sequence

$$\emptyset \to (1) \to (1^2) \to (2, 1) \to (2, 2) \to (3, 2) \to (3, 3) \to (4, 3)$$
$$\to (4, 4) \to (5, 4) \to (5, 5) \to (6, 5) \to (7, 5) \to (8, 5) \to (9, 5) \to (10, 5).$$

Identifying this sequence of partitions with the corresponding sequence of points in the Euclidean geometry, this allows us to think of a standard tableau as a path in $\widehat{\mathbb{E}}^+_n$. For example, the path corresponding to $\mathsf{t}_{(10,5,0)}$ is depicted in Figure 12.12.

Example 12.2.4. The polynomials in the $(13, 1, 1)$th column of the ∇-matrix were calculated in (12.2.3) and the corresponding tableaux were given in Figure 12.11. Compare this with the third column of the matrix in Table 8.1.

Exercise 12.2.5. Compare your answers to Exercise 12.1.32 and Exercise 8.3.4.

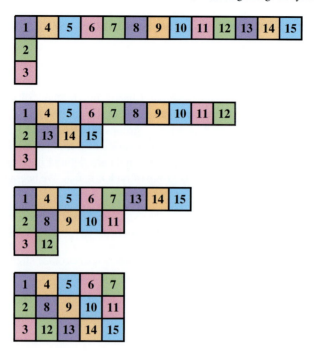

Fig. 12.11: The four tableaux belonging to $\mathrm{Std}(\lambda, \mathsf{t}_{(13,1,1)})$ for $\lambda \in \Gamma^+(3, 15)$. These tableaux are of degree 0, 1, 1, 2 respectively. Compare with Figure 12.13.

Now, with our new shifted perspective we wish to consider the elements of $\mathrm{Std}(\lambda, \mathsf{s})$ for s a reduced path (we usually set $\mathsf{s} = \mathsf{t}_\mu = \mathsf{s}_\mu$, as in Lemma 12.2.1). We notice that a partition has two addable nodes of the same residue if and only if it lies on a hyperplane in our Euclidean geometry. Thus $\mathrm{Std}(\lambda, \mathsf{s})$ consists of the paths which can be obtained from s by reflection through these hyperplanes. This is illustrated in the examples in Figures 12.11 and 12.13.

Remark 35. We believe that "definition by example" as in Figure 12.13 is easier than us defining the "reflection of a path through a hyperplane" (a rigorous definition can be found in [BCH23, Section 2.3]).

Exercise 12.2.6. Construct the 6 tableaux belonging to $\mathrm{Std}(\lambda, \mathsf{t}_{(14,1)})$ and match-them-up with the paths depicted in Figure 12.14.

Finally, we are ready to state the main theorem of this section:

Theorem 12.2.7. *Let* $\Bbbk = \mathbb{F}_p$. *Let* $\lambda, \mu \in \Gamma_{n,r}^+$ *be regular points. We have that the LLT polynomials coincide with the Kazhdan–Lusztig polynomials, that is:*

$$n_{w_\lambda, w_\mu}(q) = m_{\lambda,\mu}(q).$$

12.2 Bringing together LLT and Kazhdan–Lusztig polynomials

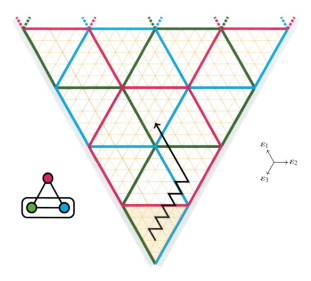

Fig. 12.12: Here we depict the first tableau $t_{(10,5,0)}$ as a path.

Sketch proof. By our definition of t_μ (for μ a regular point) we note that it only ever lies on at most one hyperplane at a time. Thus we can identify the reduced tableau t_μ with the sequence of coloured hyperplanes through which it passes in an intuitive manner. For example, the path corresponding to of $t_{(10,5,0)}$ is depicted in Figure 12.12, where it passes through the four hyperplanes labelled by σ, τ, ρ, τ in order. For another example, the path corresponding to of $t_{(13,1,1)}$ is depicted in Figure 12.13, where it passes through the four hyperplanes labelled by $\sigma, \rho, \tau, \sigma$ in order.

The sequence of hyperplanes passed through by t_μ corresponds to a reduced path in the Bruhat graph $^P W$ for $(W, P) = (\widehat{\mathfrak{S}}_n, \mathfrak{S}_n)$ and in this manner we can identify

$$\mathrm{Std}(\lambda, t_\mu) \leftrightarrow \mathrm{Path}(w_\lambda, \mathsf{T}_{w_\mu}). \tag{12.2.8}$$

In this manner we match-up $t_{(14,1)} \leftrightarrow \mathsf{U}^1_\sigma \mathsf{U}^1_\rho \mathsf{U}^1_\tau \mathsf{U}^1_\sigma \mathsf{U}^1_\rho$ and the four tableaux of Figure 12.11 with $\mathsf{U}^1_\sigma \mathsf{U}^1_\rho \mathsf{U}^1_\tau \mathsf{U}^1_\sigma$, $\mathsf{U}^1_\sigma \mathsf{U}^1_\rho \mathsf{U}^1_\tau \mathsf{U}^0_\sigma$, $\mathsf{U}^1_\sigma \mathsf{U}^0_\rho \mathsf{U}^0_\tau \mathsf{D}^0_\sigma$, and $\mathsf{U}^1_\sigma \mathsf{U}^0_\rho \mathsf{U}^0_\tau \mathsf{D}^1_\sigma$ respectively. Moreover, the map of (12.2.8) is a graded bijection. Therefore $\Delta_{\lambda,\mu} = \nabla_{\lambda,\mu}$ for any λ, μ in the principal block. The result now follows as the factorisation of this matrix

$$NB = \Delta = \nabla = MA$$

is uniquely determined by the property that the off-diagonal entries of N, M belong to $q\mathbb{Z}[q]$ and the entries of A, B belong to $\mathbb{Z}[q + q^{-1}]$. □

Exercise 12.2.9. Verify that the bijection of (12.2.8) does indeed respect the grading.

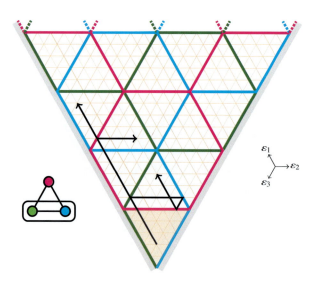

Fig. 12.13: Here we depict the first three tableaux from Figure 12.11 as paths belonging to the same W-orbit. (Notice that we do not depict the first three steps of this path, which go from $\varnothing$ to (1^3), this is because we have projected onto a 2-dimensional space where $\varnothing$ to (1^3) occupy the same position — which is ugly to picture.)

We now chase James' conjecture through the combinatorics of Theorem 12.2.7 in order to formulate its "toy version", which is entirely equivalent to the toy version of Andersen's conjecture. The only slight tweak we must do in the combinatorics is to make $r = p\binom{n}{2}$ in order that all partitions around the Steinberg partition appear (recall that the Steinberg partition is a partition of $(p-1)\binom{n}{2}$).

Conjecture 12.2.10 (The "toy version" of James' conjecture). *Let $p\binom{n}{2} = r < p^2$. We have that*

$$[\Delta_{\Bbbk \mathfrak{S}_r}((p-1)\rho_n + x\rho_n) : L_{\Bbbk \mathfrak{S}_r}((p-1)\rho_n + y\rho_n)]_{\Bbbk \mathfrak{S}_r} = n_{x,y}(q)|_{q=1}$$

where $n_{x,y}(q)$ is the Kazhdan–Lusztig polynomial associated to $x, y \in \mathfrak{S}_n$.

Remark 36. The LLT polynomials for *non-principal blocks* also have a counterpart in the setting of "singular Kazhdan–Lusztig theory", this involves looking at double cosets of Coxeter groups. This is much more complicated than "regular" or "non-singular Kazhdan–Lusztig theory" and is beyond the realms of this book. However, it is easy enough to calculate some examples of these LLT polynomials by hand. For $p = 5$ we invite the reader to calculate the LLT polynomials $m_{\lambda,\mu}$ for $\lambda, \mu \in \{(19, 4), (18, 5), (18, 4, 1), (14, 9), (13, 10), (13, 9, 1)\}$.

12.3 The quiver Hecke algebra

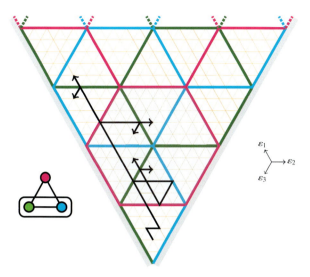

Fig. 12.14: We depict the six tableaux from $\mathrm{Std}(\lambda, \mathsf{t}_{(14,1)})$ as paths belonging to the same W-orbit. Compare this with the second column of the matrix in Table 8.1. In grey we highlight the region around the Steinberg partition $(8, 4, 0)$.

12.3 The quiver Hecke algebra

We now lift the gradings and combinatorics of coloured tableaux to the level of graded generators and bases of the group algebras of symmetric groups, $\Bbbk\mathfrak{S}_n$. We do this using the diagrammatics of quiver Hecke algebras (also known as Khovanov–Lauda–Rouquier algebras, who discovered the algebras in [KL09] and [Rou]).

Definition 12.3.1. Fix a prime $p > 0$. We define a **KLR-diagram** to be a permutation diagram in which: each strand is coloured by an element of $\mathbb{F}_p$ (its **residue**) and each solid strand can carry a finite number of dots. We further require that there are no triple points or tangencies and that no dot can lie on a crossing.

The rank of a KLR-diagram is defined to be the number of strands in the diagram. For $p = 5$ an example of a quiver Hecke diagram of rank 9 is drawn in Figure 12.15. We define the northern (respectively southern) residue sequence of a diagram to be the sequence of residues read from left to right. For example, the diagram in Figure 12.15 has southern residue sequence $(0, 4, 3, 3, 1, 2, 0, 1, 4)$.

In order to discuss these diagrams in more detail, we now fix some notation. We set $e(\underline{j})$ for $\underline{j} \in (\mathbb{Z}/p\mathbb{Z})^r$ to be the diagram with r vertical straight lines and northern/southern residue sequence given by $\underline{j} \in (\mathbb{Z}/p\mathbb{Z})^r$. We set $e(\underline{j})\psi_q$ for $1 \leqslant q < r$ to be the diagram with a single crossing of the q and $(q + 1)$th strands and with northern residue sequence given by $\underline{j} \in (\mathbb{Z}/p\mathbb{Z})^r$ (and southern residue sequence obtained from $\underline{j} \in (\mathbb{Z}/p\mathbb{Z})^r$ by swapping the residues in the q and $(q+1)$th positions). Finally, we set $e(\underline{j})y_q$ for $1 \leq q \leq r$ to be the diagram with r vertical straight lines and a single dot placed on the qth strand (and northern/southern residue sequence given by $\underline{j} \in (\mathbb{Z}/p\mathbb{Z})^r$).

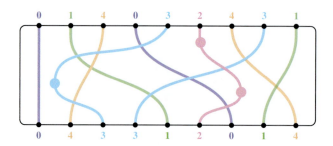

Fig. 12.15: For $p = 5$ an example of a quiver Hecke diagram of rank 9 with colouring from $\mathbb{F}_5$ as in Figure 12.1.

Given two diagrams D_1, D_2 of fixed rank $r \in \mathbb{Z}_{\geq 0}$, we define $D_1 \circ D_2$ to be the vertical concatenate of D_1 above D_2 if the southern residue sequence of D_1 is equal the northern residue sequence of D_2, and 0 otherwise.

Definition 12.3.2 ([KL09] and [Rou]). Fix $\Bbbk$ a field of characteristic $p > 2$. We define the **quiver Hecke** (or **KLR**) algebra $\mathcal{H}_r$ to be the $\Bbbk$-algebra generated by the KLR-diagrams of rank $r \in \mathbb{Z}_{\geq 0}$ with multiplication given by $\circ$ subject to the following relations together with their flips through the horizontal axis:

(R1) Any diagram may be deformed isotopically; that is, by a continuous deformation of the diagram which avoids tangencies, double points and dots on crossings.
(R2) We now consider the effect of pulling a dot through a strand. Any dot on a k-strand pass through a crossing of j- and k-strands providing $j \neq k$

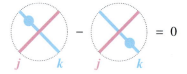

whereas in the like-residue-labelled case we get an error term as follows

12.3 The quiver Hecke algebra

(R3) Double crossings. For $|j - k| \neq 1$ we can undo double-crossings as follows

whereas for adjacent residues $|j - k| = 1$ we obtained a pair of dotted terms

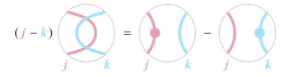

(R4) Braid relations. For adjacent residues $|j - k| = 1$ we have a deformed version of the classical braid relation

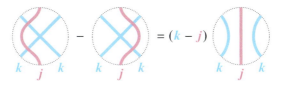

and for any other triple $i, j, k \in \mathbb{Z}/p\mathbb{Z}$, the classical braid relation holds

$$\bowtie - \bowtie = 0.$$

(R5) Finally we have the cyclotomic relations:

$$e(\underline{j}) = 0 \text{ for } \underline{j} \neq \text{res}(t) \text{ for some } t \in \text{Std}(\lambda) \text{ for } \lambda \in \Gamma_r^+ \text{ and}$$
$$y_1 e(\underline{j}) = 0 \text{ for } \underline{j} \in (\mathbb{Z}/p\mathbb{Z})^r.$$

The only way to get to grips with these relations is by getting your hands dirty with examples, which we will do momentarily. What makes this effort worth while is the huge amount of representation theoretic information encoded in this presentation, for example we have the following very important theorem.

Theorem 12.3.3 ([KL09] and [Rou]). *The algebra $\mathcal{H}_r$ is $\mathbb{Z}$-graded, with this grading defined on the local neighbourhoods of a diagram as follows*

$$\deg(e(\underline{i})) = 0 \quad \deg(y_r) = 2 \quad \deg(\psi_r e(\underline{i})) = \begin{cases} -2 & \text{if } i_r = i_{r+1} \\ 1 & \text{if } i_r = i_{r+1} \pm 1 \\ 0 & \text{otherwise.} \end{cases}$$

Example 12.3.4. In this example we consider, in great detail, the rank $r = 3$ case with the prime $p = 3$. We colour our residues using the colouring in Figure 12.16. We have seen earlier in this book that when one defines an algebra via a presentation it is not immediately obvious how "big" the algebra is; therefore it is helpful to try and derive a basis of the algebra. This is the guiding principle of this example.

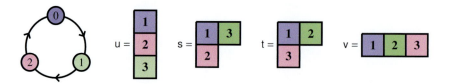

Fig. 12.16: For $p = 3$ we colour the field $\mathbb{F}_3$ and the standard tableaux of rank $n = 3$.

We begin by considering the KLR-idempotents in $\mathcal{H}_3$. The standard tableaux of rank 3 are depicted in Figure 12.16 and the residue sequences of all these tableaux are of the form $(0, 1, 2)$ or $(0, 2, 1)$. Thus, by the cyclotomic relations, we conclude that the only non-zero KLR-idempotents are the diagrams

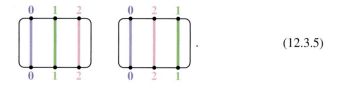

. (12.3.5)

We now ask "how many dots can I put on each strand in an idempotent?". We know that putting a single dot on the first strand kills these idempotents (by the cyclotomic relation). Can the second and third strands "support" having dots put on them? Or do they die?

If we put a single dot on the second strand of the leftmost idempotent in (12.3.5), then we have that

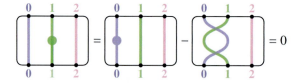

where the first equality follows by the double-crossing relations; the latter equality follows by the cyclotomic relations (note that the latter diagram has a 1-strand all the way to the left). We leave it as an exercise for the reader to verify (similarly, up to sign change) that $y_2 e(0, 2, 1) = 0$.

12.3 The quiver Hecke algebra

If we put two-dots on the third strand of the leftmost idempotent in (12.3.5), then we have that

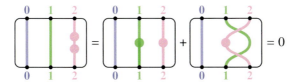

where the first equality follows from the double-crossing and dot-through-strand relations; the latter follows from our above observations on placing one dot on the middle strand. We leave it as an exercise for the reader to verify (similarly, up to sign change) that $y_3^2(0, 2, 1) = 0$.

Thus we can conclude that (for $p = 3$ and $n = 3$) the KLR-idempotents can support at most 1 dot and said dot must lie on the third strand. In fact, these elements are non-zero and so the following elements are linearly independent and represent all ways of putting dots on KLR-idempotents

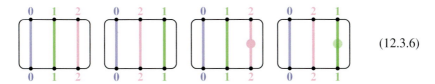

(12.3.6)

What about diagrams which contain crossings? We claim that the following diagrams, and their flips through the horizontal axis, are all zero

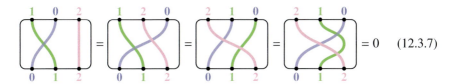

(12.3.7)

where here we have applied a permutation on top of a good idempotent (namely $e(0, 1, 2)$) and this has resulted in a *bad* idempotent (namely $e(1, 0, 2)$, $e(1, 2, 0)$, ...) which is zero. We leave it as an exercise for the reader to verify the analogous relations with $e(0, 2, 1)$.

Thus we conclude that each of these idempotents can only support having their final two strands permuted. We thus obtain

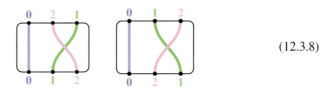

(12.3.8)

and we note that these diagrams (together with those of (12.3.9)) are linearly independent by the idempotent relations. One might ask if any further *double-crossings* might result in genuinely new diagrams. By (12.3.7) the only double-crossings left to consider are those involving the finals two strands, and we have that

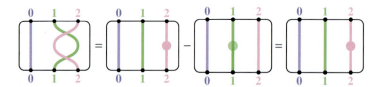

where the former equality follows by the double-crossing relation and the latter equality follows by arguments above. Thus we have expressed the (apparently new) double-crossing diagram in terms of diagrams we have already encountered. The diagram obtained by double-crossing the final two strands of $(0, 2, 1)$ can be handled similarly.

Finally, placing a dot on any strand in a diagram in (12.3.8) annihilates said diagram. For example we have that

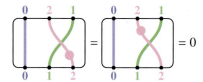

where the first equality follows from the dot-through-strand relation; the second follows since we have already shown that $y_2 e(0, 2, 1) = 0$. The other cases can be dealt with in a similar fashion.

Thus we conclude that our algebra $\mathcal{H}_3$ has graded dimension equal to

$$\dim_q(\mathcal{H}_3) = 2 + 2q + 2q^2$$

and, as an $\mathbb{F}_3$-algebra, $\mathcal{H}_3$ has basis given by the following 6 diagrams

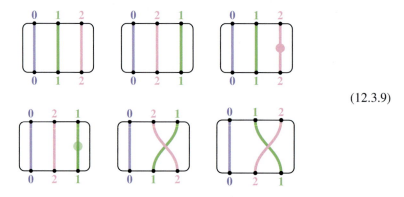

(12.3.9)

12.3 The quiver Hecke algebra

The multiplication table with respect to the basis in (12.3.9) is recorded in Figure 12.17.

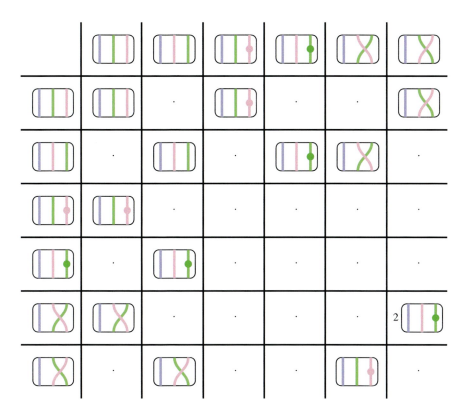

Fig. 12.17: The multiplication table for the $\mathbb{F}_3$-algebra $\mathcal{H}_3$. (Note that $2 = -1$ in $\mathbb{F}_3$.)

In the next few examples, we construct some representations of the algebra $\mathcal{H}_3$.

Example 12.3.10. The 1-dimensional "trivial" submodule $\mathbb{1} \subseteq \mathcal{H}_3$ can be constructed as follows

where every generator acts on $\mathbb{1}$ as zero with the exception of the idempotent $e(0, 1, 2)$ which acts as scalar multiplication by 1. (You can check this action using Figure 12.17.)

Example 12.3.11. From Figure 12.17 we are able to read-off the full graded Alperin diagrams of the direct summands of $\mathcal{H}_3$. Each idempotent generates a 3-dimensional submodule, the crossing diagrams each generate 2-dimensional submodules of these 3-dimensional modules, and finally the dotted elements each generate 1-dimensional submodules.

It is then clear that there are two distinct 1-dimensional simple modules concentrated in degree zero; these are generated by $e(0,1,2)$ and $e(0,2,1)$; we denote the corresponding simple modules by $L(0,1,2)$ and $L(0,2,1)$ respectively. We picture one of these two direct summands in Figure 12.18.

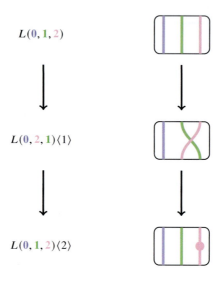

Fig. 12.18: The graded Alperin diagram of the direct summand $e(0,1,2)\mathcal{H}_3$.

Example 12.3.12. Notice that the "trivial" 1-dimensional submodule $\mathbb{1} \subseteq e(0,1,2)\mathcal{H}_3$ is isomorphic to the degree 2 graded shift of the simple $L(0,1,2)$ and that this isomorphism is given by multiplying with the degree 2 element y_3.

Exercise 12.3.13. Calculate the graded Alperin diagram of the direct summand $e(0,2,1)\mathcal{H}_3$.

Exercise 12.3.14. Calculate the action of the KLR-generators on $e(0,1,2)\mathcal{H}_3$ and $e(0,2,1)\mathcal{H}_3$.

Exercise 12.3.15. Continue with $p = 3$. Calculate the graded dimension of the $\mathcal{H}_4$-module $y_3 e(0,2,1,0)\mathcal{H}_4$.

12.4 The Brundan–Kleshchev isomorphism theorem

We want to think of the quiver Hecke algebra not as "yet another new algebra for the reader to have to cope with" but rather as a "coloured" version of $k\mathfrak{S}_r$ which brings a hidden graded presentation of $k\mathfrak{S}_r$ to the foreground. To justify this we must realise these new generators directly within $k\mathfrak{S}_r$ itself. Our first port-of-call is the idempotent generators, which are not as foreign as one might think: we set

$$E_{\underline{j}} = \sum_{\mathrm{res}(t)=\underline{j}} F_t \in k\mathfrak{S}_r$$

for $\underline{j} \in (\mathbb{Z}/p\mathbb{Z})^r$ where the righthand-side is the sum over all the Murphy operators for these tableaux, as defined in (11.1.22). Recall that these operators were defined in terms of the Jucys–Murphy elements X_q for $1 \leqslant q \leqslant r$. These Jucys–Murphy elements are key to the construction of the remaining generators, as we now illustrate.

Theorem 12.4.1 ([BK09]). *We have an isomorphism*: $\mathcal{H}_r \to k\mathfrak{S}_r$ *given by mapping*

$$e(\underline{j}) \mapsto E_{\underline{j}},$$
$$y_q e(\underline{j}) \mapsto (X_q - j_q) E_{\underline{j}},$$
$$\psi_q e(\underline{j}) \mapsto \begin{cases} (1 + s_q(X_q - X_{q+1})) E_{\underline{j}} & \text{if } j_{q+1} = j_q + 1 \\ \left((s_q - 1) \dfrac{1}{X_q - X_{q+1} + 1}\right) E_{\underline{j}} & \text{if } j_{q+1} = j_q \\ \left(1 + (s_q - 1) \dfrac{X_q - X_{q+1}}{X_q - X_{q+1} + 1}\right) E_{\underline{j}} & \text{otherwise} \end{cases}$$

We do not really expect the reader to gain a great deal of insight from the statement of Theorem 12.4.1. We have included this theorem in order to illustrate the manner in which the KLR-generators come about from very "down to Earth" (if daunting!) calculations in the group algebra.

Example 12.4.2. This already affords us new information on $k\mathfrak{S}_3 \cong \mathcal{H}_3$. Namely, we now have the graded Alperin diagrams of the direct summands of this algebra thanks to our work in Figure 12.18.

Exercise 12.4.3. Identify the unique *graded* 2-dimensional submodule of $k\mathfrak{S}_3 \cong \mathcal{H}_3$ which is isomorphic to the 2-dimensional Specht module from Section 6.4.

Exercise 12.4.4. Calculate the multiplication table of the $\mathbb{F}_2$-algebra $\mathcal{H}_3 \cong \mathbb{F}_2\mathfrak{S}_3$ for $p = 2$. Decompose $\mathcal{H}_3$ as a direct sum of three 2-dimensional right $\mathcal{H}_3$-modules (two of which are isomorphic to each other). Compare this with the result for $p = 2$ in Section 6.4.

12.5 The Hu–Mathas cellular basis

We have already seen that $\Bbbk \mathfrak{S}_r$ has a cellular basis constructed in terms of standard tableaux. We have also seen how to incorporate the grading into this tableaux-theoretic picture, using "LLT theory". We now wish to bring these two ideas together by way of the KLR-presentation and Theorem 12.4.1. We define idempotents $e_s = e(\text{res}(t))$ for $t \in \text{Std}(\lambda)$ and quasi-idempotent (nilpotent) elements

$$y_\lambda = \prod_{k=1}^{n} y_k^{|\mathcal{A}_{t_\lambda}(k)|} e_{t_\lambda}, \qquad (12.5.1)$$

these will be the generators of our chain of cell-ideals.

Example 12.5.2. Let $p = 3$. Recall that the tableau $t_{(1^3)}$ is as follows

$$t_{(1^3)} = \begin{array}{|c|} \hline 1 \\ \hline 2 \\ \hline 3 \\ \hline \end{array}.$$

We have that $\mathcal{A}_{t_\lambda}(k) = \emptyset$ for $k = 1, 2$. For $k = 3$, we observe that we could have added the green node in the first row and so $|\mathcal{A}_{t_\lambda}(k)| = 1$. Therefore

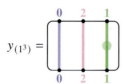

Exercise 12.5.3. Let $p = 2$. Construct the elements $y_\lambda \in \mathcal{H}_6$ for every λ a partition of 6.

Exercise 12.5.4. Provide a simple formula for relating the degree of the element $y_\lambda \in \mathcal{H}_r$ to the degree of the tableau t_λ.

Definition 12.5.5. Let $\lambda \in \Gamma_{n,r}^+$. Given a pair of paths $s, t \in \text{Std}(\lambda)$ and fix a choice $\underline{w}_t^s = s_{i_1} s_{i_2} \dots s_{i_k}$ for some minimal length word (in the Coxeter generators) for the permutation w_t^s. We define

$$\psi_t^s := e_s \psi_{i_1} \psi_{i_2} \dots \psi_{i_k} e_t$$

and we typically decorate the northern/southern vertices with the corresponding row numbers of s and t.

Example 12.5.6. For $p = 5$ we now recolour Example 11.1.10. The coloured tableaux s and t are depicted in Figure 12.19 and the corresponding coloured permutation $\psi_t^s \in \mathcal{H}_r$ is depicted in Figure 12.20.

12.5 The Hu–Mathas cellular basis

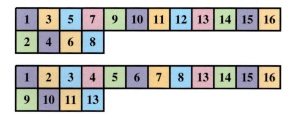

Fig. 12.19: The coloured versions of the tableaux in (11.1.11) and (11.1.12).

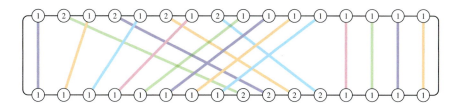

Fig. 12.20: The coloured permutation ψ_t^s for s, t as in Figure 12.19 for $p = 5$ with the colouring as in Figure 12.1.

Theorem 12.5.7 ([HM10]). *The algebra $\mathcal{H}_r$ is a graded cellular algebra with respect to the basis*
$$\{\psi_{st} = \psi_{t_\lambda}^s y_\lambda \psi_t^{t_\lambda} \mid s, t \in \mathrm{Std}(\lambda), \lambda \in \Gamma_r^+\}$$
and the anti-involution given by flipping a diagram through its horizontal axis and any total refinement of the dominance ordering on Γ_r^+.

The Hu–Mathas cellular basis for $\mathcal{H}_3 \cong \mathbb{F}_3 \mathfrak{S}_3$ is depicted in Figure 12.21.

 ▭▭▭	$\psi_{vv} = e(0, 1, 2)$	
▭▭/▭	$\psi_{ss} = e(0, 2, 1)$	$\psi_{st} = e(0, 2, 1)\psi_2$
	$\psi_{ts} = \psi_2 e(0, 2, 1)$	$\psi_{tt} = \psi_2 e(0, 2, 1)\psi_2$
 ▭/▭/▭	$\psi_{uu} = y_3 e(0, 2, 1)$	

Fig. 12.21: The Hu–Mathas graded cellular basis of $\mathcal{H}_3 \cong \Bbbk\mathfrak{S}_3$.

Definition 12.5.8. For any $t \in \mathrm{Std}(\lambda)$ we set $\psi_t := y_{t_\lambda} \psi_t^{t_\lambda} + \mathcal{H}_r^{<\lambda}$.

Example 12.5.9. We consider the $\mathbb{F}_3$-algebra $\mathcal{H}_3 \cong \mathbb{F}_3 \mathfrak{S}_3$. The graded tableaux depicted in (12.16) are of degrees 1, 0, 1, and 0 respectively. The Hu–Mathas basis for this algebra is as depicted in Figure 12.21. We note that the basis of Figure 12.21 is equal to the basis of (12.3.9), the only *apparent* difference can be reconciled as follows

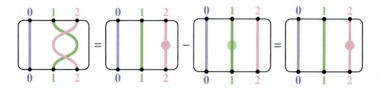

where the lefthand-side is the element ψ_{tt} and the righthand-side is the corresponding basis element in (12.3.9). The graded dimensions of the cell-modules are

$$\dim_q(\Delta_{\Bbbk\mathfrak{S}_3}(3)) = 1 \qquad \dim_q(\Delta_{\Bbbk\mathfrak{S}_3}(2,1)) = 1 + q \qquad \dim_q(\Delta_{\Bbbk\mathfrak{S}_3}(1^3)) = q$$

and we have that

$$\dim_q(\mathcal{H}_3) = 1^2 + (1+q)^2 + q^2 = 2 + 2q + 2q^2.$$

This algebra is non-negatively graded and so we can use the grading trick to immediately deduce that the decomposition matrix and the graded Alperin diagrams of the direct summands of $\mathcal{H}_3$. In particular, the simples modules have bases

$$L_{\Bbbk\mathfrak{S}_3}(\square\square\square) = \Bbbk\{\psi_s\} \qquad L_{\Bbbk\mathfrak{S}_3}(\square\!\square) = \Bbbk\{\psi_t + \mathrm{rad}\langle -, -\rangle_{\square\!\square}^{\Bbbk}\}$$

and the graded decomposition matrix is as follows.

	$L_{\Bbbk\mathfrak{S}_3}(\square\square\square)$ $L_{\Bbbk\mathfrak{S}_3}(\square\!\square)$
$\Delta_{\Bbbk\mathfrak{S}_3}(\square\square\square)$	1 0
$\Delta_{\Bbbk\mathfrak{S}_3}(\square\!\square)$	q 1
$\Delta_{\Bbbk\mathfrak{S}_3}(\square\!\square)$	0 q

The following result explains the importance of reduced tableaux in the LLT-theory of the previous section.

Proposition 12.5.10. *Let* $t \in \mathrm{Std}(\lambda)$ *be a reduced tableau. Then* $\psi_t \notin \mathrm{rad}(\Delta^{\Bbbk}(\lambda))$ *and so* $\lambda \in \Gamma_r^{\Bbbk}$. *In particular* $\dim(L^{\Bbbk}(\lambda)) \geq \#\{t \in \mathrm{Std}(\lambda) \mid t \text{ is reduced}\}$.

12.5 The Hu–Mathas cellular basis

Proof. By Proposition 6.2.22, we have $[\Delta^{\Bbbk}(\lambda) : L^{\Bbbk}(\mu)] \neq 0$ implies that $\mu \trianglerighteq \lambda$. By definition, $L^{\Bbbk}(\mu)e_t = \Delta^{\Bbbk}(\mu)e_t = 0$ for any reduced t and any $\mu \triangleright \lambda$. Now, since $\psi_t e_t = \psi_t$, this implies that ψ_t cannot belong to any simple composition factor of $\Delta^{\Bbbk}(\lambda)$ except for $L^{\Bbbk}(\lambda)$. The result follows. □

Building on our earlier work in Example 12.1.12, we now revisit the LLT polynomials we calculated for $\mathbb{F}_5\mathfrak{S}_5$, but this time we provide a structural interpretation of these polynomials in terms of the graded decomposition matrix of this algebra.

Example 12.5.11. We consider the algebra $\mathcal{H}_5 \cong \mathbb{F}_5\mathfrak{S}_5$ with the colouring as in Figure 12.1. Recall that we have already constructed the coloured tableaux necessary for running the LLT algorithm in Figures 12.5 and 12.6 and and we claimed that the (graded) decomposition matrix is as in (12.1.12).

We will see that $\mathcal{H}_5$ is positively-graded and therefore we can calculate the Gram matrices of the intersection forms (and therefore the graded decomposition matrices) using only the grading and idempotent tricks. Moreover, we have that

$$\dim_q(\Delta_{\Bbbk\mathfrak{S}_5}(5-k, 1^k)) = \binom{4}{k} + \binom{4}{k-1}q.$$

and we are able to explicitly construct the bases of the simple modules. For example, the 1-dimensional Specht module $\Delta_{\Bbbk\mathfrak{S}_5}(5)$ is simple and has basis

$$L_{\Bbbk\mathfrak{S}_5}(5) = \Bbbk\left\{\psi_s \mid s = \boxed{1\,2\,3\,4\,5}\right\} \tag{12.5.12}$$

simply because it is the maximal partition in the dominance ordering (and so the cell module is simple).

For λ an arbitrary partition of 5, the radical of $\Delta_{\Bbbk\mathfrak{S}_5}(\lambda)$ certainly contains all the basis elements of degree 1 (by the grading trick and the fact that the algebra is positively graded). Moreover, one can check that every basis element of degree zero is labelled by a reduced tableau. Thus by Proposition 12.5.10 we have that

$$\mathrm{rad}(\Delta_{\mathcal{H}_5}(\lambda)) = \Bbbk\{\psi_t \mid t \in \mathrm{Std}(\lambda), \deg(t) = 1\}$$
$$L_{\mathcal{H}_5}(\lambda) = \Bbbk\{\psi_t + \mathrm{rad}\langle -, -\rangle_\lambda^{\Bbbk} \mid t \in \mathrm{Std}(\lambda), \deg(t) = 0\}$$

(setting $L_{\mathcal{H}_5}(\lambda)$ to be the zero module). Moreover, we have a sequence of injective homomorphisms

$$\Delta_{\mathcal{H}_5}(5) \hookrightarrow \Delta_{\mathcal{H}_5}(4, 1) \hookrightarrow \Delta_{\mathcal{H}_5}(3, 1^2) \hookrightarrow \Delta_{\mathcal{H}_5}(2, 1^3) \hookrightarrow \Delta_{\mathcal{H}_5}(1^5)$$

where for each pair

$$\Delta_{\mathcal{H}_5}(\lambda) \hookrightarrow \Delta_{\mathcal{H}_5}(\mu)$$

in the sequence, the map takes each degree zero λ-tableau of a given residue sequence to the corresponding degree one μ-tableau of the same residue sequence. (Or rather, it does this for the elements ψ_t indexed by these tableaux, but let's not quibble.)

The easiest example is the homomorphism $\Delta_{\mathcal{H}_5}(5) \hookrightarrow \Delta_{\mathcal{H}_5}(4,1)$ under which the unique degree zero tableau s in (12.5.12) is mapped to the unique degree one tableau of the same residue sequence, t, below.

$$\operatorname{rad}(\Delta_{\mathcal{H}_5}(4,1)) = \Bbbk \left\{ \psi_t \mid t = \begin{array}{|c|c|c|c|} \hline 1 & 2 & 3 & 4 \\ \hline 5 \\ \cline{1-1} \end{array} \right\},$$

and the resulting quotient module, $L_{\mathcal{H}_5}(4,1)$, is as follows

$$\Bbbk \left\{ \psi_u + \operatorname{rad}\langle -,-\rangle_\lambda^{\Bbbk} \mid u = \begin{array}{|c|c|c|c|} \hline 1 & 3 & 4 & 5 \\ \hline 2 \\ \cline{1-1} \end{array}, \begin{array}{|c|c|c|c|} \hline 1 & 2 & 4 & 5 \\ \hline 3 \\ \cline{1-1} \end{array}, \begin{array}{|c|c|c|c|} \hline 1 & 2 & 3 & 5 \\ \hline 4 \\ \cline{1-1} \end{array} \right\}.$$

The next example is the homomorphism $\Delta_{\mathcal{H}_5}(4,1) \hookrightarrow \Delta_{\mathcal{H}_5}(3,1^2)$ under which each of the three tableaux above are mapped to the following respective tableaux of the same residue sequence below,

$$\operatorname{rad}(\Delta_{\mathcal{H}_5}(3,1^2)) = \Bbbk \left\{ \psi_v \mid v = \begin{array}{|c|c|c|} \hline 1 & 3 & 4 \\ \hline 2 \\ \hline 5 \\ \hline \end{array}, \begin{array}{|c|c|c|} \hline 1 & 2 & 4 \\ \hline 3 \\ \hline 5 \\ \hline \end{array}, \begin{array}{|c|c|c|} \hline 1 & 2 & 3 \\ \hline 4 \\ \hline 5 \\ \hline \end{array} \right\}.$$

Continuing in this manner, one can deduce that the simple modules and graded decomposition matrix of $\mathcal{H}_5$ are as claimed.

Remark 37. Another way to characterise the homomorphisms of Example 12.5.11 is as follows. For each tableau $t \in \operatorname{Std}(\lambda)$ with the entry 5 in the first row, let t' be the standard tableau in which we delete the 5 from the end of first row and add a new 5 at the end of the first column. In which case, all the homomorphisms of Example 12.5.11 can be characterised as the maps : $\psi_t \mapsto \psi_{t'}$.

Exercise 12.5.13. Construct the bases of all the (remaining) simple modules of $\mathcal{H}_5 \cong \mathbb{F}_5 \mathfrak{S}_5$.

Exercise 12.5.14. Construct the graded Alperin diagrams for the direct summands of the natural module of $\mathcal{H}_5 \cong \mathbb{F}_5 \mathfrak{S}_5$. (Hint: the summands $e(0,4,1,0,2)\mathcal{H}_5$ and $e(0,4,3,1,0)\mathcal{H}_5$ are each 25-dimensional and are both semisimple, and so you can ignore these; there are 4 more direct summands to consider, up to isomorphism.)

Exercise 12.5.15. Consider the $\Bbbk = \mathbb{F}_p$-algebra $\mathcal{H}_p \cong \mathbb{F}_p \mathfrak{S}_p$. Prove that

$$\dim_q(\Delta_{\Bbbk \mathfrak{S}_5}(p-k,1^k)) = \binom{p-1}{k} + \binom{p-1}{k-1}q$$

and hence generalise the ideas of Example 12.5.11 to calculate the graded decomposition matrix of $\mathcal{H}_p \cong \mathbb{F}_p \mathfrak{S}_p$ and construct the basis of $L_{\mathcal{H}_p}(\lambda)$ for $\lambda \in \Gamma_p^{\Bbbk}$.

12.6 The quiver Temperley–Lieb algebra

Finally, we conclude this chapter by considering how the above behaves with respect to the Temperley–Lieb quotients of Section 11.2. We will continue (as in Section 11.2) to consider the quotients $\Psi_n(\Bbbk\mathfrak{S}_r)$ under the restriction that $p > n$. We begin with the following observation of Plaza and Ryom–Hansen.

Proposition 12.6.1. *Let $p > n$. The generalised Temperley–Lieb algebra* $\mathrm{TL}_{n,r}$ *is isomorphic to the quotient of the $\mathbb{F}_p$-algebra $\mathcal{H}_{n,r}$ by the additional cyclotomic relation: $e(\underline{j}) = 0$ if $\underline{j} \neq \mathrm{res}(\mathsf{t})$ for some $\mathsf{t} \in \mathrm{Std}(\lambda)$ for $\lambda \in \Gamma_{n,r}^+ \subseteq \Gamma_r^+$.*

Proof. This follows because the cell-ideals of $\mathcal{H}_r$ and $\mathbb{F}_p\mathfrak{S}_r$ are identified under the Brundan–Kleshchev isomorphism and because $y_\lambda = e_\lambda$ for $\lambda \in \Gamma_{n,r}^+$ for $p > n$ (see below for more detail). □

Because of our restriction that $p > n$, we notice that the residue-sets for nodes in adjacent columns are disjoint, that is

$$\lambda \cap \{\mathrm{res}[x,y] \mid x \geq 1\} \cap \{\mathrm{res}[x,y+1] \mid x \geq 1\} = \emptyset$$

for $\lambda \in \Gamma_{n,r}^+$. This implies that every step in the tableau t_λ for $\lambda \in \Gamma_{n,r}^+$ is of degree zero. Therefore $y_\lambda = e_{\mathsf{t}_\lambda}$ and we have the following simplification of the Hu–Mathas construction:

Theorem 12.6.2 (The Hu–Mathas basis, again [HM10]). *Let $\Bbbk = \mathbb{F}_p$ for $p > n$. The $\Bbbk$-algebra $\mathcal{H}_{n,r}$ is a graded weighted cellular algebra with idempotents $1 = \sum_\mathsf{t} e_\mathsf{t}$ for t varying over the reduced tableaux and with respect to the basis*

$$\{\psi_{\mathsf{st}} = \psi_{\mathsf{t}_\lambda}^{\mathsf{s}} e_{\mathsf{t}_\lambda} \psi_{\mathsf{t}}^{\mathsf{t}_\lambda} \mid \mathsf{s}, \mathsf{t} \in \mathrm{Std}(\lambda, -), \lambda \in \Gamma_{n,r}^+\}.$$

This algebra is isomorphic to $\mathrm{TL}_{n,r} = \Psi_n(\Bbbk\mathfrak{S}_r)$.

In particular, we re-emphasise that the graded decomposition matrix of $\mathcal{H}_{n,r}$ is square uni-triangular for $p > n$.

Further Reading. The LLT-algorithm described in Definition 12.1.23 is a tableaux-theoretic re-imagining of the work of Lascoux–Lerclerc–Thibon [LLT96] (which was originally given in the setting of Fock spaces). This re-imagining of the the original LLT algorithm is due to Kleshchev–Nash [KN10]. The author learnt about much of the material on ladders covered in this section from [Mat99, Section 4].

Here we have tried to cast the quiver Hecke algebra solely in terms of "coloured permutations" which give us a "hidden grading" on the group algebra of the symmetric group. Actually, there is a general notion of quiver Hecke algebras for arbitrary quivers due to Rouquier [Rou] and Khovanov–Lauda [KL09]. Choosing the quiver in this construction to be a circle of p vertices as in Figure 12.1, we specialise to obtain $\mathbb{F}_p\mathfrak{S}_n$. We refer the reader to [Mat15] for a more thorough introduction to the KLR algebra through this lens. The connection between the KLR structure and the "seminormal form" of the symmetric groups (Theorem 11.1.24) is made explicit in [HM16] and further developed in [EvMa]. An analogue of the Jantzen sum formula for the KLR algebra is given in [Mat22]. The KLR algebras have also provided the tools for resolving other questions in the representation theory of symmetric groups, for example their "finiteness" behaviour [ALS23]. The construction of explicit KLR-graded Schur algebras by analogy with the definition considered here is given in [HM15], but this construction works only in the case of linear quivers. For non-linear quivers, the counterpart of the classical Schur algebra is provided by Webster's theory of "weighted" KLR algebras [Web19].

More exotic versions of the quiver Hecke algebras have evolved over time, including "weightings" [Web19] and "super structures" [BE17, KKT16, EKL14], which have led to applications in modular representation theory of finite groups [EK18, BK, ELV, KL], categorical knot theory [KL09, Web17a], and seems to provide the correct mathematical setting for the study of Coulomb and Higgs branches [LW23, Web12].

There is an alternative approach to the study of quiver Hecke algebras via "semi-cuspidal systems" and "imaginary Schur Weyl duality" pioneered by Kleshchev–Muth. This approach has led to some beautiful combinatorial and algebraic constructions [KM17, Mut19, KM19, KM20, KM22, ADM$^+$23] and, most famously, to the resolution of Turner's conjecture for RoCK blocks [EK18]. This theory recently led to the resolution of the beautiful Kleshchev–Martin "self-extension conjecture" for RoCK blocks [GKM22].

Chapter 13
The p-Kazhdan–Lusztig theory for Temperley–Lieb algebras

In this section we prove that the decomposition numbers of symmetric groups labelled by 2-row partitions are, indeed, equal to the p-Kazhdan–Lusztig polynomials for $(W, P) = (\widehat{\mathfrak{S}}_2, \mathfrak{S}_2)$. We thus rephrase the group theoretic conjectures of the past two chapters within the diagram algebras $\mathcal{H}_{(W,P)} = (\widehat{\mathfrak{S}}_n, \mathfrak{S}_n)$ for $n = 2$. For $n \geqslant 2$ this is a famous result of Riche–Williamson [RW18] later reproven and generalised by Elias–Losev, Bowman–Cox–Hazi, Achar–Makisumi–Riche–Williamson [EL, AMRW19, BCH23, BCHM22]. The proof here is via an explicit isomorphism theorem interrelating the quiver Hecke algebras and the diagram algebras $\mathcal{H}_{(W,P)}$ (by specialising to $n = 2$, we obtain a much more manageable version of the proof of [BCH23] — which is over 100 pages!). In our simplified "Temperley–Lieb case" we see all of the main ideas, but with dramatically simpler notation and with far fewer technicalities. We believe this is the easiest way to gain an understanding as to *why* the diagram algebras $\mathcal{H}_{(W,P)}$ are the key to understanding the original, group theoretic, statement of Lusztig's conjecture.

13.1 Graded path combinatorics of the Temperley–Lieb algebra

For this chapter we fix $(W, P) = (\widehat{\mathfrak{S}}_2, \mathfrak{S}_2)$ with $W = \langle \sigma, \tau \mid \sigma^2 = 1 = \tau^2 \rangle$ and $P = \langle \tau \mid \tau^2 = 1 \rangle$. We now delve a little deeper into the combinatorics of 2-part partitions within a plane $\mathbb{R}^2$ equipped with an action of W. We equip a 2-dimensional real vector space $\mathbb{E}_2 = \mathbb{R}\{\varepsilon_1, \varepsilon_2\}$ with a family of parallel τ-hyperplanes through the points

$$\mathbb{E}_2(\tau, rp) = \{(x_1, x_2) \in \mathbb{R}\{\varepsilon_1, \varepsilon_2\} \mid x_1 - x_2 = -1 + 2rp, r \in \mathbb{Z}\}$$

and parallel σ-hyperplanes through the points

$$\mathbb{E}_2(\sigma, rp) = \{(x_1, x_2) \in \mathbb{R}\{\varepsilon_1, \varepsilon_2\} \mid x_1 - x_2 = (p-1) + 2rp, r \in \mathbb{Z}\}.$$

Reflection through these hyperplanes defines the action of $W = \widehat{\mathfrak{S}}_2$ on $\mathbb{R}^2$. We define the fundamental region to be the set of points

$$\mathcal{F}_n^p = \{(x_1, x_2) \in \mathbb{E} \mid -1 \leq x_1 - x_2 \leq p - 1\}.$$

By repeatedly reflecting this region, we can tile the whole of $\mathbb{E}_2 = \cup_{w \in W} \mathcal{F}_n^p w$. We will be primarily concerned with the subregion labelled by elements $w \in {}^P W$. We will grey out the region of $\mathbb{E}_2$ tiled by the $\mathcal{F}_n^p w$ labelled by $w \notin {}^P W$.

We regard $\mathsf{s} \in \mathrm{Std}(\lambda)$ as a path in the alcove geometry by taking a $+\varepsilon_1$ step (respectively $+\varepsilon_2$ step) every time we add a node in the first (respectively second) row of $\mathsf{s} \in \mathrm{Std}(\lambda)$. Two tableaux s, t have the same residue sequence if and only if they are obtained by reflecting one another through a sequence of hyperplanes. This is best illustrated via an example, see Figures 13.1 and 13.2).

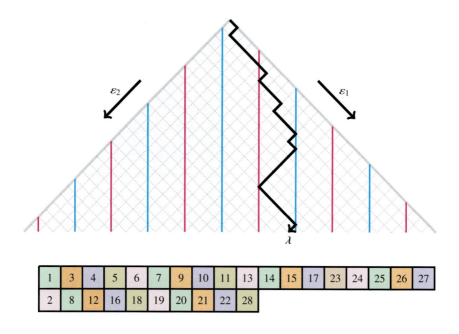

Fig. 13.1: A tableau $\mathsf{s} \in \mathrm{Std}((18, 10))$ as a path in $\mathbb{E}_2$. We indicate the directions ε_1 and ε_2 and the placement of hyperplanes for $p = 5$. Each entry in row 1 of the tableau corresponds to a step in the ε_1-direction and each entry in row 2 of the tableau corresponds to a step in the ε_2-direction. The greyed-out region depicts the points which do *not* label partitions. Ignoring this greyed-out region corresponds to taking cosets $P \leq W$ (and so removes "half" the region, since $|P| = 2$).

Lemma 13.1.1. *Let $\lambda = (\lambda_1, \lambda_2)$ lie on a hyperplane in $\widehat{\mathbb{B}}_2$. Then the nodes $[1, \lambda_1 + 1]$ and $[2, \lambda_2 + 1]$ are of equal residue, that is $\mathrm{ct}[1, \lambda_1 + 1] - \mathrm{ct}[2, \lambda_2 + 1] \in p\mathbb{Z}$.*

13.1 Graded path combinatorics of the Temperley–Lieb algebra

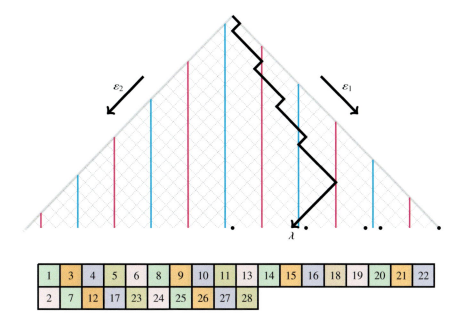

Fig. 13.2: A tableau $t \in \mathrm{Std}((18, 10))$ obtained by reflecting $s \in \mathrm{Std}((18, 10))$ through a sequence of hyperplanes.

Proposition 13.1.2. *The degree of a tableau* $t \in \mathrm{Std}(\lambda)$ *can be calculated by summing over the regions of the corresponding path which are of the form*

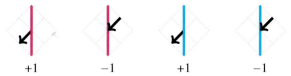

with corresponding degree contributions ± 1.

Proof. By Lemma 13.1.1 a partition λ lies on a hyperplane if and only if has two addable nodes with the *same* residue $r \in \mathbb{Z}/p\mathbb{Z}$. Adding a box in the first (respectively second) row of a partition corresponds to a step towards (away from) the origin. Therefore for λ on a hyperplane, the steps

$$\lambda \xrightarrow{+\varepsilon_1} \mu \qquad \mu \xrightarrow{+\varepsilon_1} \lambda \qquad \lambda \xrightarrow{+\varepsilon_2} \mu \qquad \mu \xrightarrow{+\varepsilon_2} \lambda$$

correspond to adding boxes boxes in a tableau of degrees 0, 0, 1 and −1 respectively. All steps not of the above form have degree 0 (as the relevant residues are "far apart"). The result follows. □

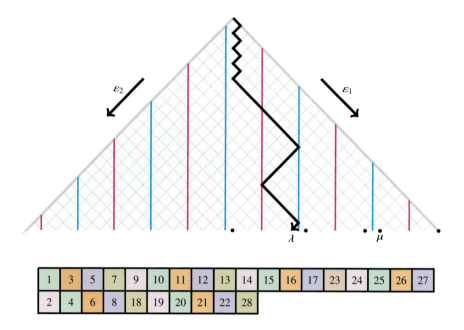

Fig. 13.3: An example of a tableau u ∈ Std($(18, 10), t_{(24,4)}$) as a path in 2-dimensional Euclidean space. The partition $\mu = (24, 4)$ is also highlighted. The reader is invited to check that the tableaux u and $t_{(24,4)}$ have the same residue sequences, and to apply repeated reflections to pass between the two paths.

Exercise 13.1.3. Calculate the degree of the path s ∈ Std($(18, 10)$) depicted in Figure 13.1. Compare this with the degree of the tableau s in Figure 13.1 as defined in Section 12.1.

Example 13.1.4. The path t ∈ Std($(18, 10), t_{(24,4)}$) depicted in Figure 13.3 has degree 1. To see this, we note that the 18th step, 22nd step, and 28th step are of the form
$$\lambda^{(1)} \xrightarrow{+\varepsilon_2} \mu^{(1)} \qquad \lambda^{(2)} \xrightarrow{+\varepsilon_2} \mu^{(2)} \qquad \lambda^{(3)} \xrightarrow{+\varepsilon_2} \mu^{(3)}$$
respectively, for $\lambda^{(1)}$ and $\lambda^{(3)}$ lying on the $x_1 - x_2 = 9$ hyperplane and $\mu^{(2)}$ lying on the $x_1 - x_2 = 4$ hyperplane. Therefore these steps have degree +1, −1, and +1 respectively. All other steps have degree 0 and so the total degree of the path is 1.

Exercise 13.1.5. Let s ∈ Std$(18, 10)$ be as in Figure 13.1. Calculate the total number of elements of Std$((18, 10)$ with the same residue sequence as s ∈ Std$((18, 10)$. (Hint: the answer is a power of 2).

Exercise 13.1.6. Let t ∈ Std$((18, 10)$ be as in Figure 13.1. Calculate the total number of elements of Std(λ) for $\lambda \in \Gamma_{2,28}$ with the same residue sequence as s ∈ Std$((18, 10)$.

13.2 Hyperplane-coloured residue sequences

Given λ a 2-row partition, we define the residue-pair, $\text{res}(\lambda) \in \mathbb{Z}/p\mathbb{Z}^2$ to be the unordered pair

$$\text{res}(\lambda) := \{\text{res}[x,y] \mid \text{for } [x,y] \notin \lambda, [x,y-1] \in \lambda, \text{ for } 1 \leq x \leq 2\}$$

We define the hyperplane-coloured residue-pair to be the pair obtained by colouring the elements of $\text{res}_\kappa(\lambda)$ pink if $\lambda \in \mathbb{E}_2(\sigma, rp)$ or blue if $\lambda \in \mathbb{E}_2(\tau, rp)$ for some $r \in \mathbb{Z}$, or simply black otherwise.

Example 13.2.1. For $p = 5$, the partitions $(1,1)$, $(5,1)$, and $(12,3) \in \Gamma^+_{2,15}$ have coloured residue pairs $\{0,1\}$, $\{0,0\}$, $\{2,2\}$ respectively.

For $t \in \text{Std}_2(\lambda)$ of the form

$$\varnothing \to \text{Shape}(t\!\downarrow_{\leq 1}) \to \text{Shape}(t\!\downarrow_{\leq 2}) \to \cdots \to \text{Shape}(t\!\downarrow_{\leq n}) = \lambda$$

we define its coloured residue sequence to be the sequence of pairs for the partitions $\text{Shape}(t\!\downarrow_{\leq k})$ for $1 \leq k \leq n$. We note that the residue of the boxes added at the kth step can be calculated by considering the set difference

$$\text{res}(\text{Shape}(t\!\downarrow_{\leq k})) \setminus (\text{res}(\text{Shape}(t\!\downarrow_{\leq k})) \cap \text{res}(\text{Shape}(t\!\downarrow_{<k}))).$$

Example 13.2.2. The tableau in Figure 13.1 has coloured residue sequence

$$\{0,4\}, \{1,4\}, \{0,1\}, \{0,2\}, \{0,3\}, \{0,4\}, \{0,0\}, \{0,1\}, \{1,1\}, \{1,2\}, \{1,3\}, \ldots$$

and we remark that the sequence of a tableau

$$\text{res}(\text{Shape}(t\!\downarrow_{\leq k})) \setminus (\text{res}(\text{Shape}(t\!\downarrow_{\leq k})) \cap \text{res}(\text{Shape}(t\!\downarrow_{<k}))) \quad (13.2.3)$$

is equal to the classical notion of residue sequence

$$0, 4, 1, 2, 3, 4, 0, 1, 1, 2, 3 \ldots$$

Remark 38. If s, t are obtained from one another by a series of reflections through hyperplanes then they have the exact same coloured residue sequences.

Example 13.2.4. Let $p = 5$. Calculate the coloured residue sequence of the following tableau

$$s = \begin{array}{|c|c|c|c|c|c|c|c|c|} \hline 1 & 2 & 3 & 4 & 5 & 6 & 7 & 8 & 9 \\ \hline 10 \\ \cline{1-1} \end{array}$$

has coloured-residue sequence as follows

$$\{0,4\}, \{1,4\}, \{2,4\}, \{3,4\}, \{4,4\}, \{0,4\}, \{1,4\}, \{2,4\}, \{3,4\}, \{4,4\}, \{0,4\}.$$

13.3 Recolouring the quiver Temperley–Lieb algebra

In this section we "refine" our KLR-diagrams into local regions, indexed by paths in the alcove geometry of length 1 or 2. We equip the diagrams with hyperplane-colourings and rename the generators so as to suggest the underlying isomorphism with the Soergel diagrammatic algebras. We first consider the single strand idempotent generators. Let μ be a partition lying on a σ-hyperplane, so that the addable nodes of μ are both of equal residue $i \in \mathbb{Z}/p\mathbb{Z}$, say. We can step onto $(\lambda \to \mu)$ or off-of $(\mu \to \nu)$ the σ-hyperplane like so

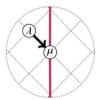 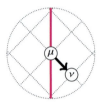

and we have coloured KLR-idempotents associated to these paths as follows

$$e_{\lambda \to \mu} := e_{\varnothing\sigma}(i-1) := \qquad\qquad e_{\mu \to \nu} := e_{\sigma\varnothing}(i) =$$

where we record the pair of addable residues at each point (λ, μ in the former case and μ, ν in the latter case) along the northern and southern edges. Now consider a pairs of points λ, μ not lying on a hyperplane with $\lambda \to \mu$. The addable nodes of λ are of residue $i, j \in \mathbb{Z}/p\mathbb{Z}$ such that $|i - j| > 1$. We can take a step $\lambda \to \mu$ like so

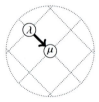

and we have an idempotent associated to this path as follows

$$e_{\varnothing\varnothing}(j, i \to i+1) =$$

where we have likewise recorded the pair of addable residues at each point (λ, μ).

13.3 Recolouring the quiver Temperley–Lieb algebra

We now consider the two-strand KLR generators ψ_r. Let λ, μ be partitions both lying on a σ-hyperplane, and suppose that the addable nodes of λ are both of equal residue $i \in \mathbb{Z}/p\mathbb{Z}$. Associated to the pair of paths

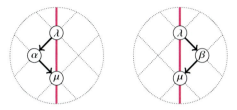

we have the like-labelled-residue crossing, which we refer to as the "double-fork diagram", as follows

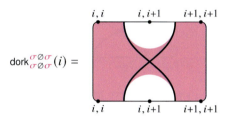

where we note that the pairs of residues along the northern and southern edges of this diagram are the residues of the addable nodes of λ, then α (respectively β) and then μ. Let λ, μ be partitions both lying *near* a σ-hyperplane, that is the addable nodes of λ are of residue $i, i-1 \in \mathbb{Z}/p\mathbb{Z}$. Associated to the pair of paths

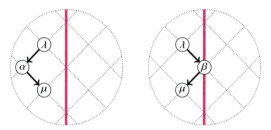

we have the adjacent-residue-crossing, which we refer to as the "spot diagram", as follows

$$\text{spot}_{\emptyset\sigma\emptyset}^{\emptyset\emptyset\emptyset}(i) = \begin{array}{c} i-1,i \quad i-1,i+1 \quad i,i+1 \\ \text{[diagram]} \\ i-1,i \quad i,i \quad i,i+1 \end{array}$$

where we note that the pairs of residues along the northern and southern edges of this diagram are the residues of the addable nodes of λ, then α (respectively β) and then μ. The τ-versions of the above are defined in a similar fashion.

Let λ, μ be partitions both lying *far* from a σ-hyperplane, (the addable nodes of λ are of residue $i, j \in \mathbb{Z}/p\mathbb{Z}$ such that $|i - j| > 1$). Associated to the pair of paths

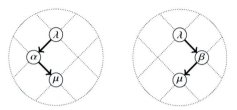

we have the distant-residue-crossing diagram, which we refer to as the "transposition of alcove walks diagram", as follows

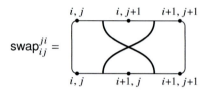

where the pairs of residues along the northern and southern edges of this diagram are the residues of the addable nodes of λ, then α (respectively β) and then μ.

Remark 39. For any skew tableau $\mathsf{s} \in \mathrm{Std}(\nu \setminus \lambda)$ of the form

$$\mathsf{s} = (\emptyset = \lambda^0 \to \lambda^1 \to \cdots \to \lambda^r = \nu)$$

we let e_s denote the diagram obtained from horizontal concatenation of the corresponding idempotents for each step. Any horizontal concatenate of idempotents what is not of this form is set to be zero. For example

$$e_{\emptyset\sigma}(i) \otimes e_{\tau\emptyset}(j) = 0 \qquad e_{\emptyset\tau}(i) \otimes e_{\sigma\emptyset}(j) = 0 \qquad e_{\emptyset\tau}(i) \otimes e_{\emptyset\emptyset}(j) = 0$$

for $\sigma \neq \tau$ and any $i, j \in \mathbb{Z}/p\mathbb{Z}$.

Example 13.3.1. Let $p = 5$ and s as in Example 13.2.4. we have an associated idempotent e_s constructed via concatenation as follows

$$e_\mathsf{s} = e_{\emptyset\emptyset}(4, 0 \to 1) \otimes e_{\emptyset\emptyset}(4, 1 \to 2) \otimes e_{\emptyset\emptyset}(4, 2 \to 3) \otimes e_{\emptyset\sigma}(4) \otimes e_{\sigma\emptyset}(0) \otimes \ldots$$

which can be pictured as a hyperplane-coloured KLR diagram as follows

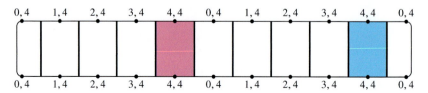

13.3 Recolouring the quiver Temperley–Lieb algebra

This is merely a recolouring of the familiar (residue-coloured) KLR diagram

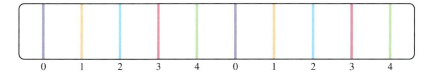

where we notice that the residue-subscripts in the latter diagram are obtained from those of the former diagram as in (13.2.3).

What have we done with this recolouring and why have we done it? The answer to "what have we done?" is that we have simply tweaked our perspective somewhat. We can readily identify the old and new diagrams and we now restate this identification for the purposes of clarity. We have that

$$\psi_1 e(i,i) = \sum_{\pi \in \{\sigma,\tau\}} \text{dork}_{\pi\emptyset\pi}^{\pi\emptyset\pi}(i) \qquad \psi_1 e(i,i+1) = \sum_{\pi \in \{\sigma,\tau\}} \text{spot}_{\emptyset\pi\emptyset}^{\emptyset\emptyset\emptyset}(i)$$

$$\psi_1 e(i,j) = \text{swap}_{ij}^{ji} \qquad e(i) = e_{\emptyset\sigma}(i) + e_{\sigma\emptyset}(i) + \sum_{|i-j|>1} e_{\emptyset\emptyset}(j, i \to i+1)$$

where the pairs of addable residues forming the subscripts of the diagrams on the righthand-side of these equalities uniquely determine the residues of the strands of the diagrams on the lefthand-side of these equalities as in (13.2.3). Of course, it would appear that we have introduced many new idempotents and massively refined the existing KLR idempotents according to the hyperplane-colourings... however, this is not the case at all! First of all, we recall that the quiver Hecke algebra has a basis indexed by pairs of tableaux as in Theorem 12.6.2; this can be interpreted in terms of paths in the alcove geometry as in Section 13.1. The hyperplane colouring of a path is uniquely determined by its residue sequence (see Remark 38) and so the hyperplane-colouring is uniquely determined by horizontal concatenation (and Remark 39). The answer to "why have we done it?" is to make connections to the Soergel diagrams more intuitive, as (hopefully!) illustrated in the following example.

Example 13.3.2. Given $i \in \mathbb{Z}/p\mathbb{Z}$ we define the "fork diagram" via horizontal and vertical concatenation of dork, spot, and idempotent diagrams as follows

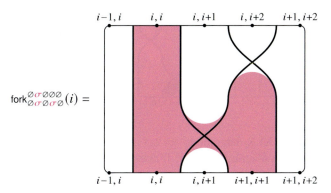

13.4 The isomorphism theorem

We are now ready to outline the isomorphism theorem interrelating the Soergel and KLR diagrammatic worlds (in the Temperley–Lieb case). The purpose of this final section of the book is to sketch a proof of this result. We define a "forgetful" map from (recoloured) KLR diagrams to Soergel diagrams as follows

$$f(e_{\emptyset\pi}(i-1) \otimes e_{\pi\emptyset}(i)) = 1_\pi \quad f(e_{\emptyset\emptyset}(j, i \to i+1)) = 1_\emptyset \quad f(\mathsf{swap}_{ij}^{ji}) = 1_\emptyset$$
$$f(\mathsf{spot}_{\emptyset\pi\emptyset}^{\emptyset\emptyset\emptyset}(i)) = \mathsf{spot}_\pi^\emptyset \quad f(\mathsf{fork}_{\emptyset\pi\emptyset\pi\emptyset}^{\emptyset\pi\emptyset\emptyset\emptyset}(i)) = \mathsf{fork}_{\pi\pi}^\pi$$

for all admissible $i, j \in \mathbb{Z}/p\mathbb{Z}$ and $\pi \in \{\sigma, \tau\}$. We extend this map by $\otimes$ and $\circ$ and duality in the obvious fashion. Diagrammatically, the map f simply deletes the black strands in the KLR diagram, leaving only the colouring behind (thus obtaining a Soergel diagram, up to "squinting"). For example, we apply the forgetful map to the (recoloured) KLR diagram in Figure 13.4 to obtain the diagram in Figure 13.5.

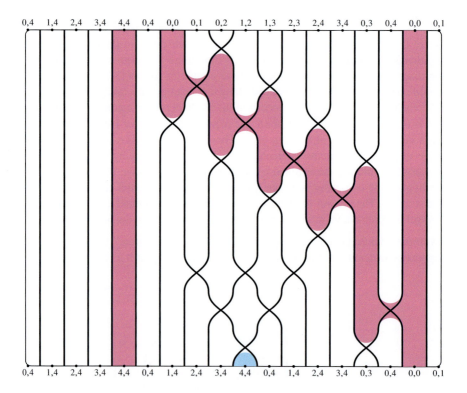

Fig. 13.4: A recoloured KLR diagram $\psi_\mathsf{t}^\mathsf{s}$ for s, t as pictured in Figure 13.6. This diagram is mapped to the diagram in Figure 13.5 under the forgetful map.

13.4 The isomorphism theorem

Remark 40. Notice that we consider horizontal concatenations of the (recoloured) KLR generators as they are somehow "skinnier" than their Soergel-theoretic counterparts. This is okay because we will only consider KLR diagrams whose left and right edges are both white.

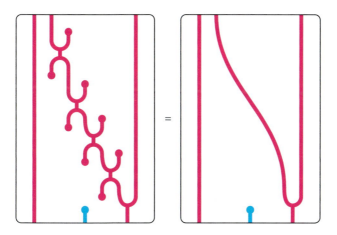

Fig. 13.5: The Soergel diagram obtained from the diagram in Figure 13.4 under the forgetful map. The lefthand diagram is obtained from the diagram in Figure 13.4 by deleting the black strands and "squinting". The righthand diagram is obtained by simplifying the picture using the fork-spot relation.

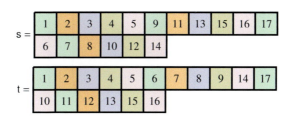

Fig. 13.6: The tableaux $\mathsf{s}, \mathsf{t} \in \mathrm{Std}_{17}(11,6)$ corresponding to the top and bottom of the diagram in Figure 13.4.

We define a "prototype inverse" map g from Soergel diagrams to (recoloured) KLR diagrams as follows

$$g(1_\pi) = \sum_{i \in \mathbb{Z}/p\mathbb{Z}} e_{\emptyset\pi}(i-1) \otimes e_{\pi\emptyset}(i) \qquad g(\mathrm{spot}_\pi^\emptyset) = \sum_{i \in \mathbb{Z}/p\mathbb{Z}} \mathrm{spot}_{\emptyset\pi\emptyset}^{\emptyset\emptyset\emptyset}(i)$$

$$g(1_\emptyset) = \sum_{|i-j|>1}(e_{\emptyset\emptyset}(j,i \to i+1) + \mathrm{swap}_{ij}^{ji}) \qquad g(\mathrm{fork}_{\pi\pi}^\pi) = \sum_{i \in \mathbb{Z}/p\mathbb{Z}} \mathrm{fork}_{\emptyset\pi\emptyset\pi\emptyset}^{\emptyset\pi\emptyset\emptyset\emptyset}(i)$$

for $\pi \in \{\sigma, \tau\}$ and extending this by $\otimes$ and $\circ$ and duals in the usual manner.

13.4.1 Fork-annihilation in the KLR algebra

This is the first of a series of subsections in which we provide analogues of the monochrome Soergel relations within the recoloured KLR algebra. The analogue of the fork-annihilation relation is that

$$\text{fork}^{\emptyset\sigma\emptyset\emptyset\emptyset}_{\emptyset\sigma\emptyset\sigma\emptyset}(i) \circ \text{fork}^{\emptyset\sigma\emptyset\sigma\emptyset}_{\emptyset\sigma\emptyset\emptyset\emptyset}(i) = 0$$

for all $i \in \mathbb{Z}/p\mathbb{Z}$. This is pictured in Figure 13.7. We will show that the local relation depicted in Figure 13.7 does indeed follow from the classical quiver Hecke relations (that is, upon forgetting the colouring). We observe that the lefthand diagram in Figure 13.7 is equal to

$$\psi_3\psi_2\psi_2\psi_3 e(i-1,i,i+1,i) = \psi_3\psi_2 e(i-1,i,i,i+1)\psi_2\psi_3 = 0$$

where the first equality follows by the idempotent relations and the second follows from the like-residue-labelled double-crossing relation (R3) of Definition 12.3.2.

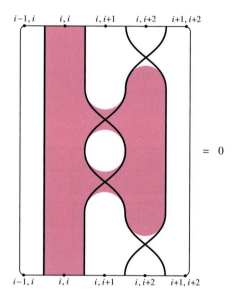

Fig. 13.7: The analogue of the fork-annihilation relation.

Exercise 13.4.1. Write down the standard tableaux corresponding to the top and bottom of the element $\text{fork}^{\emptyset\sigma\emptyset\emptyset\emptyset}_{\emptyset\sigma\emptyset\sigma\emptyset}(i)$.

13.4.2 Fork-spot contraction in the KLR algebra

The analogue of fork-spot contraction in the (recoloured) KLR algebra is as follows:

$$\text{fork}^{\emptyset\sigma\emptyset\emptyset\emptyset}_{\emptyset\sigma\emptyset\sigma\emptyset}(i) \circ \left(e_{\emptyset\sigma}(i-1) \otimes e_{\sigma\emptyset}(i) \otimes \text{spot}^{\emptyset\sigma\emptyset}_{\emptyset\emptyset\emptyset}(i+1)\right)$$
$$= e_{\emptyset\sigma}(i-1) \otimes e_{\sigma\emptyset}(i) \otimes e_{\emptyset\emptyset}(i-1, i \to i+1) \otimes e_{\emptyset\emptyset}(i+1, i-1 \to i)$$

for all $i \in \mathbb{Z}/p\mathbb{Z}$. This is pictured in Figure 13.8.

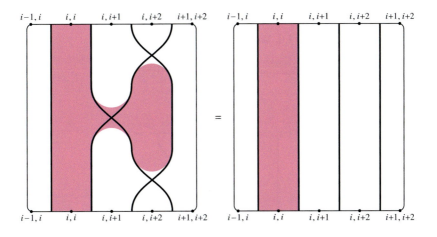

Fig. 13.8: The analogue of fork-spot contraction.

We will show that the local relation depicted in Figure 13.8 does indeed follow from the classical quiver Hecke relations (that is, upon forgetting the colouring). We observe that the lefthand diagram in Figure 13.8 is a braid to which we can apply the KLR-braid relation $(R4)$ of Definition 12.3.2 to obtain

$$\psi_3\psi_2\psi_3 e(i-1, i, i+1, i) = (\psi_2\psi_3\psi_2 + 1)e(i-1, i, i+1, i)$$

and we observe that the braided diagram on the righthand-side factors through an idempotent of the form $e(i-1, i+1, i, i)$ and this is not the residue sequence of a skew-standard tableau and so this term is zero by the cyclotomic relation, thus

$$\psi_3\psi_2\psi_3 e(i-1, i, i+1, i) = (\psi_2\psi_3\psi_2 + 1)e(i-1, i, i+1, i) = e(i-1, i, i+1, i),$$

this is the idempotent depicted on the righthand-side of Figure 13.8, as claimed.

13.4.3 Isotopy in the KLR algebra

We have already observed in Figures 13.4 and 13.5 that drawing a "wiggly Soergel strand" in the recoloured KLR algebra is more arduous than in the Soergel case. In fact, the only way to define such a wiggle/isotopy is via a product of spots and forks/dorks as follows

$$\text{iso}_{\varnothing\varnothing\varnothing\sigma\varnothing}^{\varnothing\sigma\varnothing\varnothing\varnothing}(i) = (1_{\varnothing\sigma}(i-1) \otimes 1_{\sigma\varnothing}(i) \otimes \text{spot}_{\varnothing\sigma\varnothing}^{\varnothing\varnothing\varnothing}(i+1)) \circ$$
$$(1_{\varnothing\sigma}(i-1) \otimes \text{dork}_{\sigma\varnothing\sigma}^{\sigma\varnothing\sigma}(i) \otimes 1_{\sigma\varnothing}(i+1)) \circ$$
$$(\text{spot}_{\varnothing\varnothing\varnothing}^{\varnothing\sigma\varnothing}(i) \otimes 1_{\varnothing\sigma}(i) \otimes 1_{\sigma\varnothing}(i+1))$$

this is pictured in Figure 13.9. We claim that

$$\text{iso}_{\varnothing\varnothing\varnothing\sigma\varnothing}^{\varnothing\sigma\varnothing\varnothing\varnothing}(i)\text{iso}_{\varnothing\sigma\varnothing\varnothing\varnothing}^{\varnothing\varnothing\varnothing\sigma\varnothing}(i) = e_{\varnothing\sigma\varnothing\varnothing\varnothing}(i-1, i, i+1, i). \tag{13.4.2}$$

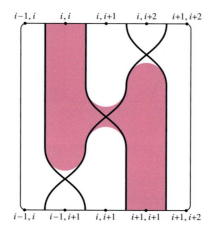

Fig. 13.9: The isotopy diagram $\text{iso}_{\varnothing\varnothing\varnothing\sigma\varnothing}^{\varnothing\sigma\varnothing\varnothing\varnothing}(i)$ for $i \in \mathbb{Z}/p\mathbb{Z}$.

We now verify the claim (13.4.2). We have that

$$\begin{aligned}
\text{iso}_{\varnothing\varnothing\varnothing\sigma\varnothing}^{\varnothing\sigma\varnothing\varnothing\varnothing}(i)\text{iso}_{\varnothing\sigma\varnothing\varnothing\varnothing}^{\varnothing\varnothing\varnothing\sigma\varnothing}(i) &= \psi_3\psi_2\psi_1\psi_1\psi_2\psi_3 e(i-1, i, i+1, i) \\
&= \psi_3\psi_2 e(i-1, i, i, i+1)(y_2 - y_1)\psi_2\psi_3 \\
&= \psi_3\psi_2 e(i-1, i, i, i+1) y_2 \psi_2 \psi_3 \\
&= (\psi_3\psi_2\psi_3 + \psi_3 y_3 \psi_2 \psi_2 \psi_3) e(i-1, i, i+1, i) \\
&= \psi_3\psi_2\psi_3 e(i-1, i, i+1, i) \\
&= e(i-1, i, i+1, i)
\end{aligned}$$

13.4 The isomorphism theorem

where the first equality follows by definition, the second by the adjacent-residue double-crossing relation (R3), the third and fifth follow by the like-residue double-crossing relation (R3), the fourth by the dot-through-a-crossing relation (R2), and the final equality we have already proven, see Figure 13.8.

Exercise 13.4.3. Draw the recoloured KLR diagram version of (13.4.2).

Exercise 13.4.4. Verify that the "other" isotopy relation

$$\mathsf{iso}_{\varnothing\sigma\varnothing\varnothing\varnothing}^{\varnothing\varnothing\varnothing\sigma\varnothing}(i)\mathsf{iso}_{\varnothing\varnothing\varnothing\sigma\varnothing}^{\varnothing\sigma\varnothing\varnothing\varnothing}(i) = e_{\varnothing\varnothing\varnothing\sigma\varnothing}(i, i-1, i, i+1). \tag{13.4.5}$$

holds in the KLR algebra. (Note that we can now wiggle a vertical strand to the right using (13.4.2) or to the left using (13.4.5).)

Remark 41. For ease of notation, we set $\mathsf{iso}_{\varnothing\varnothing\varnothing\sigma\varnothing}^{\varnothing\sigma\varnothing\varnothing\varnothing} = \sum_{i\in\mathbb{Z}/p\mathbb{Z}} \mathsf{iso}_{\varnothing\varnothing\varnothing\sigma\varnothing}^{\varnothing\sigma\varnothing\varnothing\varnothing}(i)$.

13.4.4 The monochrome barbell relation in the KLR algebra

In order to state the analogue of the barbell relation, we must first fix some notation

$$\sigma\mathsf{bar}(i) = \mathsf{spot}_{\varnothing\sigma\varnothing}^{\varnothing\varnothing\varnothing}(i)\mathsf{spot}_{\varnothing\varnothing\varnothing}^{\varnothing\sigma\varnothing}(i) \quad \sigma\mathsf{gap}(i) = \mathsf{spot}_{\varnothing\varnothing\varnothing}^{\varnothing\sigma\varnothing}(i) \circ \mathsf{spot}_{\varnothing\sigma\varnothing}^{\varnothing\varnothing\varnothing}(i).$$

With this notation in place, the analogue of the one-colour barbell relation is

$$e_{\varnothing\sigma\varnothing}(i-1,i) \otimes \sigma\mathsf{bar}(i+1) + \mathsf{iso}_{\varnothing\varnothing\varnothing\sigma\varnothing}^{\varnothing\sigma\varnothing\varnothing\varnothing}(\sigma\mathsf{bar}(i) \otimes e_{\varnothing\sigma\varnothing}(i,i+1))\mathsf{iso}_{\varnothing\sigma\varnothing\varnothing\varnothing}^{\varnothing\varnothing\varnothing\sigma\varnothing}$$
$$=2\sigma\mathsf{gap}(i) \otimes e_{\varnothing\varnothing\varnothing}(i+1,i).$$

This relation is depicted (with $i = 1$) in Figure 13.10.

Exercise 13.4.6. Construct the standard tableaux for the top/bottom and middle of each diagram in Figure 13.10.

Lemma 13.4.7. *We have that*

$$\psi_1 y_1^2 \psi_1 e(i,i) = (y_2\psi_1 + y_1\psi_1)e(i,i) \tag{13.4.8}$$
$$\psi_2\psi_1 y_1\psi_1\psi_2 e(i,i,i+1) = \psi_2\psi_1\psi_2 e(i,i,i+1) \tag{13.4.9}$$
$$y_2 e(i-1,i,i+1,i) = y_3 e(i-1,i,i+1,i) \tag{13.4.10}$$

Proof. The first equation follows from multiple applications of (R2) and (R3), the second equation is similar. For the third equation, we have that

$$y_2 e(i-1,i,i+1,i) = \psi_2 e(i-1,i+1,i,i)\psi_2 + y_3 e(i-1,i,i+1,i) = y_3 e(i-1,i,i+1,i)$$

where the first equality holds by (R2) and the second equality follows by the cyclotomic relation (R5). □

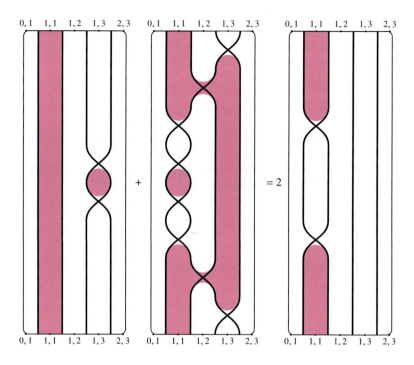

Fig. 13.10: The KLR analogue of the monochrome barbell relation with $i=1$.

We now verify the KLR-analogue of the one-colour barbell. We have that the lefthand-side of the relation is equal to

$$e_{\emptyset\sigma\emptyset}(i-1,i) \otimes \sigma\mathsf{bar}(i+1) + \mathsf{iso}^{\emptyset\sigma\emptyset\emptyset}_{\emptyset\emptyset\sigma\emptyset}(\sigma\mathsf{bar}(i) \otimes e_{\emptyset\sigma\emptyset}(i,i+1))\mathsf{iso}^{\emptyset\emptyset\emptyset\sigma\emptyset}_{\emptyset\sigma\emptyset\emptyset}$$
$$=(\psi_3^2 + \psi_3\psi_2\psi_1^4\psi_2\psi_3)e(i-1,i,i+1,i)$$
$$=(\psi_3^2 + \psi_3\psi_2(y_1^2 + y_2^2 - 2y_1y_2)\psi_2\psi_3)e(i-1,i,i+1,i)$$
$$=(\psi_3^2 + 0 + \psi_3(y_3\psi_2 + y_2\psi_2)\psi_3 - 2y_1\psi_3\psi_2\psi_3)e(i-1,i,i+1,i)$$
$$=((y_3 - y_4) + (y_4 + y_2) - 2y_1)e(i-1,i,i+1,i)$$
$$=(-2y_1 + y_2 + y_3)e(i-1,i,i+1,i)$$
$$=(-2y_1 + 2y_2)e(i-1,i,i+1,i)$$
$$=2\sigma\mathsf{gap}(i) \otimes e_{\emptyset\emptyset\emptyset}(i+1,i).$$

where the first and final equalities hold by definition; the second follows by applying the like-residue-double-crossing relation (R3) twice; the third follows from (13.4.8) and (13.4.9); the fourth follows from the braid-relation (R4) and the cyclotomic relation (R5) in the usual manner; the fifth equation follows (13.4.10).

13.4.5 Bijectivity and bases

The reader should now be quite convinced that there will be a sense in which $\mathcal{H}_{(W,P)}$ and $\mathcal{H}_{2,r}$ are "the same". Both algebras have bases given in terms of paths in a 2-dimensional alcove geometry, and we have already verified most of the defining relations of the former algebra have natural analogues in the latter algebra. While we have defined maps f and a "candidate inverse map" g on diagrams, we have been a bit sloppy about precisely defining the domains and codomains of these maps.

We now come to the fiddliest part of this chapter, explicitly stating the isomorphism on the level of bases. For $\ell \in \mathbb{Z}_{\geq 0}$ we let $\Pi \subseteq {}^P W$ denote the saturated subset consisting of the cosets

$$\Pi = \{\emptyset, \sigma, \sigma\tau, \sigma\tau\sigma,, \sigma\tau\sigma\tau, \ldots, \underbrace{\sigma\tau\sigma\tau\ldots}_{\text{length } \ell}\}.$$

These cosets are in bijection with the 2-row partitions of ℓp in the W-orbit of $(\ell p, 0) \in \Gamma^+_{2,\ell p}$. Abusing notation we write $\Pi \subseteq {}^P W$ and $\Pi \subseteq \Gamma^+_{2,\ell p}$. There is a single reduced path t_μ in the Bruhat graph for any $\mu \in \Pi \subseteq {}^P W$. On the other hand, there are many reduced tableaux of shape μ for any $\mu \in \Pi \subseteq \Gamma^+_{2,\ell p}$. For each $\mu \in \Pi \subseteq \Gamma^+_{2,\ell p}$, we fix t_μ to be the least dominant tableau of shape μ.

Theorem 13.4.11. *For $\ell \in \mathbb{Z}_{\geq 0}$, we have that*

$$f : (\textstyle\sum_{\mu \in \Pi} e_{\mathsf{t}_\mu}) \mathcal{H}_{2,\ell p} (\textstyle\sum_{\mu \in \Pi} e_{\mathsf{t}_\mu}) \to 1_\Pi \mathcal{H}_{(W,P)} 1_\Pi$$

is a $\mathbb{Z}$-graded algebra isomorphism.

Remark 42. At the beginning of Part 5 of this book we spoke of "chipping" away at the group algebra $\Bbbk\mathfrak{S}_r \cong \mathcal{H}_r$ until we obtain a "beautiful core" that is isomorphic to $\mathcal{H}_{(W,P)}$. Here we have made good on our promise in two steps: firstly, the quotient $\mathcal{H}_r \to \mathcal{H}_{2,r}$ was defined via quotienting by the 2-sided ideal generated by a KLR idempotent (see Proposition 12.6.1 and Theorem 12.6.2); secondly, we have truncated by the idempotent $\sum_{\mu \in \Pi} e_{\mathsf{t}_\mu}$. Thus we have "chipped away" the excess to find a subquotient isomorphic to $\mathcal{H}_{(W,P)}$, as promised!

Sketch proof of Theorem 13.4.11. We define an inverse map

$$f^{-1} : 1_\Pi \mathcal{H}_{(W,P)} 1_\Pi \to (\textstyle\sum_{\mu \in \Pi} e_{\mathsf{t}_\mu}) \mathcal{H}_{2,\ell p} (\textstyle\sum_{\mu \in \Pi} e_{\mathsf{t}_\mu})$$
$$f^{-1}(d) = (\textstyle\sum_{\mu \in \Pi} e_{\mathsf{t}_\mu}) \circ g(d) \circ (\textstyle\sum_{\mu \in \Pi} e_{\mathsf{t}_\mu})$$

for $d \in \mathcal{H}_{(W,P)}$. We have already seen that map g respects most of the local relations of $\mathcal{H}_{(W,P)}$ and we leave it as an exercise for the reader to check that the last couple of relations hold.

The algebra $\mathcal{H}_{2,\ell p}$ is generated by the $\psi^\mathsf{s}_\mathsf{t}$ for pairs of tableaux $\mathsf{s}, \mathsf{t} \in \mathrm{Std}(\lambda)$ (by Theorem 12.6.2) and every such element is in the image of f^{-1}, by definition. Thus the map f^{-1} is surjective; bijectivity then follows from a dimension count. □

Exercise 13.4.12. Verify the analogues of the double-fork and two-coloured barbell relation hold in the KLR algebra.

Exercise 13.4.13. Write down the effect of the map g on the basis of $1_\Pi \mathcal{H}_{(W,P)} 1_\Pi$ for $\Pi = \{\emptyset, \sigma\}$ and $\Pi = \{\emptyset, \sigma, \sigma\tau\}$.

Further Reading. The penultimate chapter of this book showed that the Kazhdan–Lusztig polynomials are examples of a much broader class of LLT polynomials. One can generalise the Kazhdan–Lusztig polynomials we consider here to a *singular Kazhdan–Lusztig theory* which is broad enough to encapsulate all of the LLT polynomials. The study of singular analogues of the diagrammatic Hecke categories is an area in relative infancy, we refer to [Eli17, EMTW20, EK23] for more details. All of this is simply to say that the final chapter of this part of the book implicitly borrowed ideas from the singular theory for Hecke categories.

References

ADL03. I. Ágoston, V. Dlab, and E. Lukács, *Quasi-hereditary extension algebras*, Algebr. Represent. Theory **6** (2003), no. 1, 97–117.

ADM+23. D. Abbasian, L. Difulvio, R. Muth, G. Pasternak, I. Sholtes, and F. Sinclair, *Cuspidal ribbon tableaux in affine type A*, Algebr. Comb. **6** (2023), no. 2, 285–319.

ABG04. S. Arkhipov, R. Bezrukavnikov, and V. Ginzburg. Quantum groups, the loop Grassmannian, and the Springer resolution. *J. Amer. Math. Soc.*, 17(3):595–678, 2004.

AJS94. H. Andersen, J. Jantzen, and W. Soergel, *Representations of quantum groups at a pth root of unity and of semisimple groups in characteristic p: independence of p*, Astérisque (1994), no. 220, 321.

AK08. M. Altunbulak and A. Klyachko, *The Pauli principle revisited*, Comm. Math. Phys. **282** (2008), no. 2, 287–322.

AMRW19. P. Achar, S. Makisumi, S. Riche, and G. Williamson, *Koszul duality for Kac-Moody groups and characters of tilting modules*, J. Amer. Math. Soc. **32** (2019), no. 1, 261–310.

And98. H. H. Andersen, *Tilting modules for algebraic groups*, Algebraic groups and their representations (Cambridge, 1997), NATO Adv. Sci. Inst. Ser. C Math. Phys. Sci., vol. 517, Kluwer Acad. Publ., Dordrecht, 1998, pp. 25–42.

Ari96. S. Ariki, *On the decomposition numbers of the Hecke algebra of $G(m, 1, n)$*, J. Math. Kyoto Univ. **36** (1996), no. 4, 789–808.

ALS23. S. Ariki, S. Lyle, and L. Speyer, *Schurian-finiteness of blocks of type A Hecke algebras*, J. Lond. Math. Soc. (2) **108** (2023), no. 6, 2333–2376.

Arm88. M. A. Armstrong, *Groups and symmetry*, Undergraduate Texts in Mathematics, Springer-Verlag, New York, 1988.

Bai. J. Baine, *On the coefficients in the Jones–Wenzl idempotent*, arXiv:2406.06333.

BDEA+20. M. Balagović, Z. Daugherty, I. Entova-Aizenbud, I. Halacheva, J. Hennig, M. S. Im, G. Letzter, E. Norton, V. Serganova, and C. Stroppel, *The affine VW supercategory*, Selecta Math. (N.S.) **26** (2020), no. 2, Paper No. 20, 42.

BCD19. S. Barbier, A. Cox, and M. De Visscher, *The blocks of the periplectic Brauer algebra in positive characteristic*, J. Algebra **534** (2019), 289–312.

Ben98. D. J. Benson, *Representations and cohomology. I*, second ed., Cambridge Studies in Advanced Mathematics, vol. 30, Cambridge University Press, Cambridge, 1998, Basic representation theory of finite groups and associative algebras.

BGS96. A. Beilinson, V. Ginzburg, and W. Soergel, *Koszul duality patterns in representation theory*, J. Amer. Math. Soc. **9** (1996), no. 2, 473–527.

BBM04. A. Beilinson, R. Bezrukavnikov, and I. Mirković. Tilting exercises. *Mosc. Math. J.*, 4(3):547–557, 782, 2004.

BFK99. J. Bernstein, I. Frenkel, and M. Khovanov, *A categorification of the Temperley-Lieb algebra and Schur quotients of $U(\mathfrak{sl}_2)$ via projective and Zuckerman functors*, Selecta Math. (N.S.) **5** (1999), no. 2, 199–241.

BM13. R. Bezrukavnikov and I. Mirković. Representations of semisimple Lie algebras in prime characteristic and the noncommutative Springer resolution. *Ann. of Math. (2)*, 178(3):835–919, 2013.

BMR08. R. Bezrukavnikov, I. Mirković, and D. Rumynin. Localization of modules for a semisimple Lie algebra in prime characteristic. *Ann. of Math. (2)*, 167(3):945–991, 2008. With an appendix by Bezrukavnikov and Simon Riche.

Bes18. C. Bessenrodt, *Critical classes, Kronecker products of spin characters, and the Saxl conjecture*, Algebr. Comb. **1** (2018), no. 3, 353–369.

BB17. C. Bessenrodt and C. Bowman, *Multiplicity-free Kronecker products of characters of the symmetric groups*, Adv. Math. **322** (2017), 473–529.

BB23. C. Bessenrodt and C. Bowman, *Splitting Kronecker squares, 2-decomposition numbers, Catalan combinatorics, and the Saxl conjecture*, Algebr. Comb. **6** (2023), no. 4, 863–899.

BBP22.	C. Bessenrodt, C. Bowman, and R. Paget, *The classification of multiplicity-free plethysms of Schur functions*, Trans. Amer. Math. Soc. **375** (2022), no. 7, 5151–5194.
BBS21.	C. Bessenrodt, C. Bowman, and L. Sutton, *Kronecker positivity and 2-modular representation theory*, Trans. Amer. Math. Soc. Ser. B **8** (2021), 1024–1055.
BBD[+]22.	C. Blundell, L. Buesing, A. Davies, P. Veličković, and G. Williamson, *Towards combinatorial invariance for Kazhdan-Lusztig polynomials*, Represent. Theory **26** (2022), 1145–1191.
BC14.	C. Bowman and A. Cox, *Decomposition numbers for Brauer algebras of type $G(m, p, n)$ in characteristic zero*, J. Pure Appl. Algebra **218** (2014), no. 6, 992–1002.
BC18.	C. Bowman and A. Cox, *Modular decomposition numbers of cyclotomic Hecke and diagrammatic Cherednik algebras: a path theoretic approach*, Forum Math. Sigma **6** (2018), Paper No. e11, 66.
BCD13.	C. Bowman, A. Cox, and M. De Visscher, *Decomposition numbers for the cyclotomic Brauer algebras in characteristic zero*, J. Algebra **378** (2013), 80–102.
BCH23.	C. Bowman, A. Cox, and A. Hazi, *Path isomorphisms between quiver Hecke and diagrammatic Bott-Samelson endomorphism algebras*, Adv. Math. **429** (2023), Paper No. 109185, 106.
BCHM22.	C. Bowman, A. Cox, A. Hazi, and D. Michailidis, *Path combinatorics and light leaves for quiver Hecke algebras*, Math. Z. **300** (2022), no. 3, 2167–2203.
BCS17.	C. Bowman, A. Cox, and L. Speyer, *A family of graded decomposition numbers for diagrammatic Cherednik algebras*, Int. Math. Res. Not. IMRN (2017), no. 9, 2686–2734.
BDD[+]a.	C. Bowman, A. Dell'Arciprete, M. De Visscher, A. Hazi, R. Muth, and C. Stroppel, *Faithful covers of Khovanov arc algebras*, arXiv:2411.15788.
BDE19.	C. Bowman, M. De Visscher, and J. Enyang, *The lattice permutation condition for Kronecker tableaux*, Sém. Lothar. Combin. **82B** (2019).
BDF[+].	C. Bowman, M. De Visscher, N. Farrell, A. Hazi, and E. Norton, *Oriented Temperley–Lieb algebras and combinatorial Kazhdan–Lusztig theory*, arXiv:2212.09402.
BDM11.	C. Bowman, S. Doty, and S. Martin, *Decomposition of tensor products of modular irreducible representations for* SL_3, Int. Electron. J. Algebra **9** (2011), 177–219, With an appendix by C. M. Ringel.
BDM24.	C. Bowman, S. Doty and S. Martin, Canonical bases and new applications of increasing and decreasing subsequences to invariant theory, J. Algebra **659** (2024), 23–43.
BDM15.	C. Bowman, S. Doty, and S. Martin, *Decomposition of tensor products of modular irreducible representations for* SL_3: *the* $p \geq 5$ *case*, Int. Electron. J. Algebra **17** (2015), 105–138.
BDM22a.	C. Bowman, S. Doty, and S. Martin, *Integral Schur-Weyl duality for partition algebras*, Algebr. Comb. **5** (2022), no. 2, 371–399.
BDM22b.	C. Bowman, S. Doty, and S. Martin, *An integral second fundamental theorem of invariant theory for partition algebras*, Represent. Theory **26** (2022), 437–454.
BDO15.	C. Bowman, M. De Visscher, and R. Orellana, *The partition algebra and the Kronecker coefficients*, Trans. Amer. Math. Soc. **367** (2015), no. 5, 3647–3667.
BDD[+]b.	C. Bowman, A. Dell'Arciprete, M. De Visscher, A. Hazi, R. Muth, and C. Stroppel, *Quiver presentations and Schur–Weyl duality for Khovanov arc algebras*, arXiv:2411.15520.
BDE19.	C. Bowman, M. De Visscher, and J. Enyang, *Simple modules for the partition algebra and monotone convergence of Kronecker coefficients*, Int. Math. Res. Not. IMRN (2019), no. 4, 1059–1097.
BDVE21.	C. Bowman, M. De Visscher, and J. Enyang, *The co-Pieri rule for stable Kronecker coefficients*, J. Combin. Theory Ser. A **177** (2021), Paper No. 105297, 71.
BDK15.	C. Bowman, M. De Visscher, and O. King, *The blocks of the partition algebra in positive characteristic*, Algebr. Represent. Theory **18** (2015), no. 5, 1357–1388.
BEG18a.	C. Bowman, J. Enyang, and F. Goodman, *Diagram algebras, dominance triangularity and skew cell modules*, J. Aust. Math. Soc. **104** (2018), no. 1, 13–36.

References

BEG18b. C. Bowman, J. Enyang, and F. M. Goodman, *The cellular second fundamental theorem of invariant theory for classical groups*, Int. Math. Res. Not. IMRN https://doi.org/10.1093 /imrn/rny079 (2018).

BHDS. C. Bowman, A. Hazi, M. De Visscher, and C. Stroppel, *Quiver presentations and isomorphisms of Hecke categories and Khovanov arc algebras*, arXiv:2309.13695.

BHDS. C. Bowman, A. Hazi, M. De Visscher, and E. Norton, *The anti-spherical Hecke categories for Hermitian symmetric pairs*, arXiv:2208.02584.

BHN22. C. Bowman, A. Hazi, and E. Norton, *The modular Weyl-Kac character formula*, Math. Z. **302** (2022), no. 4, 2207–2232.

BG19. C. Bowman and E. Giannelli, The integral isomorphism behind row removal phenomena for Schur algebras, Math. Proc. Cambridge Philos. Soc. **167** (2019), no. 2, 209–228.

Bow22. C. Bowman, *The many integral graded cellular bases of Hecke algebras of complex reflection groups*, Amer. J. Math. **144** (2022), no. 2, 437–504.

B13. C. Bowman, Bases of quasi-hereditary covers of diagram algebras, Math. Proc. Cambridge Philos. Soc. **154** (2013), no. 3, 393–418.

B12. C. Bowman, Brauer algebras of type C are cellularly stratified, Math. Proc. Cambridge Philos. Soc. **153** (2012), no. 1, 1–7.

BM12. C. Bowman and S. Martin, A reciprocity result for projective indecomposable modules of cellular algebras and BGG algebras, J. Lie Theory **22** (2012), no. 4, 1065–1073.

BP20. C. Bowman and R. Paget, *The uniqueness of plethystic factorisation*, Trans. Amer. Math. Soc. **373** (2020), no. 3, 1653–1666.

BP24. C. Bowman and R. Paget, *The partition algebra and the plethysm coefficients I: Stability and Foulkes' conjecture*, J. Algebra **655** (2024), 110–138.

BPW25. C. Bowman, R. Paget, and M. Wildon, *The partition algebra and the plethysm coefficients II: Ramified plethysm*, Adv. Math. **462** (2025), Paper No. 110090.

BS18. C. Bowman and L. Speyer, *Kleshchev's decomposition numbers for diagrammatic Cherednik algebras*, Trans. Amer. Math. Soc. **370** (2018), no. 5, 3551–3590.

BS19. C. Bowman and L. Speyer, *An analogue of row removal for diagrammatic Cherednik algebras*, Math. Z. **293** (2019), no. 3-4, 935–955.

BM01. T. Braden and R. MacPherson, *From moment graphs to intersection cohomology*, Math. Ann. **321** (2001), no. 3, 533–551.

BE17. J. Brundan and A. Ellis, *Super Kac-Moody 2-categories*, Proc. Lond. Math. Soc. (3) **115** (2017), no. 5, 925–973.

BCI11a. P. Bürgisser, M. Christandl, and C. Ikenmeyer, *Even partitions in plethysms*, J. Algebra **328** (2011), 322–329.

BCI11b. P. Bürgisser, M. Christandl, and C. Ikenmeyer, *Nonvanishing of Kronecker coefficients for rectangular shapes*, Adv. Math. **227** (2011), no. 5, 2082–2091.

BCM06. F. Brenti, F. Caselli, and M. Marietti, *Special matchings and Kazhdan-Lusztig polynomials*, Adv. Math. **202** (2006), no. 2, 555–601.

BK. J. Brundan and A. Kleshchev, *Odd Grassmannian bimodules and derived equivalences for spin symmetric groups*, arXiv:2203.14149, preprint.

BK99. C. Bessenrodt and A. Kleshchev, *On Kronecker products of complex representations of the symmetric and alternating groups*, Pacific Journal of Mathematics **190** (1999), no. 201-223.

BK09. J. Brundan and A. Kleshchev, *Blocks of cyclotomic Hecke algebras and Khovanov-Lauda algebras*, Invent. Math. **178** (2009), no. 3, 451–484.

Bla16. J. Blasiak, *Kronecker coefficients for one hook shape*, Sém. Lothar. Combin. **77** (2016), Art. B77c, 40.

BLMW11. P. Bürgisser, J. Landsberg, L. Manivel, and J. Weyman, *An overview of mathematical issues arising in the geometric complexity theory approach to* VP $\neq$ VNP, SIAM J. Comput. **40** (2011), no. 4, 1179–1209.

BLP23. G. Burrull, N. Libedinsky, and D. Plaza, *Combinatorial invariance conjecture for $\tilde{A}_2$*, Int. Math. Res. Not. IMRN (2023), no. 10, 8903–8933.

BN05.	D. Bar-Natan, *Khovanov's homology for tangles and cobordisms*, Geom. Topol. **9** (2005), 1443–1499.
Boe88.	B. Boe, *Kazhdan-Lusztig polynomials for Hermitian symmetric spaces*, Trans. Amer. Math. Soc. **309** (1988), no. 1, 279–294.
BOR10.	E. Briand, R. Orellana, and M. Rosas, *The stability of the Kronecker product of Schur functions*, 22nd International Conference on Formal Power Series and Algebraic Combinatorics (FPSAC 2010), Discrete Math. Theor. Comput. Sci. Proc., AN, Assoc. Discrete Math. Theor. Comput. Sci., Nancy, 2010, pp. 557–567.
Bra37.	R. Brauer, *On algebras which are connected with the semisimple continuous groups*, Ann. of Math. (2) **38** (1937), no. 4, 857–872.
Bre07.	F. Brenti, *Parabolic Kazhdan-Lusztig R-polynomials for Hermitian symmetric pairs*, J. Algebra **318** (2007), no. 1, 412–429.
BS10.	J. Brundan and C. Stroppel, *Highest weight categories arising from Khovanov's diagram algebra. II. Koszulity*, Transform. Groups **15** (2010), no. 1, 1–45.
BS11a.	J. Brundan and C. Stroppel, *Highest weight categories arising from Khovanov's diagram algebra I: cellularity*, Mosc. Math. J. **11** (2011), no. 4, 685–722, 821–822.
BS11b.	J. Brundan and C. Stroppel, *Highest weight categories arising from Khovanov's diagram algebra III: category O*, Represent. Theory **15** (2011), 170–243.
BS12a.	J. Brundan and C. Stroppel, *Gradings on walled Brauer algebras and Khovanov's arc algebra*, Adv. Math. **231** (2012), no. 2, 709–773.
BS12b.	J. Brundan and C. Stroppel, *Highest weight categories arising from Khovanov's diagram algebra IV: the general linear supergroup*, J. Eur. Math. Soc. (JEMS) **14** (2012), no. 2, 373–419.
BS17.	J. Brundan and C. Stroppel, *Semi-infinite highest weight categories*, 1808.08022 (2017).
BSW20a.	J. Brundan, A. Savage, and B. Webster, *Heisenberg and Kac-Moody categorification*, Selecta Math. (N.S.) **26** (2020), no. 5, Paper No. 74, 62.
BSW20b.	J. Brundan, A. Savage, and B. Webster, *On the definition of quantum Heisenberg category*, Algebra Number Theory **14** (2020), no. 2, 275–321.
BSW21.	J. Brundan, A. Savage, and B. Webster, *Foundations of Frobenius Heisenberg categories*, J. Algebra **578** (2021), 115–185.
BSW22.	J. Brundan, A. Savage, and B. Webster, *Quantum Frobenius Heisenberg categorification*, J. Pure Appl. Algebra **226** (2022), no. 1, Paper No. 106792, 50.
BB81.	A. Beilinson and J. Bernstein, *Localisation de g-modules*, C. R. Acad. Sci. Paris Sér. I Math. **292** (1981), no. 1, 15–18.
BK81.	J.-L. Brylinski and M. Kashiwara, *Kazhdan-Lusztig conjecture and holonomic systems*, Invent. Math. **64** (1981), no. 3, 387–410.
BW.	S. Barmeier and Z. Wang, A_∞ *deformations of extended Khovanov arc algebras and Stroppel's conjecture*, arXiv:2211.03354.
Car14.	R. D. Carmichael, *On the numerical factors of the arithmetic forms $\alpha^n \pm \beta^n$*, Ann. of Math. (2) **15** (1913/14), no. 1-4, 49–70.
CC76.	R. Carter and E. Cline, *The submodule structure of Weyl modules for groups of type A_1*, Proceedings of the Conference on Finite Groups (Univ. Utah, Park City, Utah, 1975), Academic Press, New York-London, 1976, pp. 303–311.
CD11.	A. Cox and M. De Visscher, *Diagrammatic Kazhdan-Lusztig theory for the (walled) Brauer algebra*, J. Algebra **340** (2011), 151–181.
CDDM08.	A. Cox, M. De Visscher, S. Doty, and P. Martin, *On the blocks of the walled Brauer algebra*, J. Algebra **320** (2008), no. 1, 169–212.
CDM09a.	A. Cox, M. De Visscher, and P. Martin, *The blocks of the Brauer algebra in characteristic zero*, Represent. Theory **13** (2009), 272–308.
CDM09b.	A. Cox, M. De Visscher, and P. Martin, *A geometric characterisation of the blocks of the Brauer algebra*, J. Lond. Math. Soc. (2) **80** (2009), no. 2, 471–494.
CDM11.	A. Cox, M. De Visscher, and P. Martin, *Alcove geometry and a translation principle for the Brauer algebra*, J. Pure Appl. Algebra **215** (2011), no. 4, 335–367.

References

CE00.	A. Cox and K. Erdmann, *On Ext^2 between Weyl modules for quantum GL_n*, Math. Proc. Cambridge Philos. Soc. **128** (2000), no. 3, 441–463.
CGG12.	M. Chlouveraki, I. Gordon, and S. Griffeth, *Cell modules and canonical basic sets for Hecke algebras from Cherednik algebras*, New trends in noncommutative algebra, Contemp. Math., vol. 562, Amer. Math. Soc., Providence, RI, 2012, pp. 77–89.
CHM07.	M. Christandl, A. W. Harrow, and G. Mitchison, *Nonzero Kronecker coefficients and what they tell us about spectra*, Comm. Math. Phys. **270** (2007), no. 3, 575–585.
CKM14.	S. Cautis, J. Kamnitzer, and S. Morrison, *Webs and quantum skew Howe duality*, Math. Ann. **360** (2014), no. 1-2, 351–390.
Cli79.	E. Cline, Ext^1 *for* SL_2, Comm. Algebra **7** (1979), no. 1, 107–111.
CPS88.	E. Cline, B. Parshall and L. Scott, Finite-dimensional algebras and highest weight categories, J. Reine Angew. Math. **391** (1988), 85–99.
CM18.	E. Carlsson and A. Mellit, *A proof of the shuffle conjecture*, J. Amer. Math. Soc. **31** (2018), no. 3, 661–697.
CGM03.	A. Cox, J. Graham, and P. Martin, *The blob algebra in positive characteristic*, J. Algebra **266** (2003), no. 2, 584–635.
CMPX06.	A. Cox, P. Martin, A. Parker, and C. Xi, *Representation theory of towers of recollement: theory, notes, and examples*, J. Algebra **302** (2006), no. 1, 340–360.
CMT17.	J. Chuang, H. Miyachi, and K. Tan, *Parallelotope tilings and q-decomposition numbers*, Adv. Math. **321** (2017), 80–159.
Cou18.	K. Coulembier, *The periplectic Brauer algebra*, Proc. Lond. Math. Soc. (3) **117** (2018), no. 3, 441–482.
CR81.	C. W. Curtis and I. Reiner, *Methods of representation theory. Vol. I*, Pure and Applied Mathematics, J. Wiley & Sons, Inc., New York, 1981, With applications to finite groups and orders, A Wiley-Interscience Publication.
CR08.	J. Chuang and R. Rouquier, *Derived equivalences for symmetric groups and $\mathfrak{sl}_2$-categorification*, Ann. of Math. (2) **167** (2008), no. 1, 245–298.
Cra95.	L. Crane, *Clock and category: is quantum gravity algebraic?*, J. Math. Phys. **36** (1995), no. 11, 6180–6193.
DDH08.	R. Dipper, S. Doty, and J. Hu, *Brauer algebras, symplectic Schur algebras and Schur-Weyl duality*, Trans. Amer. Math. Soc. **360** (2008), no. 1, 189–213.
Deo87.	V. Deodhar, *On some geometric aspects of Bruhat orderings II. The parabolic analogue of Kazhdan–Lusztig polynomials*, J. Algebra **111** (1987), no. 2, 483–506.
Deo90.	V. Deodhar, *A combinatorial setting for questions in Kazhdan-Lusztig theory*, Geom. Dedicata **36** (1990), no. 1, 95–119.
Der81.	D. I. Deriziotis, *The submodule structure of Weyl modules for groups of type A_1*, Comm. Algebra **9** (1981), no. 3, 247–265.
DG02.	S. Doty and A. Giaquinto, *Presenting Schur algebras*, Int. Math. Res. Not. (2002), no. 36, 1907–1944.
DH05.	S. Doty and A. Henke, *Decomposition of tensor products of modular irreducibles for* SL_2, Q. J. Math. **56** (2005), no. 2, 189–207.
DH09.	S. Doty and J. Hu, *Schur-Weyl duality for orthogonal groups*, Proc. Lond. Math. Soc. (3) **98** (2009), no. 3, 679–713.
DJM98.	R. Dipper, G. D. James, and A. Mathas, *Cyclotomic q-Schur algebras*, Math. Z. **229** (1998), no. 3, 385–416.
DM17.	M. De Visscher and P. Martin, *On Brauer algebra simple modules over the complex field*, Trans. Amer. Math. Soc. **369** (2017), no. 3, 1579–1609.
DM20.	S. Donkin and S. Martin, *A Clebsch-Gordan decomposition in positive characteristic*, J. Algebra **560** (2020), 680–699.
Don93.	S. Donkin, *On tilting modules for algebraic groups*, Math. Z. **212** (1993), no. 1, 39–60.
Don98.	S. Donkin, *The q-Schur algebra*, London Mathematical Society Lecture Note Series, vol. 253, Cambridge University Press, Cambridge, 1998.
Dot85.	S. Doty, *The submodule structure of certain Weyl modules for groups of type A_n*, J. Algebra **95** (1985), no. 2, 373–383.

Dot89. S. Doty, *Submodules of symmetric powers of the natural module for* GL_n, Invariant theory (Denton, TX, 1986), Contemp. Math., vol. 88, Amer. Math. Soc., Providence, RI, 1989, pp. 185–191.

Dot03. S. Doty, *Presenting generalized q-Schur algebras*, Represent. Theory **7** (2003), 196–213.

DVB+21. A. Davies, P. Velickoviċ, L. Buesing, S. Blackwell, D. Zheng, N. Tomašev, R. Tanburn, P. Battaglia, C. Blundell, A. Juhász, M. Lackenby, G. Williamson, D. Hassabis, and P. Kohli, *Advancing mathematics by guiding human intuition with AI*, Nature **600** (2021), 70–74.

Dvi93. Y. Dvir, *On the Kronecker product of* S_n *characters*, J. Algebra **154** (1993), no. 1, 125–140.

EG17. J. Enyang and F. M. Goodman, *Cellular bases for algebras with a Jones basic construction*, Algebr. Represent. Theory **20** (2017), no. 1, 71–121.

EP19. J. Espinoza and D. Plaza, *Blob algebra and two-color Soergel calculus*, J. Pure Appl. Algebra **223** (2019), no. 11, 4708–4745.

EH02. K. Erdmann and A. Henke, *On Schur algebras, Ringel duality and symmetric groups*, J. Pure Appl. Algebra **169** (2002), no. 2-3, 175–199.

EH18. K. Erdmann and T. Holm, *Algebras and representation theory*, Springer Undergraduate Mathematics Series, Springer, Cham, 2018.

EK10. B. Elias and M. Khovanov, *Diagrammatics for Soergel categories*, Int. J. Math. Math. Sci. (2010), Art. ID 978635, 58.

EK18. A. Evseev and A. Kleshchev, *Blocks of symmetric groups, semicuspidal KLR algebras and zigzag Schur-Weyl duality*, Ann. of Math. (2) **188** (2018), no. 2, 453–512.

EvMa. A. Evseev and A. Mathas *Content systems and deformations of cyclotomic KLR algebras of type A and C*, arXiv:2209.00134.

EK23. B. Elias and H. Ko, *A singular Coxeter presentation*, Proc. Lond. Math. Soc. (3) **126** (2023), no. 3, 923–996.

EKL14. A. Ellis, M. Khovanov, and A. Lauda, *The odd nilHecke algebra and its diagrammatics*, Int. Math. Res. Not. IMRN (2014), no. 4, 991–1062.

EL. B. Elias and I. Losev, *Modular representation theory in type A via Soergel bimodules*, arXiv:1701.00560, preprint.

EL16. B. Elias and A. Lauda, *Trace decategorification of the Hecke category*, J. Algebra **449** (2016), 615–634.

EL17. T. Etgü and Y. Lekili, *Koszul duality patterns in Floer theory*, Geom. Topol. **21** (2017), no. 6, 3313–3389.

Eli10. B. Elias, *A diagrammatic Temperley-Lieb categorification*, Int. J. Math. Math. Sci. (2010), Art. ID 530808, 47.

Eli16. B. Elias, *The two-color Soergel calculus*, Compos. Math. **152** (2016), no. 2, 327–398.

Eli17. B. Elias, *Quantum Satake in type A. Part I*, J. Comb. Algebra **1** (2017), no. 1, 63–125.

ELV. M. Ebert, A. Lauda, and L. Vera, *Derived superequivalence for spin symmetric groups and odd* sl_2-*categorifications*, arXiv:2203.14153, preprint.

EMTW20. B. Elias, S. Makisumi, U. Thiel, and G. Williamson, *Introduction to Soergel bimodules*, RSME Springer Series, vol. 5, Springer, Cham, [2020] ©2020.

Erd93. K. Erdmann, *Schur algebras of finite type*, Quart. J. Math. Oxford Ser. (2) **44** (1993), no. 173, 17–41.

Erd94. K. Erdmann, *Symmetric groups and quasi-hereditary algebras*, Finite-dimensional algebras and related topics (Ottawa, ON, 1992), NATO Adv. Sci. Inst. Ser. C: Math. Phys. Sci., vol. 424, Kluwer Acad. Publ., Dordrecht, 1994, pp. 123–161.

Erd95. K. Erdmann, Ext^1 *for Weyl modules of* $SL_2(K)$, Math. Z. **218** (1995), no. 3, 447–459.

Erd97. K. Erdmann, *Representations of* $GL_n(K)$ *and symmetric groups*, Representation theory of finite groups (Columbus, OH, 1995), Ohio State Univ. Math. Res. Inst. Publ., vol. 6, de Gruyter, Berlin, 1997, pp. 67–84.

ES16a. M. Ehrig and C. Stroppel, *2-row Springer fibres and Khovanov diagram algebras for type D*, Canad. J. Math. **68** (2016), no. 6, 1285–1333.

ES16b.	M. Ehrig and C. Stroppel, *Diagrammatic description for the categories of perverse sheaves on isotropic Grassmannians*, Selecta Math. (N.S.) **22** (2016), no. 3, 1455–1536.
ES16c.	M. Ehrig and C. Stroppel, *Schur-Weyl duality for the Brauer algebra and the orthosymplectic Lie superalgebra*, Math. Z. **284** (2016), no. 1-2, 595–613.
ES17.	M. Ehrig and C. Stroppel, *On the category of finite-dimensional representations of* $OSp(r|2n)$: *Part I*, Representation theory—current trends and perspectives, EMS Ser. Congr. Rep., Eur. Math. Soc., Zürich, 2017, pp. 109–170.
ET21.	M. Ehrig and D. Tubbenhauer, *Relative cellular algebras*, Transform. Groups **26** (2021), no. 1, 229–277.
EW06.	K. Erdmann and M. Wildon, *Introduction to Lie algebras*, Springer Undergraduate Mathematics Series, Springer-Verlag London, Ltd., London, 2006.
EL17.	B. Elias and N. Libedinsky, Indecomposable Soergel bimodules for universal Coxeter groups, Trans. Amer. Math. Soc. **369** (2017), no. 6, 3883–3910.
EW13.	B. Elias and G. Williamson, Kazhdan-Lusztig conjectures and shadows of Hodge theory, in *Arbeitstagung Bonn 2013*, 105–126, Progr. Math., 319, Birkhäuser/Springer, Cham.
EW14.	B. Elias and G. Williamson, *The Hodge theory of Soergel bimodules*, Ann. of Math. (2) **180** (2014), no. 3, 1089–1136.
EW16.	B. Elias and G. Williamson, *Soergel calculus*, Represent. Theory **20** (2016), 295–374.
EW21.	B. Elias and G. Williamson, Relative hard Lefschetz for Soergel bimodules, J. Eur. Math. Soc. (JEMS) **23** (2021), no. 8, 2549–2581.
EW23.	B. Elias and G. Williamson, Localized calculus for the Hecke category, Ann. Math. Blaise Pascal **30** (2023), no. 1, 1–73;
FH91.	W. Fulton and J. Harris, *Representation theory*, Graduate Texts in Mathematics, vol. 129, Springer-Verlag, New York, 1991, A first course, Readings in Mathematics.
FI20.	N. Fischer and C. Ikenmeyer, *The computational complexity of plethysm coefficients*, Comput. Complexity **29** (2020), no. 2, Paper No. 8, 43.
Fie12.	P. Fiebig, *An upper bound on the exceptional characteristics for Lusztig's character formula*, J. Reine Angew. Math. **673** (2012), 1–31.
Gec06.	M. Geck, *Kazhdan-Lusztig cells and the Murphy basis*, Proc. London Math. Soc. (3) **93** (2006), no. 3, 635–665.
Gec07.	M. Geck, *Hecke algebras of finite type are cellular*, Invent. Math. **169** (2007), no. 3, 501–517.
GG13.	T. Geetha and F. M. Goodman, *Cellularity of wreath product algebras and A-Brauer algebras*, J. Algebra **389** (2013), 151–190.
GJW23.	J. Gibson, L. T. Jensen, and G. Williamson, *Calculating the p-canonical basis of Hecke algebras*, Transform. Groups **28** (2023), no. 3, 1121–1148.
GKM22.	H. Geranios, A. Kleshchev, and L. Morotti, *On self-extensions of irreducible modules over symmetric groups*, Trans. Amer. Math. Soc. **375** (2022), no. 4, 2627–2676.
GL96.	J. Graham and G. Lehrer, *Cellular algebras*, Invent. Math. **123** (1996), no. 1, 1–34.
Gre93.	J. A. Green, *Combinatorics and the Schur algebra*, J. Pure Appl. Algebra **88** (1993), no. 1-3, 89–106.
Gre07.	J. A. Green, *Polynomial representations of* GL_n, second ed., Lecture Notes in Mathematics, vol. 830, Springer, Berlin, 2007, With an appendix on Schensted correspondence and Littelmann paths by K. Erdmann, J. A. Green and M. Schocker.
Hö99.	M. Härterich, *Murphy bases of generalized Temperley-Lieb algebras*, Arch. Math. (Basel) **72** (1999), no. 5, 337–345.
Haz17.	A. Hazi, *Balanced semisimple filtrations for tilting modules*, Represent. Theory **21** (2017), 4–19.
Haz.	A. Hazi, *Matrix recursion for positive characteristic diagrammatic Soergel bimodules for affine Weyl groups*, to appear in Represent. Theory, arXiv:1708.07072.
HM10.	J. Hu and A. Mathas, *Graded cellular bases for the cyclotomic Khovanov-Lauda-Rouquier algebras of type A*, Adv. Math. **225** (2010), no. 2, 598–642.

HM15.	J. Hu and A. Mathas, *Quiver Schur algebras for linear quivers*, Proc. Lond. Math. Soc. (3) **110** (2015), no. 6, 1315–1386.
HM16.	J. Hu and A. Mathas, *Seminormal forms and cyclotomic quiver Hecke algebras of type A*, Math. Ann. **364** (2016), no. 3-4, 1189–1254.
HO01.	R. Häring-Oldenburg, *Cyclotomic Birman-Murakami-Wenzl algebras*, J. Pure Appl. Algebra **161** (2001), no. 1-2, 113–144.
Hum78.	J. Humphreys, *Introduction to Lie algebras and representation theory*, Graduate Texts in Mathematics, vol. 9, Springer-Verlag, New York-Berlin, 1978, Second printing, revised.
Hum90.	J. Humphreys, *Reflection groups and Coxeter groups*, Cambridge Studies in Advanced Mathematics, vol. 29, Cambridge University Press, Cambridge, 1990.
IP17.	C. Ikenmeyer and G. Panova, *Rectangular Kronecker coefficients and plethysms in geometric complexity theory*, Adv. Math. **319** (2017), 40–66.
Irv88.	R. Irving, *The socle filtration of a Verma module*, Ann. Sci. École Norm. Sup. (4) **21** (1988), no. 1, 47–65.
Jam76.	G. D. James, *On the decomposition matrices of the symmetric groups. II*, J. Algebra **43** (1976), no. 1, 45–54.
Jam78.	G. D. James, *The representation theory of the symmetric groups*, Lecture Notes in Mathematics 682, Springer, 1978.
Jam90.	G. D. James, *The decomposition matrices of* $GL_n(q)$ *for* $n \leq 10$, Proc. London Math. Soc. (3) **60** (1990), no. 2, 225–265.
Jam94.	G. D. James, *The representation theory for Buckminsterfullerene*, J. Algebra **167** (1994), no. 3, 803–820.
Jan03.	J. C. Jantzen, *Representations of algebraic groups*, second ed., Mathematical Surveys and Monographs, vol. 107, American Mathematical Society, Providence, RI, 2003.
Jen21.	L. Jensen, *Correction of the Lusztig–Williamson billiards conjecture*, arXiv:2105.04665.
JBS.	J. Morse, A. Pun, J. Blasiak, M. Haiman and G. H. Seelinger, *Dens, nests and the Loehr–Warrington conjecture*, arXiv:2112.07070, preprint.
Jon85.	V. Jones, *A polynomial invariant for knots via von Neumann algebras*, Bull. Amer. Math. Soc. (N.S.) **12** (1985), no. 1, 103–111.
Jon87.	V. Jones, *Hecke algebra representations of braid groups and link polynomials*, Ann. of Math. (2) **126** (1987), no. 2, 335–388.
JW17.	L. Jensen and G. Williamson, *The p-canonical basis for Hecke algebras*, Categorification and higher representation theory, Contemp. Math., vol. 683, Amer. Math. Soc., Providence, RI, 2017, pp. 333–361.
Kau90.	L. Kauffman, *An invariant of regular isotopy*, Trans. Amer. Math. Soc. **318** (1990), no. 2, 417–471.
Kho00.	M. Khovanov, *A categorification of the Jones polynomial*, Duke Math. J. **101** (2000), no. 3, 359–426.
KKT16.	S. Kang, M. Kashiwara, and S. Tsuchioka, *Quiver Hecke superalgebras*, J. Reine Angew. Math. **711** (2016), 1–54.
KT95.	M. Kashiwara and T. Tanisaki. Kazhdan-Lusztig conjecture for affine Lie algebras with negative level. *Duke Math. J.*, 77(1):21–62, 1995.
KT96.	M. Kashiwara and T. Tanisaki. Kazhdan-Lusztig conjecture for affine Lie algebras with negative level. II. Nonintegral case. *Duke Math. J.*, 84(3):771–813, 1996.
KL.	A. Kleshchev and M. Livesey, *RoCK blocks for double covers of symmetric groups and quiver Hecke superalgebras*, arXiv:2201.06870, preprint.
KL79.	D. Kazhdan and G. Lusztig, *Representations of Coxeter groups and Hecke algebras*, Invent. Math. **53** (1979), no. 2, 165–184.
KL80.	D. Kazhdan and G. Lusztig, *Schubert varieties and Poincaré duality*, Geometry of the Laplace operator (Proc. Sympos. Pure Math., Univ. Hawaii, Honolulu, Hawaii, 1979), Proc. Sympos. Pure Math., vol. XXXVI, Amer. Math. Soc., Providence, RI, 1980, pp. 185–203.

KL93.	D. Kazhdan and G. Lusztig, *Tensor structures arising from affine Lie algebras. I, II*, J. Amer. Math. Soc. **6** (1993), no. 4, 905–947, 949–1011.
KL94a.	D. Kazhdan and G. Lusztig, *Tensor structures arising from affine Lie algebras. III*, J. Amer. Math. Soc. **7** (1994), no. 2, 335–381.
KL94b.	D. Kazhdan and G. Lusztig, *Tensor structures arising from affine Lie algebras. IV*, J. Amer. Math. Soc. **7** (1994), no. 2, 383–453.
KL09.	M. Khovanov and A. Lauda, *A diagrammatic approach to categorification of quantum groups. I*, Represent. Theory **13** (2009), 309–347.
KM11.	P. Kronheimer and T. Mrowka, *Khovanov homology is an unknot-detector*, Publ. Math. Inst. Hautes Études Sci. (2011), no. 113, 97–208.
KM17.	A. Kleshchev and R. Muth, *Imaginary Schur-Weyl duality*, Mem. Amer. Math. Soc. **245** (2017), no. 1157, xvii+83.
KM19.	A. Kleshchev and R. Muth, *Affine zigzag algebras and imaginary strata for KLR algebras*, Trans. Amer. Math. Soc. **371** (2019), no. 7, 4535–4583.
KM20.	A. Kleshchev and R. Muth, *Based quasi-hereditary algebras*, J. Algebra **558** (2020), 504–522.
KM22.	A. Kleshchev and R. Muth, *Schurifying quasi-hereditary algebras*, Proc. Lond. Math. Soc. (3) **125** (2022), no. 3, 626–680.
KN10.	A. Kleshchev and D. Nash, *An interpretation of the Lascoux–Leclerc–Thibon algorithm and graded representation theory*, Comm. Algebra **38** (2010), no. 12, 4489–4500.
Koi89.	K. Koike, *On the decomposition of tensor products of the representations of the classical groups: by means of the universal characters*, Adv. Math. **74** (1989), no. 1, 57–86.
KS02.	M. Khovanov and P. Seidel, *Quivers, Floer cohomology, and braid group actions*, J. Amer. Math. Soc. **15** (2002), no. 1, 203–271.
Kup96.	G. Kuperberg, *Spiders for rank 2 Lie algebras*, Comm. Math. Phys. **180** (1996), no. 1, 109–151.
Lan17.	J. Landsberg, *Geometry and complexity theory*, Cambridge Studies in Advanced Mathematics, vol. 169, Cambridge University Press, Cambridge, 2017.
Lib08.	N. Libedinsky, *Sur la catégorie des bimodules de Soergel*, J. Algebra **320** (2008), no. 7, 2675–2694.
Liu17.	R. Liu, *A simplified Kronecker rule for one hook shape*, Proc. Amer. Math. Soc. **145** (2017), no. 9, 3657–3664.
LLT96.	A. Lascoux, B. Leclerc, and J.-Y. Thibon, *Hecke algebras at roots of unity and crystal bases of quantum affine algebras*, Comm. Math. Phys. **181** (1996), no. 1, 205–263.
Los16.	I. Losev, *Proof of Varagnolo–Vasserot conjecture on cyclotomic categories O*, Selecta Math. **22** (2016), no. 2, 631–668.
LS81.	A. Lascoux and M. Schützenberger, *Polynômes de Kazhdan & Lusztig pour les grassmanniennes*, Young tableaux and Schur functors in algebra and geometry (Toruń, 1980), Astérisque, vol. 87, Soc. Math. France, Paris, 1981, pp. 249–266.
LS12.	A. Licata and A. Savage, *A survey of Heisenberg categorification via graphical calculus*, Bull. Inst. Math. Acad. Sin. (N.S.) **7** (2012), no. 2, 291–321.
LT00.	B. Leclerc and J.-Y. Thibon, *Littlewood-Richardson coefficients and Kazhdan-Lusztig polynomials*, Combinatorial methods in representation theory (Kyoto, 1998), Adv. Stud. Pure Math., vol. 28, Kinokuniya, Tokyo, 2000, pp. 155–220.
LP20.	N. Libedinsky and D. Plaza, *Blob algebra approach to modular representation theory*, Proc. Lond. Math. Soc. (3) **121** (2020), no. 3, 656–701.
LPRH21.	D. Lobos, D. Plaza, and S. Ryom-Hansen, *The nil-blob algebra: an incarnation of type $\tilde{A}_1$ Soergel calculus and of the truncated blob algebra*, J. Algebra **570** (2021), 297–365.
Lus80.	G. Lusztig, *Some problems in the representation theory of finite Chevalley groups*, The Santa Cruz Conference on Finite Groups (Univ. California, Santa Cruz, Calif., 1979), Proc. Sympos. Pure Math., vol. 37, Amer. Math. Soc., Providence, RI, 1980, pp. 313–317.

Lus80a.	G. Lusztig, *Hecke algebras and Jantzen's generic decomposition patterns*, Adv. in Math. **37** (1980), no. 2, 121–164.
Lus93.	G. Lusztig, *Introduction to quantum groups*, Progress in Mathematics, 110, Birkhäuser Boston, Boston, MA, 1993.
LW18a.	G. Lusztig and G. Williamson, *Billiards and tilting characters for* SL_3, SIGMA Symmetry Integrability Geom. Methods Appl. **14** (2018), 015, 22 pages.
LW22.	N. Libedinsky and G. Williamson, *The anti-spherical category*, Adv. Math. **405** (2022), Paper No. 108509, 34.
LW23.	E. LePage and B. Webster, *Rational Cherednik algebras of $G(\ell, p, n)$ from the Coulomb perspective*, Adv. Math. **433** (2023), Paper No. 109295, 49.
Mac14.	M. Mackaay, *The sl_N-web algebras and dual canonical bases*, J. Algebra **409** (2014), 54–100.
Mar91.	P. Martin, *Potts models and related problems in statistical mechanics*, Series on Advances in Statistical Mechanics, 5, World Scientific Publishing Co., Inc., Teaneck, NJ, 1991.
Mar93.	S. Martin, *Schur algebras and representation theory*, Cambridge Tracts in Mathematics, vol. 112, Cambridge University Press, Cambridge, 1993.
Mar96.	P. Martin, *The structure of the partition algebras*, J. Algebra **183** (1996), 319–358.
Mar15.	P. Martin, *The decomposition matrices of the Brauer algebra over the complex field*, Trans. Amer. Math. Soc. **367** (2015), no. 3, 1797–1825.
Mar18a.	M. Marietti, *The combinatorial invariance conjecture for parabolic Kazhdan-Lusztig polynomials of lower intervals*, Adv. Math. **335** (2018), 180–210.
Mar18b.	S. Martin, *On certain tilting modules for SL_2*, J. Algebra **506** (2018), 397–408.
Mat99.	A. Mathas, *Iwahori–Hecke Algebras and Schur Algebras of the Symmetric Group*, University Lecture Series, vol. 15, American Mathematical Society, Providence, RI, 1999.
Mat15.	A. Mathas, *Cyclotomic quiver Hecke algebras of type A*, Modular representation theory of finite and p-adic groups, Lect. Notes Ser. Inst. Math. Sci. Natl. Univ. Singap., vol. 30, World Sci. Publ., Hackensack, NJ, 2015, pp. 165–266.
Mat22.	A. Mathas, *Positive Jantzen sum formulas for cyclotomic Hecke algebras*, Math. Z. **301** (2022), no. 3, 2617–2658.
MT24.	A. Mathas and D. Tubbenhauer, *Cellularity and subdivision of KLR and weighted KLRW algebras*, Math. Ann. **389** (2024), no. 3, 3043–3122.
Moo03.	D. Moon, *Tensor product representations of the Lie superalgebra $\mathfrak{p}(n)$ and their centralizers*, Comm. Algebra **31** (2003), no. 5, 2095–2140.
MPT14.	M. Mackaay, W. Pan, and D. Tubbenhauer, *The sl_3-web algebra*, Math. Z. **277** (2014), no. 1-2, 401–479.
JBS.	J. Morse, A. Pun, J. Blasiak, M. Haiman and G. H. Seelinger, *Dens, nests and the Loehr–Warrington conjecture*, arXiv:2112.07070, preprint.
MS22.	C. Yu Mak and I. Smith, *Fukaya-Seidel categories of Hilbert schemes and parabolic category O*, J. Eur. Math. Soc. (JEMS) **24** (2022), no. 9, 3215–3332.
MST05.	T. McLeish, P. Stockley, and R. Twarock, *Mathematical virology*, J. Theor. Med. **6** (2005), no. 2, 67–68.
MT19.	M. Mackaaij and D. Tubbenhauer, *Two-color Soergel calculus and simple transitive 2-representations*, Canad. J. Math. **71** (2019), no. 6, 1523–1566.
Mur92.	G. Murphy, *On the representation theory of the symmetric groups and associated Hecke algebras*, J. Algebra **152** (1992), no. 2, 492–513.
Mur95.	G. Murphy, *The representations of Hecke algebras of type A_n*, J. Algebra **173** (1995), no. 1, 97–121.
Mut19.	R. Muth, *Graded skew Specht modules and cuspidal modules for Khovanov-Lauda-Rouquier algebras of affine type A*, Algebr. Represent. Theory **22** (2019), no. 4, 977–1015.
MW00b.	P. Martin and David Woodcock, *On the structure of the blob algebra*, J. Algebra **225** (2000), no. 2, 957–988.

References

MW22. E. McDowell and M. Wildon, *Modular plethystic isomorphisms for two-dimensional linear groups*, J. Algebra **602** (2022), 441–483.

Nat15. S. Natale, *Jordan-Hölder theorem for finite dimensional Hopf algebras*, Proc. Amer. Math. Soc. **143** (2015), no. 12, 5195–5211.

NV18. G. Naisse and P. Vaz, *Odd Khovanov's arc algebra*, Fund. Math. **241** (2018), no. 2, 143–178.

Par07. A. E. Parker, *Higher extensions between modules for* SL_2, Adv. Math. **209** (2007), no. 1, 381–405.

Pic20. L. Piccirillo, *The Conway knot is not slice*, Ann. of Math. (2) **191** (2020), no. 2, 581–591.

Pla13. D. Plaza, *Graded decomposition numbers for the blob algebra*, J. Algebra **394** (2013), 182–206.

Pla17. D. Plaza, *Graded cellularity and the monotonicity conjecture*, J. Algebra **473** (2017), 324–351.

PRH14. D. Plaza and S. Ryom-Hansen, *Graded cellular bases for Temperley-Lieb algebras of type A and B*, J. Algebraic Combin. **40** (2014), no. 1, 137–177.

PW21. R. Paget and M. Wildon, *Plethysms of symmetric functions and representations of* $SL_2(\mathbf{C})$, Algebr. Comb. **4** (2021), no. 1, 27–68.

Ras10. J. Rasmussen, *Khovanov homology and the slice genus*, Invent. Math. **182** (2010), no. 2, 419–447.

Rib90. K. Ribet, *From the Taniyama-Shimura conjecture to Fermat's last theorem*, Ann. Fac. Sci. Toulouse Math. (5) **11** (1990), no. 1, 116–139.

Rou. R. Rouquier, *2-Kac–Moody algebras*, arXiv:0812.5023, preprint.

RSA14. D. Ridout and Y. Saint-Aubin, *Standard modules, induction and the structure of the Temperley-Lieb algebra*, Adv. Theor. Math. Phys. **18** (2014), no. 5, 957–1041.

RSVV16. R. Rouquier, P. Shan, M. Varagnolo, and E. Vasserot, *Categorifications and cyclotomic rational double affine Hecke algebras*, Invent. Math. **204** (2016), no. 3, 671–786.

RW18. S. Riche and G. Williamson, *Tilting modules and the p-canonical basis*, Astérisque (2018), no. 397, ix+184.

Sar16. A. Sartori, *Categorification of tensor powers of the vector representation of* $U_q(\mathfrak{gl}(1|1))$, Selecta Math. (N.S.) **22** (2016), no. 2, 669–734.

Sha12. P. Shan, *Graded decomposition matrices of v-Schur algebras via Jantzen filtration*, Represent. Theory **16** (2012), 212–269.

Soe97. W. Soergel, *Kazhdan–Lusztig polynomials and a combinatoric for tilting modules*, Represent. Theory **1** (1997), 83–114 (electronic).

Soe92. W. Soergel, *The combinatorics of Harish-Chandra bimodules*, J. Reine Angew. Math. **429** (1992), 49–74.

Soe97. W. Soergel, *Kazhdan–Lusztig polynomials and a combinatoric for tilting modules*, Represent. Theory **1** (1997), 83–114 (electronic).

Soe00. W. Soergel, *On the relation between intersection cohomology and representation theory in positive characteristic*, vol. 152, 2000, Commutative algebra, homological algebra and representation theory (Catania/Genoa/Rome, 1998), pp. 311–335.

Soe07. W. Soergel, *Kazhdan-Lusztig-Polynome und unzerlegbare Bimoduln über Polynomringen*, J. Inst. Math. Jussieu **6** (2007), no. 3, 501–525.

SS16. S. Sam and A. Snowden, *Proof of Stembridge's conjecture on stability of Kronecker coefficients*, J. Algebraic Combin. **43** (2016), no. 1, 1–10.

Sta80. R. Stanley, *Unimodal sequences arising from Lie algebras*, Combinatorics, representation theory and statistical methods in groups, Lect. Notes Pure Appl. Math., vol. 57, Dekker, New York, 1980, pp. 127–136.

Sta90. D. Stanton, *Unimodality and Young's lattice*, J. Combin. Theory Ser. A **54** (1990), no. 1, 41–53.

Sta15. R. Stanley, *Catalan numbers*, Cambridge University Press, New York, 2015.

Ste96. J. Stembridge, *On the fully commutative elements of Coxeter groups*, J. Algebraic Combin. **5** (1996), no. 4, 353–385.

Ste12.	J. Stembridge, *Generalized stability of kronecker coefficients*, , 2012, preprint.
Str09.	C. Stroppel, *Parabolic category $\mathcal{O}$, perverse sheaves on Grassmannians, Springer fibres and Khovanov homology*, Compos. Math. **145** (2009), no. 4, 954–992.
Str23.	C. Stroppel, *Categorification: tangle invariants and TQFTs*, ICM—International Congress of Mathematicians. Vol. 2. Plenary lectures, EMS Press, Berlin, [2023] ©2023, pp. 1312–1353.
TL71.	H. Temperley and E. Lieb, *Relations between the "percolation" and "colouring" problem and other graph-theoretical problems associated with regular planar lattices: some exact results for the "percolation" problem*, Proc. Roy. Soc. London Ser. A **322** (1971), no. 1549, 251–280.
Tub14.	D. Tubbenhauer, sl_3-*web bases, intermediate crystal bases and categorification*, J. Algebraic Combin. **40** (2014), no. 4, 1001–1076.
Tub20.	D. Tubbenhauer, gl_n-*webs, categorification and Khovanov-Rozansky homologies*, J. Knot Theory Ramifications **29** (2020), no. 11, 2050074, 96.
Tub24.	D. Tubbenhauer, *Sandwich cellularity and a version of cell theory*, Rocky Mountain J. Math. **54** (2024), no. 6, 1733–1773.
Tur89.	V. G. Turaev, *Operator invariants of tangles, and R-matrices*, Izv. Akad. Nauk SSSR Ser. Mat. **53** (1989), no. 5, 1073–1107, 1135.
TW21.	D. Tubbenhauer and P. Wedrich, *Quivers for* SL_2 *tilting modules*, Represent. Theory **25** (2021), 440–480.
TW22.	D. Tubbenhauer and P. Wedrich, *The center of* SL_2 *tilting modules*, Glasg. Math. J. **64** (2022), no. 1, 165–184.
Web12.	B. Webster, *Koszul duality between Higgs and Coulomb categories O*, arXiv:1611.06541.
Web17a.	B. Webster, *Knot invariants and higher representation theory*, Mem. Amer. Math. Soc. **250** (2017), no. 1191, v+141.
Web17b.	B. Webster, *Rouquier's conjecture and diagrammatic algebra*, Forum Math. Sigma **5** (2017), e27, 71.
Web19.	B. Webster, *Weighted Khovanov-Lauda-Rouquier algebras*, Doc. Math. **24** (2019), 209–250.
Web20.	B. Webster, *On graded presentations of Hecke algebras and their generalizations*, Algebr. Comb. **3** (2020), no. 1, 1–38.
Wes09.	B. Westbury, *Invariant tensors and cellular categories*, J. Algebra **321** (2009), no. 11, 3563–3567.
Wil95.	A. Wiles, *Modular elliptic curves and Fermat's last theorem*, Ann. of Math. (2) **141** (1995), no. 3, 443–551.
Wil16.	G. Williamson, *Local Hodge theory of Soergel bimodules*, Acta Math. **217** (2016), no. 2, 341–404.
Wil17.	G. Williamson, *Schubert calculus and torsion explosion*, J. Amer. Math. Soc. **30** (2017), no. 4, 1023–1046, With a joint appendix with A. Kontorovich and P. J. McNamara.
Wil18.	G. Williamson, *Parity sheaves and the Hecke category*, Proceedings of the International Congress of Mathematicians—Rio de Janeiro 2018. Vol. I. Plenary lectures, World Sci. Publ., Hackensack, NJ, 2018, pp. 979–1015.

Index

$L^{\Bbbk}(\lambda)$, 119
$\Bbbk$-algebra, 81, 83
$\Delta^{\Bbbk}(\lambda)$, 118
$\Gamma_{n,r}$, 295
Λ^+, 117
$\Lambda^0_{\Bbbk}$, 119
Λ_r, 96
Λ^+_r, 137
$\Lambda_{m,n}$, 94
$\Lambda_{m,n}$, 169
$\mathbb{T}_r$, 96
$\mathbb{T}_{n,r}$, 295
$\mathrm{rad}\langle -,-\rangle^{\Bbbk}_\lambda$, 119

Alperin diagram, 144, 148
Anti-involution, 89
Anti-spherical Kazhdan–Lusztig polynomials, 180, 203

Binary Schur algebra, 100
Bruhat graph, 169, 181
Bruhat order, 191
Burnside's counting theorem, 59

Cell module, 118
Cellular algebra, 117, 137
Cellular basis, 117
Composition series, 51, 122
Coxeter group, 37

Decomposition numbers, 122, 142

Determinant homomorphism, 101

Fibonacci numbers, ix, x, 268
First isomorphism theorem, 49
Formal character, 131

Grading trick, 134
Group, 7, 11
Group actions, 57
Group cosets, 39

Homomorphism, 47, 84, 92, 110, 132

Ideal, 106
Idempotent trick, 134
Idempotent truncation, 150, 153
Idempotents, 87
Intersection form, 251
Isomorphism, 15, 84, 92, 110, 132

Kazhdan–Lusztig polynomial, 167, 173, 175, 188
Kazhdan–Lusztig positivity, 188
Kernel, 47
KLR algebra, 330

Lagrange's theorem, 42
Length function, 26, 27
Light leaves, 212, 230, 244, 250
LLT polynomial, 323

© The Author(s), under exclusive license to Springer Nature Switzerland AG 2025
C. Bowman, *Diagrammatic Algebra*, Universitext,
https://doi.org/10.1007/978-3-031-88801-4

Lusztig's conjecture, 283, 309, 310

Module, 106
Morita equivalent, 154
Multiplicity-free module, 144, 148

Normal subgroup, 43

Orbit stabiliser theorem, 59
Oriented Temperley–Lieb algebra, 95

Palindromic polynomial, 165
Partition, 163, 295
Presentations, 14, 83, 90

Quantum binomials, 162
Quiver Hecke algebra, 330
Quotient module, 107

Radical, 129
Rank-generating function, 164

Saturated set, 153, 192

Second isomorphism theorem, 50
Simple module, 119–121
Simple group, 45
Simple module, 107, 140
Spherical Kazhdan–Lusztig
   polynomials, 180, 203
Steinberg, 195
Submodule, 107
Symmetric group, 21
Symmetry groups of solids, 62, 65,
   67, 70

Temperley–Lieb algebras, 85
The Jordan–Hölder Theorem, 52

Uni-modal polynomial, 165
Unipotent element, 89

Weighted cellular algebra, 131

Zig-zag algebras, 90

Printed in the United States
by Baker & Taylor Publisher Services